P9-DGV-369

About Island Press

Island Press is the only nonprofit organization in the United States whose principal purpose is the publication of books on environmental issues and natural resource management. We provide solutions-oriented information to professionals, public officials, business and community leaders, and concerned citizens who are shaping responses to environmental problems.

In 2003, Island Press celebrates its nineteenth anniversary as the leading provider of timely and practical books that take a multidisciplinary approach to critical environmental concerns. Our growing list of titles reflects our commitment to bringing the best of an expanding body of literature to the environmental community throughout North America and the world.

Support for Island Press is provided by The Nathan Cummings Foundation, Geraldine R. Dodge Foundation, Doris Duke Charitable Foundation, Educational Foundation of America, The Charles Engelhard Foundation, The Ford Foundation, The George Gund Foundation, The Vira I. Heinz Endowment, The William and Flora Hewlett Foundation, Henry Luce Foundation, The John D. and Catherine T. MacArthur Foundation, The Andrew W. Mellon Foundation, The Moriah Fund, The Curtis and Edith Munson Foundation, National Fish and Wildlife Foundation, The New-Land Foundation, Oak Foundation, The Overbrook Foundation, The David and Lucile Packard Foundation, The Pew Charitable Trusts, The Rockefeller Foundation, The Winslow Foundation, and other generous donors.

The opinions expressed in this book are those of the author(s) and do not necessarily reflect the views of these foundations.

The Death of Our Planet's Species

A Challenge to Ecology and Ethics

MARTIN GORKE

Translated from the German by Patricia Nevers

ISLAND PRESS
Washington • Covelo • London

Copyright © 2003 Klett-Cotta and Martin Gorke

All rights reserved under International and Pan-American Copyright Conventions. No part of this book may be reproduced in any form or by any means without permission in writing from the publisher: Island Press, 1718 Connecticut Avenue, N.W., Suite 300, Washington, DC 20009.

ISLAND PRESS is a trademark of The Center for Resource Economics.

The translation of this book was made possible through a grant from Goethe-Institut Inter Nationes, Bonn.

Klett-Cotta
© 1999 J. G. Cotta'sche Buchhandlung Nachfolger GmbH, Stuttgart
For the English translation
© 2003 Martin Gorke, Greifswald

Library of Congress Cataloging-in-Publication Data

Gorke, Martin, 1958–
    [Artensterben. English]
    The death of our planet's species : a challenge to ecology and ethics / Martin Gorke ; translated
from German by Patricia Nevers.
        p. cm.
Includes bibliographical references and index.
    ISBN 1-55963-957-1 (hard cover : alk. paper)
    1. Ecology—Philosophy. I. Title.
    QH540.5.G6713 2003
    333.95—dc21

2003006059

British Cataloguing-in-Publication Data available

Printed on recycled, acid-free paper ♻
Manufactured in the United States of America
09 08 07 06 05 04 03   10 9 8 7 6 5 4 3 2 1

For our fellow travelers
on this grand journey
through space and time

CONTENTS

PREFACE

Even though human-induced species extinction presently seems to rank low on peoples' attention scale compared to other political and societal topics, this does not mean that its significance in earth history or its ecological consequences have diminished in any way. It must repeatedly be made clear that if current trends continue, within the next one hundred years half of all our planet's species will most likely have become extinct. Thus, members of today's generation are witnesses and also perpetrators of the greatest catastrophe in the history of life since the disappearance of the dinosaurs 65 million years ago.

The irrevocable loss of species, which is probably the most disturbing symptom of our ecological crisis, is not only a challenge to ecology, politics, economics, law, and nature conservation. It is also a challenge to *ethics*, because among ethicists there is still controversy about why extirpating a species is morally reprehensible. In this book, I shall attempt to provide an answer to this question. My intention is not simply to provide another example of a series of publications in which facts about the extinction of species that elicit concern are documented, but rather to create a philosophical foundation that will enable all those interested in protecting nature to *evaluate* such facts. The aim of this book is to outline the *ethical dimensions* of species extinction.

This will be carried out in two steps. The first step is to clearly demonstrate that the ecological crisis and the disappearance of species accompanying it does indeed represent an ethical problem and that this problem cannot be resolved merely by scientific and technological means. Restrictions to such an "ecological solution" are posed by the basic limits of ecological knowledge as well as by the fact that it is logically and factually erroneous to try to deduce norms for "the right way" to deal with nature *directly* from ecological theory. In a second step, I shall develop the thesis that the ethical dimensions of species loss are not solely due to the damage that may be incurred for the interests of future generations. I shall argue instead that the most important reason for protecting species is their *intrinsic value*.

Now, it is quite easy to postulate that nature has intrinsic value, but it is considerably more difficult to conclusively justify this argument. It is virtually impossible to achieve this in the context of traditional ethical theories, which are in essence more or less *anthropocentric*. Justifying species protection by referring to the intrinsic value of nature requires a more comprehensive understanding of ethics. It requires extending ethical theory from anthropocentrism to *pluralistic holism*. This book outlines a scheme of justification for such a concept of ethics, one which grants intrinsic value not only to humans but also to all natural entities and entire systems, including species.

Thus, first it addresses all those who are adamantly convinced that we humans in principle have no right to destroy other species, but who are unable to explain exactly why. For these readers, especially those actively involved in nature conservation, this book aims to provide basic ethical arguments for rationally justifying this position both personally and in discussions with others. Second, it is directed at those who continue to reject the concept of intrinsic value of nature as "an irrational construct." I hope to be able to show these people, in particular, philosophers and ethicists, that contrary to what is commonly held, more reasonable arguments *in support* of such a thesis and a corresponding concept of ethics can be found than ones that oppose it.

A book such as this one, which attempts to address quite a diverse audience, including philosophers, biologists, conservationists, and interested lay people, and moreover operates in the mined area that lies between natural science and the humanities, is subject to special risks. Depending upon previous knowledge and personal interests, various readers will tend to consider some of the argumentative steps presented here to be superfluous. For example, someone actively involved in species protection may be primarily interested in arguments for justifying holistic ethics, the limits of utilitarian arguments, and sociopsychological considerations, while discussions of ecological and philosophical theory may be of lesser interest. On the other hand, the latter topic may be the very one that particularly captures the attention of a philosopher, while he or she might regard the pragmatic question of the most "effective" ethics for achieving species protection merely of secondary importance. Therefore, allow me to give the following advice to both groups of readers: For a profound understanding of the theme of this book, it is definitely useful to read it in order from cover to cover. However, many of the chapters are as such more or less complete so that it may also be worthwhile to read one or the other of them individually. In particular, the two main divisions, Part A and Part B, can be understood independently of one another. Someone who is familiar with ecological theory, understands its limits and possibilities, and wants to progress as quickly as possible to the ethical dimensions of species extinction, can move immediately from the introduction to Part B. Likewise, philosophical novices can skip over Chapters 25 and 29, in which various different objections to holism are presented, or read them sometime later.

However, critical readers should not forego the opportunity to pursue the thoughts presented in this book from the beginning to the end, including those parts that appear to be "difficult." Justifiably, they expect a *philosophical* treatise to not only present propositions but also to make them as "watertight" as possible. Obviously, this cannot be achieved without a certain degree of argumentative effort, or rather, as Kant ([1783] 1976, 6) puts it: "A person may very well use a hammer and chisel to produce a piece of furniture, but it takes an etching needle to make a copperplate engraving."

Greifswald, May 2003

ACKNOWLEDGMENTS

Numerous people have contributed to this book by providing encouragement, reviewing sections of it, or suggesting valuable literature sources, and I am very grateful to all of them. First and foremost I wish to thank Wilhelm Vossenkuhl, who supervised my Ph.D. thesis in philosophy, upon which much of this book is based, and further supported it in various different ways. Thanks are also due to Helmut Zwölfer, whose detailed criticism served to greatly improve the ecological sections of the book; furthermore to Jürgen Gerdes, Norbert Niclauss, Kai Grosch, Klaus Looft, Wolfgang Völkl, Mark Frenzel, Eike Hartwig, Ludwig Trepl, and last but not least my mother, Hanna Gorke. I wish to thank Uli Seizinger and Gerhard Dörfler of the science section of the University of Bayreuth library for their help in locating many important books and articles from journals, and I gratefully acknowledge the contributions of Johannes Czaja and Heidrun Kochmann from Klett-Cotta, who edited the German manuscript and made stylistic improvements.

I certainly would never have considered having the book translated into English if Dieter Birnbacher had not recommended this in his evaluation of my thesis, a suggestion for which I am most grateful. Publication of the book by Island Press, one of the most well-known publishing companies in the field of environmental studies, was made possible by the friendly support of Roland Knappe, Edward O. Wilson, and Michael Nelson, whose help I wish to acknowledge. I also wish to extend my thanks to Barbara Dean, Barbara Youngblood, Chace Caven, and Erin Johnson from Island Press for their professional and kind assistance in editing the English version of the text.

Like so many things in life, the dimensions of having a book translated only became truly obvious after the work had commenced. One of the things I underestimated was the effort and difficulties involved in substituting original English quotations for those in the German version that were derived from German translations. Fortunately, Jeannine Bohn from the University of Hamburg was able to assist me in finding the German quotes in their text context. In a most laudable undertaking, Craig Buttke from the University of Wisconsin–Stevens Point then used these passages to locate the original quotations in English. I wish to further acknowledge Rainer Schimming and Jürgen Bolik for helping me find the right expressions in matters of mathematics and physics. Christian Bartholomaeus was a great help by converting the illustrations of this book to an electronic database. A special word of thanks is due to Thomas Seiler, who accompanied the translation project steadfastly and provided many valuable suggestions and recommendations. I am further grateful to Konrad Ott, Michael Succow, and Hans Joosten for supporting my work on the translation during my employment as an assistant professor in the Department of Environmental Ethics at the University of Greifswald. I also gratefully acknowledge the support of the Michael-Otto-Foundation for Environmental Protection, which financed this position and assumed part of the translation costs.

Most of all, I wish to thank Patricia Nevers, who translated this book into English over a period of almost two years on weekends and holidays parallel to her work as a professor of biology education at the University of Hamburg. She performed this task with remarkable persistence and exceptional care. I particularly appreciate the patience and friendliness with which she took pains to consider special wishes on my part concerning certain expressions. The pleasure of working together was surpassed only by the pleasure of kayaking together on the Baltic Sea.

1. Introduction: The Basic Problem and Possible Solutions

Almost inaudibly, but yet nonetheless real, a life-destructing process is currently taking place that is unprecedented in the history of humanity. A large part of the biological diversity of our planet is on the brink of extinction. According to estimates of the biologist Wilson (1992, 280), species are dying worldwide at a rate of about three per hour, or more than seventy per day, and 27,000 per year, each a unique specimen of life that has gradually come to be over hundreds of thousands of years. Extrapolating from present trends, we can expect an even greater increase in the loss of species (Figure 1). Pimm and Raven (2000, 844) estimate that the extinction rate in the middle of this century may be about 50,000 per million species and decade. If we assume that approximately one species per year disappeared before the coming of humankind (Markl 1989, 31), this translates to a rate of species extinction that is more than one thousandfold greater than the natural one.

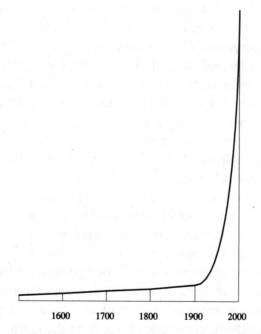

Figure 1. The growing rate of species extinction. The diagram shows the estimated annual rate of extinction based on current investigations of habitats and potential threats to their existence (according to Durrell 1986, 29).

1

Just how many species have already disappeared and how many will be lost in the future if the current global trends continue cannot be determined exactly. We still do not know exactly how many living species exist on earth (Wilson 1985, 700; May 1988, 1448). So far about 1.7 million animal and plant species have been recognized and described scientifically, but extrapolations about the total number of species vary within the enormous range of 2 to 50 million (in some cases even 100 million) depending upon estimates of the yet unknown number of invertebrate organisms in tropical rain forests (May 1988, 1441; Adis 1990, 115). Most biologists consider an estimate of between 5 and 15 million to be realistic, but concede that it will not be possible to verify this number in the near future (Stork 1993, 218, 228).[1]

Even though we are presently unable to quote figures about biological diversity and the degree to which it is endangered with absolute certainty, there seems to be considerable agreement among biologists and paleontologists that if current trends continue, the loss of species we are now experiencing could attain the dimensions of the five greatest instances of mass species extinction that life on earth has had to sustain in its 3.5-billion-year history. From the time that the first multicellular organisms began to develop around 670 million years ago, the number of species increased continually, and all in all the process by which new species arise (*speciation*) predominated over the process of so-called background extinction. Nevertheless, this process of constantly increasing biological diversity was interrupted five times by dramatic instances of mass extinction caused by meteorites hitting the earth and/or changes in climate (Benton 1986; Jablonski 1991; Eldredge 1991). Each time the "evolutionary clock" was reset, so to speak (Eldredge 1991, 216). The worst of all such resetting events occurred 245 million years ago at the end of the Permian Period when possibly 96 percent of all the species that existed at that time became extinct, and life on earth just barely managed to escape total destruction (Erwin 1989, 225). Although the causes of this case of mass extinction, the greatest in earth history, are still not yet well understood, there seems to be little doubt about the explanatory model proposed for the fifth and so far last instance of mass species death 65 million years ago. A meteorite with a diameter of 10 kilometers struck the earth on the north coast of what is now called the Yucatán Peninsula of Mexico and the fiery explosion that followed brought forth catastrophic atmospheric destruction and worldwide climate changes (Alvarez et al. 1980; Keller 1992, 108). Approximately half of all the existing species, including the dinosaurs, died as a result of the complex geochemical and biological chain reactions that ensued in the 50,000 years that followed (Hsü et al. 1982, 255).

Right now an equally serious instance is in the offing that might well develop into the sixth great case of mass extinction. Within the brief period of barely one hundred years, an extremely short time span from a geological standpoint, one-fourth to one-half of all biological species are once again on the verge of destruction (Roberts 1988, 1759; Smith et al. 1993a, 375; Pimm 2001, 231). This time the main cause is a singular one in the history of the planet, an individual species, Homo sapiens. Since the first small population of humans arose in Africa somewhat more than 2 million years ago, this very successful species has spread out over all the continents of the planet. Today it utilizes about 40 percent of the land surface and has depleted the land vegetation by one-third (see Hannah et al. 1994, 248; Vitousek et al. 1997, 495). Since 1850, the human population has increased from one billion to more than six billion. Both the high population number and the numerous achievements of science and technology have resulted in humans gaining "power beyond precedent to influence natural environmental systems" (Ehrlich et al. 1973, 4)—something that no doubt is occurring increasingly to the disadvantage of other species.

The various ways in which the activities of humans have directly or indirectly caused other species to be endangered or become extinct are so manifold and complex that it would be beyond the scope of this book to describe them in detail. For this purpose the reader is referred to many scientific publications.[2] These indicate that in general the death of our planet's species can be attributed to eight different complex causes: (1) direct destruction, (2) overexploitation of stocks, (3) introduction of exotic species, (4) the burden of chemical pollution, (5) intensive agriculture, (6) habitat loss (especially in the course of destroying tropical rain forests), (7) mass tourism, and (8) the greenhouse effect. In this connection, Wehnert (1988, 75) has shown that the significance of these various individual causes has shifted in the last one hundred years. While in the past direct destruction (by hunting and trapping) and the introduction of exotic species were the primary causes of species death, today habitat loss and intensive agriculture are mainly responsible. Due to these factors, in Germany one-third of all the ferns and flowering plants and one-half of the approximately five hundred indigenous species of vertebrates have now either been lost or are on the verge of becoming extinct (Markl 1989, 32). However, in the future another factor may become a major one in endangering species: the green-house effect. If we do indeed experience a 3-degree Celsius increase in global temperature, as many predict, then the climate zones will shift so rapidly that many plant and animal species will not be able to adapt and will subsequently die (McKibbben 1990, 58). As the paleontologist Eldredge (1991, 218) points out, it was mostly changes in the size and location of habitats that led to extinction in the past. But the rate at which human-induced changes occur surpasses anything known in the history of the planet. "We seem to be able to effect more environmental change per unit of time than any other factor ever proposed as a cause for serious bouts of extinction, with the sole exception of the most catastrophic of the bolide impact scenarios" (Eldredge 1991, 274). Many species are no longer able to keep up with these changes.

The main question addressed in this book is *why we should even bother about all of this*? What is there to lament about the current loss of species and reduction in biodiversity, if extinction processes and species replacement are basic facets of nature?

In order to respond to this question more accurately, it is necessary to make two things clear. First, this time, contrary to a meteorite striking the earth at the end of the Cretaceous Period, we are not dealing with a fateful natural event but rather with a complex web of human actions for which humans are collectively and (to a certain extent) also individually responsible. The basic source of these actions consists of human desires, motives, convictions, attitudes, and worldviews, which could be subjected to both factual and ethical criticism. Thus an ethical dimension is present from the very beginning. Second, it should be recalled that relevant convictions, attitudes, and worldviews not only determine the actions that have led to species extinction but also affect other aspects of the way we deal with nature and other humans. Thus, it is not surprising that the loss of species we are currently experiencing is not an isolated phenomenon but rather one of many symptoms of a more comprehensive context in which life itself is being threatened, commonly referred to as our *ecological crisis*. This term is usually employed to summarize all the ecological consequences of human activities that have been regarded since the 1970s as "endangering the foundations of life" (whereby, of course, the exact meaning of the latter expression is left open). Since many symptoms of the ecological crisis are well-known by now and have been described in detail in publications (Global 2000 1981, E. U. von Weizsäcker 1992), they do not have to be listed separately here. Catchwords such as the greenhouse effect, the ozone hole, population explosion, soil erosion, forest depletion, increased waste, DDT in human milk (and even in penguins in the Antarctic) should suffice to remind us how comprehensive and complex the threat to life through human activities has become in the past years. No wonder species extinction is often thought to be one of the most disturbing of all these symptoms, in addition to the greenhouse effect. It is *irreversible*, and it indicates that if destruction of nature continues, a point will eventually be reached at which even our own species may cease to exist or at least may forfeit all hopes of leading a *good* life.

In light of this general perspective, people seldom openly question the thought that our scientifically and technologically oriented civilization is in the middle of a major crisis, but there is a great deal of controversy about the measures that must be taken to master it. While considering the various different solutions that have been proposed as possible ways out of the crisis, it is important to differentiate between *specific* solutions for particular aspects and more *fundamental* solutions that would require a change in philosophical outlook or ethical assessment of the entire problem. In this book, I will primarily be dealing with the latter. That is, my main goal is to address fundamental questions that arise in connection with considerations of the ecological crisis and the loss of our planet's species. It would be important and interesting as well to broaden the discussion to include things that conservation or even politics could do to solve the problem, but that would be well beyond the scope of this book. Furthermore, I feel that it is fundamental normative clarification that is particularly lacking in current discussions among proponents of nature conservation or species protection, much more so than detailed factual analysis. I have gathered this from reading publications on nature conservation[3] as well as through numerous discussions with people involved in conservation in which concern is repeatedly expressed about being unable to properly justify and ethically assess actions and about the lack of a convincing ethical foundation for them. Many conservationists and environmentalists feel that academic philosophy has provided little support in these matters (see Hartkopf and Bohne 1983, 64). And indeed, if you look for systematic philosophical treatises on how to justify nature and species protection, you will not find many except for a few notable American studies (e.g., Norton 1986, 1987; Taylor 1986; Rolston 1985, 1988) and a couple of German ones (e.g., von der Pfordten 1996). However, assuming that the appropriateness of practical solutions always depends upon the appropriateness of the normative premises and the view of nature upon which they are based, the significance of this deficit is evident. This is probably the reason why two estimates concerning the ecological crisis and species loss are usually accepted without further questioning and believed to provide sufficient grounds for conserving nature and species, a view that I question. These are (1) ecological scientism, and (2) the position of ethical anthropocentrism, which can be outlined as follows:

Characteristic of the position of *ecological scientism* is the idea that the ecological crisis is not an ethical crisis but merely a matter of facts. It is accompanied by the conviction that the problems associated with the ecological crisis can be resolved simply by *scientific and technological means*. Symptoms such as species extinction are regretted, but only because they reflect ignorance and shortsightedness in our dealings with nature. Advocates of this position believe that in order to avoid such "accidents" and other ecological problems in the future and secure proper functioning of spaceship earth, we must continue on the course of scientific and technological mastery of nature we have pursued so far without wavering. Technological control of nature, which has been initiated but is still unfinished, must be perfected by a different kind of planning, one that takes ecological consequences into account that have been ignored so far. In particular, ecological research must be promoted in order to be able to better evaluate ecological risks and sound out the limits of what is technologically possible. If changing the course of progress proves to be necessary, this can best be achieved within the context of the existing economic and industrial system and the rationality upon which it is based, as the proponents of this position maintain. As exemplified by a brochure from the German Federal Department of the Interior from the year 1985, "environmental protection" is first and foremost a factual matter that requires no new "ideology" but "rational action, drive and persistence" instead (Bundesminister des Innern 1985, 7).

Contrary to the position of ecological scientism, the position of *ethical anthropocentrism*[4] does not consider the ecological crisis and species extinction to be merely a factual matter but an *ethical* problem as well. However, the moral problem of species loss is thought to derive solely from the fact that reducing biodiversity might cause people living today as well as future generations to lose useful resources and perhaps even suffer irreversible damage. Advocates of this position either refuse to consider or explicitly reject the idea of direct ethical responsibility for nature and protecting species for their own sake. This strictly anthropocentric view of the ecological crisis is illustrated by a standard scientific text on species and biotope protection, which contains the following introductory passage: "Ethics provides no additional criteria in support of arguments for protecting nature beyond those of the usefulness of nature, the quantitative significance of species and ecosystems for nature's economy, the beauty of nature or its importance for future generations" (Kaule 1986, 16). If we eliminate species and restrict their distribution significantly, then according to Kaule this way of dealing with nature is immoral only to the extent that "we pass on the earth to future generations in a reduced state and thus limit their possibilities. In so doing, we thrive on the capital instead of the interest." Mohr (1987, 170) believes that with the exception of "a call for establishing a contract between generations" there is "no rigorous reason" why biodiversity should be maintained. "Species protection remains a postulate for which no further justification can be provided."

I consider both viewpoints to be incorrect. Therefore, I intend to subject them to detailed analysis and criticism in the two major sections of my book and then present an opposing standpoint based on considerations of theory of science and natural philosophy as well as on a broader understanding of morality. More precisely, I will proceed as follows:

In Part A, the section of the book that deals with scientific theory, I will first examine the claims of ecological scientism that the ecological crisis is merely a matter of facts and that resolution can be expected from the efforts of science and technology. Since the science of ecology is often assigned a key position in the context of such a view, this section will focus on whether or not and to what extent ecology is really capable of fulfilling the many hopes that have been attached to it. If you take a closer look at the expectations connected with ecology, you will find that they actually can be divided into two different groups depending upon their source and aims. One such group, consisting primarily of representatives of political, economic, and industrial management, regards ecology as an ultimately reliable database for making predictions to ensure more environmentally amenable and thus also more economically efficient management of habitats and resources. The other, which includes mostly representatives of the environmental movement and active conservationists, thinks that ecology is capable of providing norms for the "right way" to deal with nature. However, in both cases I believe that too much is expected of ecology.

I shall attempt to demonstrate this in Part A, Section I, with respect to *technical optimism*, a position which assumes that ecological processes and relationships in nature can be understood well enough to be controlled completely (Chapters 2 and 3). Both epistemological arguments and specific aspects of ecology will be discussed that contradict this kind of optimism, including such topics as complexity, nonlinearity, and problems of generalization, boundaries, measurement distortion, and quantification (Chapters 4 and 5). Analysis of these aspects leads to fundamental considerations of scientific theory centered around two questions, one concerning the possibility of "alternative" science (Chapter 6) and the other addressing the relationship between science and worldviews (Chapter 7).

In Part A, Section II, I will subsequently argue that for logical and factual reasons it is just as mistaken to attempt to deduce ethical principles directly from ecological knowledge (Chapters 8–10). On the basis of ecological slogans (ecological balance, stability, biodiversity, cycles, ecological health and nature's economy), I hope to demonstrate that such instances of naturalistic fallacy are widespread and have found their way unnoticed into ecological discussions (Chapters 11 and 12). As a result, a kind of "ecologism" is arisen, which will be criticized in Chapter 13.

What often fails to be addressed is the opposite of the naturalistic fallacy, namely the so-called *normativistic fallacy* (Chapter 14). A normativistic fallacy consists of the erroneous assumption that one can derive specific obligations purely on the basis of normative considerations. In distinct opposition to the latter position, in Part A, Section III, I will outline two *positive* contributions ecology can make toward solving the ecological crisis. First, by acknowledging its limits ecology can help us to develop a more cautious and modest attitude in our dealings with nature (Chapters 15 and 16). In addition, it can provide knowledge that enables us to formulate questions to be directed at ethics (Chapter 17).

Subsequent to Part A, the part of the book that deals with theory of science in which an attempt is made to "deconstruct" ecological metaphysics as it is often found in publications on environmental ethics, in Part B, which addresses ethics, the problem of our relationship to nature comes sharply into focus. It will now become apparent that many "ecological problems" are in essence ethical ones. Therefore, in order to resolve them, it is necessary to resort to considerations of environmental ethics. Since this still relatively young field of ethics is quite heterogeneous, in Chapter 18 I will describe four basic types that have arisen in the course of the last three or four decades. Afterward, the history and current state of the discussion will be briefly reviewed (Chapter 19). I intend to interject my own thoughts into this discussion by confronting the positions outlined in Chapter 18 with the main topic of my book, the phenomenon of species extinction. The attempt to find the "right" ethical response to this problem will be conducted at two different levels. The first is a *pragmatic* approach (Part B Section I), involving examination of the scope and motivational power of the different schools of ethical thought. The second consists of a *theoretical* approach (Part B Section II) aimed at analyzing the conclusiveness of the theoretical justification these positions have to offer.

Since protecting species seems to be a postulate firmly rooted in intuition and generally quite well accepted, regardless of the way people justify it (Chapter 20), in Part B, Section I, I shall first of all address the practical question of just which type of environmental ethics is best suited to achieve this end. In the context of a detailed analysis of the most common economic, ecological, and aesthetic reasons offered for protecting species (Chapters 21 and 22), the so-called *convergence hypothesis* will be shown to be untenable. According to this hypothesis, anthropocentric and non-anthropocentric ethical positions are in effect practically indistinguishable. Contrary to this hypothesis, I propose that only holistic ethics is capable of guaranteeing general species protection (i.e., a kind of protection that in principle includes *all* species) in a scientifically convincing and psychologically coherent manner (Chapters 23 and 24).

In the following chapters in Part B, Section II, I shall present theoretical justification for a holistic and pluralistic concept of morality and examine three fundamental objections to this more comprehensive concept of human responsibility that are commonly debated in the literature (Chapter 25). After reviewing a few general considerations regarding the possibilities of justification itself (Chapter 26), I will turn to a discussion of different worldviews and views of humanity, in the course of which I will argue in favor of a conceptual foundation that accounts for both the ecological and evolutionary contingencies of humanity as well as the unique position humans hold in nature (Chapter 27). The heart of my theoretical arguments, my *formal* justification scheme, will be presented in Chapter 28, followed by a discussion of objections to certain aspects of its content (Chapter 29). Then in Chapter 30, following the sections on justification, I will finally take up some of the theoretical and practical consequences of a holistic concept of environmental ethics for species protection. This leads to a major quandary of any kind of ethics, namely how to balance conflicting duties and how to deal with moral dilemmas (Chapter 31). The treatise will conclude with a brief summary and outlook (Chapter 32), in which a few unanswered questions will be outlined.

In the course of weighing different positions discussed in Parts A and B the following basic proposal of this book will be developed: The ecological crisis is not simply a problem of facts and science and therefore cannot be mastered merely by scientific and technological means. While ecological knowledge is undoubtedly indispensable for reaching a solution, the science of ecology is neither in a position to justify "environmentally adequate" behavior in and of itself, nor is it capable of providing instructions for how to manage the biosphere in the future. Hopes of achieving perfect control of nature through ecology are futile. Furthermore, ways of dealing with nature based on criteria that take only human desires and wishes into consideration are not only unreasonable (in view of all the things involved) but also ethically flawed. A concept of environmental ethics that seriously takes into account both our current knowledge of the position of humans in the cosmos and the universal character of morality has no other recourse than to abandon an anthropocentric perspective and grant all nature surrounding us intrinsic value. In the context of such a concept of ethics the extinction of innumerable species of animals and plants is not only an injustice toward future generations of humans, it is morally reprehensible in and of itself.

The American philosopher Rolston (1982, 150) contends that when future historians look back on this century, they will note an enormous breadth of knowledge coupled with extreme narrowness in value judgment. "Never have humans known so much about, and valued so little in, the great chain of being." Therefore, Rolston says, "the ecological crisis is really not surprising." With this book I would like to make a contribution toward correcting the discrepancy Rolston criticizes. By deconstructing both ecological scientism and ethical anthropocentrism I hope to open readers' eyes to the *entire ethical dimension* of the death of our planet's species, which in the end means presenting a perspective of *the intrinsic value of nature*.

A. Hopes for an "Ecological Solution"

I. Ecology as the Epitome of Controlling Nature?

2. Technical Optimism

In view of such unequivocal symptoms as the disappearance of species, the ozone hole, or the greenhouse effect, it is hardly possible to deny the existence of an ecological crisis. But an attitude of skepticism concerning its scope and fundamental significance is still widespread, particularly among advocates of *technical optimism*. To put it more succinctly, people with this attitude believe that for any problem, including ecological ones, a technical solution can be found sooner or later (Rifkin 1981). When new problems crop up, this simply reflects a temporary gap in our knowledge about the world and our ability to control nature, one that will eventually be closed by science and technology. The technical optimist is convinced that increasingly efficient expertise will not only help us to analyze the causes and effects of problems that already exist and find solutions for them but that it will also permit us to prevent problems by prediction and risk assessment. The expectations often attached to modern medicine and its role in health care are applied to the science of ecology[5] and environmental problems in an analogous manner and can be summed up in the terms diagnosis, therapy, and prevention.

Although I shall discuss later on whether or not it is legitimate to draw an analogy between medicine and ecology on the basis of these three terms (Chapter 12), a first impression suggests that at least with respect to the diagnosis of environmental threats the relatively young discipline of ecology has been very successful (Heinrich and Hergt 1990). It is not unreasonable to maintain that as far as basic tenets are concerned, our understanding of many of our most significant global ecological problems such as the greenhouse effect is so good that we could readily move on to therapy. The reason that this doesn't happen at all or only inadequately is usually of a political or economical nature, even though the explanation often advanced involves insufficient knowledge or intolerable differences of opinion among experts. Exacting demands on scientific proof as a prerequisite for taking action are often coupled with an antiquated but still widespread understanding of empirical science as an undertaking that is capable of providing absolutely certain knowledge and infallible evidence. However, as various studies in epistemology and the theory of science have shown, demands of this kind on the certainty of scientific knowledge are not justified.[6] In reality, all theories based on empirical evidence, and in the terms of formal logics, all synthetic statements about the world, are merely hypothetical or preliminary. The inevitable question that then arises is just how much diagnostic certainty it takes to declare the scientific understanding of an ecological problem to be *sufficient* to warrant initiating a therapy that may be economically painful. As we shall see later on regarding the possible ecological consequences of massive species extinction, ethical aspects are of primary importance in addition to risk assessment and epistemological considerations (Chapters 15 and 22.b).

When the issue is the *therapy* of environmental problems rather than their diagnosis, the science of ecology is not really in a position to take on the very important part often assigned to it. A decision about which therapy should be selected among all those available cannot be reached solely on the basis of ecological expertise. In principle, any number of therapy types are imaginable depending upon the kind and degree of control of nature involved. These range between two different extremes. At one end of the scale there is the therapy of *reversal*. This means that once the consequences of intervening in nature have been found to be detrimental and the most important causal relationships have been analyzed, then the causes are stopped or at least reduced. The other end of the scale is defined by the therapy of *technological correction*. In this case, only certain undesirable consequences are eliminated, preferably without altering the cause itself. A good illustration of these positions is the case of forest damage due to acid rain. Since emissions from motor vehicles, households, and industry have been identified as the most likely causes of this problem, a reversal therapy would require sufficiently reducing the use of fossil fuel. An example of technological correction, on the other hand, is the catalyzer, with which the emissions thought to be the most damaging ones are reduced but the use of fossil fuel remains unaltered. One step further on the scale toward technological correction would be to apply lime in order to neutralize soil that has been badly damaged by acid rain. The most severe symptoms of acid rain can be temporarily alleviated by this measure without doing anything at all about the emission problem. The most extreme example of technological correction finally is the attempt to "reconstruct" trees with the help of gene technology in order to make them less sensitive to pollution. This constitutes the bizarre peak of technical optimism, which aims at arbitrarily adjusting nature to the interests of human beings.

Despite apparently unwavering belief in the possibilities of technological correction, even a technical optimist will agree that *prevention* is better than therapy. Therefore, the greatest hopes attached to the science of ecology have to do with its supposed ability to predict environmental problems and with possible preventive strategies. If we succeed in sufficiently analyzing and quantifying all the ecological rules and relationships that exist in nature, then we will be able to prevent serious damage to the environment from the very start—such are the expectations of technical optimists. The idea is that with the help of theoretical models and computer simulations all interventions in nature and their consequences will be able to be calculated ahead of time, and thus specific predictions can be incorporated into technological planning. Along the same line of thought, politicians, legal representatives, and administrative bodies responsible for planning and testing the ecological compatibility of measures are particularly interested in establishing well-defined threshold levels for determining exactly when serious damage to humans and the environment can be excluded with reasonable certainty (see Peine 1990). According to these groups, it is ecology's job to analyze ecosystems from A to Z in order to determine their stress capacity in the event of human intervention.[7]

But just how realistic are these in part rather broad demands on the science of ecology and the attendant hopes that we will eventually be able to control all environmental problems on the basis of science? How well do the premises of technical optimism hold up under more exact examination?

3. Supposed and Temporary Limits

When a technical optimist is confronted with the possible limits of scientific knowledge and its predictive ability, he often refers to cases of mistaken judgment in the past that clearly show that what was once thought to be an insurmountable barrier sooner or later was indeed mastered. For example, Comte (1798–1857) was convinced that we would never be able to figure out the chemical composition of the stars, and in keeping with his positivistic philosophy, he therefore found it worthless to even think about the matter. But as early as 1863, Higgins succeeded in solving the problem with the help of spectral analysis and discovered that the same elements exist on stars as we have on earth. Similar examples for the temporary nature of supposedly absolute limits can be found throughout the history of science (Vollmer 1989, 387).

In ecology, which must deal with highly complex systems, it was the enormous amount of data and the problems of mathematical computation that first presented a serious hurdle and stymied quantitative approaches for a long time. But since the discovery of computer technology and the development of increasingly fast processors and greater storage capacities, this barrier no longer seems to be a fundamental one. Thus with the help of mathematical models and information technology such complex systemic relationships as the effects of deep sea fishing on fish populations have become accessible to computational analysis (May et al. 1978).

After so many limits have toppled in the history of science and technology, it is no wonder that an unlimited optimism has arisen that regards almost every problem as capable of being grasped by science and mastered by technology. Might not the causal relationships involved in ecological problems also be one day understood well enough to be able to include nature in technological developments and expose what we currently imagine to be "limits of growth" (Meadows et al. 1972) as only temporary limits?

At this point it is important to emphasize that optimism of this kind is not completely refutable by rational means. From the standpoint of an opponent of such optimism, be it that of a hypothetical realist, a skeptic, or an epistemological pessimist, to claim to be able to make any absolute predictions about the *impossibility* of future discoveries and inventions would require contradicting one's own point of view. But the reverse, of course, is also true. The conclusions that technical optimists tend to draw from various instances of false predictions in the history of science are just as untenable. The empirically based fact that *many* limits once supposed to be insurmountable proved to be manageable after all does not mean that there are no real limits *at all*. As in any area outside of mathematics, what is at stake in discussions between optimists and skeptics is not absolutely certain evidence but rather plausibilities and probabilities. In the context of the ecological crisis, however, it is really not important whether something thought to be an absolute limit might someday prove to have been only a temporary one. Since we are pressed for time, it would be irresponsible to place all hopes on such a distant possibility instead of accepting supposedly fundamental limits as being *valid for the time being*.

It would be just as unreasonable to operate on premises that are incompatible with what are currently considered valid (that is, not yet falsified) laws of nature. Of course it is theoretically possible that these laws may be refuted or that they at least may have to be modified some day in the future. But when we are dealing with practical matters, the practical reality of the validity of scientific theory provides a more reliable argumentative basis than the theoretical possibility of future refutation. In this sense, it would, for example, be mistaken to support a position that ignores the principle of entropy, as most economic theories practically do, in hopes that this fundamental law of thermodynamics may someday prove to be wrong.[8] A basic rule of thumb such as that discussed above, which gives priority to well-established empirical knowledge over vague optimism, appears all the more convincing when a practical decision must be reached about something that involves large risks or ones that are difficult to estimate.

In the following section I will show that there are not only temporary and practical limits to the science of ecology but also fundamental ones that make many expectations of technical optimists appear to be wishful thinking. The claim that these are truly *fundamental* limits is based on epistemological arguments and theory of science on the one hand and on empirical evidence generated by science itself in recent years on the other. In the sense of the qualified view of empirical knowledge presented above, this evidence appears to provide a sufficiently certain basis for further discussion.[9]

4. Fundamental Limits of Ecology

4.a. Complexity

One of the defining properties of life is its tremendous diversity and complexity. "Regardless of how one defines the concept of complexity, at any rate all living systems are considerably more complex than all inanimate ones (although they may not be more complex than all man-made objects)" (Vollmer 1990, 3). In order to appreciate the significance of complexity for our ability to describe and govern living systems, it is necessary to briefly define the term *system*. But we then run into the difficulty that in spite of ongoing attempts to embed this concept in general system theory, the beginnings of which date back to the 1960s (see Bertalanffy 1973), many different definitions are still used simultaneously in publications, "the special and often limited applicability of which is confusing." However, what they have in common is that "they all are more or less based on a 'naïve' concept of quantity" (Kornwachs and von Lucadou 1984, 111). Thus Wuketits (1981, 87, 88) favors a broad concept of a system as "a complex of reciprocally related elements" characterized by "fundamental *invariance* with respect to the fluctuations of the individual elements." Another characteristic of systems he describes, which is relevant for the arguments presented in this book, is the *hierarchical structure* of systems. "Every system is made up of subordinate subsystems and is itself integrated in a larger super system. A complex is the same as a system when it can be separated into smaller complexes which . . . interact with one another." In a system, interactions exist not only between elements on the same level but also between elements on one level with those on higher or lower systemic levels (see Weiss 1969).

If we assume the perspective of system theory and regard the cosmos as a hierarchically organized and layered structure embracing many different levels from elementary particles to clusters of galaxies, then the working field of ecology includes levels extending from organisms to populations, communities, ecosystems, complexes of ecosystems, the system of human society, and its environment and finally the ecosphere (Haber 1984, 193). If we consider higher systemic levels, it becomes quite obvious that these systemic terms do not apply to "objects" with clearly defined outer boundaries. Instead, systems consist of "descriptions of sectors of reality" (Kornwachs and von Lucadou 1984, 112) and therefore vary depending upon the particular interests of the author who describes the system, his scientific research topic, and the means of description available. Quine (1977, 26–68) speaks of "ontological relativity" in connection with this kind of relationship between description and frame of reference.

In the long run, just how many layers of subsystems can be grasped all at once within such a frame of reference depends upon the efficiency of the human mind and the possibilities of electronic data processing. Since cumulative complexity from level to level poses definite limits, from a certain level downward the properties of the subordinate systems must be regarded as axiomatic givens, that is, as "black boxes" (Ott 1985, 50). In doing this, one both rightfully and unrightfully (see Chapter 4.b) assumes that because of the systemic properties that emerge at higher systemic levels, beyond a certain point the properties of lower levels are no longer important for explaining those of higher levels.

When talking about "cumulative complexity," it must be conceded that we still have no generally accepted *quantitative* measure of complexity (see Pippenger 1978). All that mathematics has to offer is the proposal that we define complexity on the basis of the minimum number of bits required to describe a system (Chaitin 1975, 49). But a minimal description of this kind is not necessarily unambiguous. Thus the term complexity is usually employed as a comparative term, or in other words "we can normally agree quite readily on which of two systems is the more complicated one" (Vollmer 1986b, 166). According to this intuitive understanding of complexity there is, for example, hardly any question that the central nervous system of humans is more complex than an amoeba. What is decisive about the complexity of these systems is not so much the number of "building blocks" involved (i.e., atoms, biomolecules, organelles, and cells) but rather the number and kind of *relationships* between them, that is, the degree of networking.

It is completely beyond the power of our imagination to conceive of how the number of possible relationships increases with the number of building blocks. While only two different combinations are conceivable with two building blocks (either one or none) and 8 with three, 64 with four and 1024 with five, the number of possible structural relationships among only 24 building blocks is as great as $1,2 \times 10^{83}$! This "mega-astronomical" number is more than a thousandfold greater than the total number of all the atoms of the visible universe (Kafka 1989, 23, 24).[10] If the number of possible relationships among twenty-four building blocks is greater than the number of all the particles in the world, just imagine how great the number of possible interactions within one of the most comprehensive kinds of systems might be, an ecosystem with all its manifold communities, hundreds of plant and animal species, thousands of populations, billions of higher organisms, and even more microorganisms, not to mention the abiotic factors (climate, weather, soil properties, etc.). Even if each and every one of all the possible combinations of interactions may not exist in reality, it is obvious that it is in principle impossible to grasp more than a fraction of all these relational structures, let alone computationally grasp an ecosystem. No fashion of progress in electronic data processing can alter this very fundamental fact. In the end the results of any kind of computation depend upon the unavoidably spotty database and knowledge of the programmer. Ecologists, who gratefully accept the new possibilities for quantitative analysis provided by the computer, still have no illusions about its limitations. "We will never achieve perfect ecosystem analysis capable of revealing all the functions that exist" (Kreeb 1979, 127).

Of course a possible response to this view is that there are many problems that can be solved without *perfect* ecosystem analysis, and that for practical purposes we can easily get along with appropriate and more simple measures. For example, a higher order system made up of two complex subsystems, each of which comprises one thousand elements, can be regarded as a very simple system consisting of only two elements. This objection draws attention to a very common procedure, one that is fundamental to science, namely, the process of reducing complexity by deliberately excluding higher or lower systemic levels from causal analysis.[11] For example, in his treatise titled *Discours de la Méthode* Descartes ([1637] 1956, 15) proposes that to deal with a problem that is too complicated to be solved in one fell swoop, one should "divide each of the difficulties . . . encountered into as many parts as possible, and as might be required for an easier solution." Ott (1985, 50) compares this *analytical* procedure of a scientist with the use of "a magnifying glass that is moved up and down a ruler in order to magnify a particular area for measurement." There is no question that impressive results can be obtained in this manner with certain ecological problems involving only a few parameters. Any ecology textbook contains numerous examples of such procedures. However, when it comes to more sophisticated problems or, for that matter, comprehensive ecological "management" such as a technical optimist might envision, skepticism is due. The more systemic levels and parameters of the total system we exclude from consideration, the less certain we can be about really sufficiently understanding the entire system.

Errors can be expected with both "external" and "internal" reduction. Thus with *internal* reductionism (by which higher systemic levels are excluded from consideration) the implicit assumption is that once all individual problems have been solved at lower levels of the system, these solutions just have to be put together like pieces of a mosaic, and eventually they will permit us to understand the greater, more complex system in its entirety. In the meantime, however, with respect to highly complex systems "this assumption, or rather hope" has proven to be "fundamentally wrong" (Cramer 1986, 1153). When more complex systems are involved, the well-known saying holds true that a whole is more than the sum of its parts. Cramer (1979) calls systems that cannot be successfully broken down into individual processes and that therefore are not completely accessible to analysis *fundamentally complex* systems.

A different kind of error may ensue from the process of *external* reductionism, by which lower systemic levels are excluded from scrutiny and only the properties of higher systemic levels are investigated, as in the example mentioned above of two subsystems that form a higher order system. In this case the very decisive effects of lower systemic levels on higher ones may be underestimated. In ecosystems, for example, not only do the systemic properties of higher levels, that is, the environment, affect those of lower levels, the organism. The reverse is also true. A few individuals of a species can operate over several systemic levels to exercise a surprising effect on the entire ecosystem. This is exemplified by the spreading of epidemics, the release of predatory mammals on islands or the introduction of exotic new plant species into foreign ecosystems.[12] Cases such as these clearly show how severely the structure and function of ecosystems can be altered by unpredictable external or internal conditions.

A change in the structure of an ecosystem is usually accompanied by a change in the *degree of complexity* of the system. In this respect Kornwachs and von Lucadou (1984, 127) have shown that "under certain circumstances systems that exhibit a variable degree of complexity . . . may be inaccessible to complete description in the sense of determination." However, a system that cannot be completely described can also not be completely controlled. From this observation the authors arrive at the recommendation not to treat ecological systems as "reliable," manageable entities but rather to always reckon with a certain amount of autonomy. Thus the complexity of life processes places *fundamental* limitations on ecological knowledge and management possibilities. Reducing complexity does not cause them to simply disappear.

The fact that completely exaggerated expectations are still attached to ecology in spite of these reservations and that politics continues to demand "total analysis of ecosystems" (Borchardt et al., 1989) may be rooted in human psychology. Thus dealing with the super-exponential character of complexity may simply be beyond the conceptual capacities of humans, which have evolved in and are adapted to other, intermediate dimensions, those of the human "mesocosmos" (see Vollmer 1986c, 161). The problem is aggravated by the fact that linear thinking is enhanced by everyday experience, which in turn is strongly influenced by the use of technical apparatus, and that the kind of linear causality found in the technical world is unwittingly applied to the domain of life processes as well. As Dörner (1993, 137) has shown, everyday experience is often "a bad teacher for dealing with complex, dynamic systems." Although it is usually appropriate in everyday life to treat similar things in a similar manner and to predict the near future in a "linear" manner, this often proves to be wrong for activities involving complex ecological reality. This is because in recent years it has become evident that many important life processes that can be quantified have proven to be *nonlinear* ones (Eilenberger 1989).

4.b. Nonlinearity

A linear relationship between cause and effect is characterized as follows: When the factors to be measured are wisely chosen, a twofold increase in the causal factor will lead to an equally large increase in effect. Linear cause-and-effect relationships can be described by differential equations, and these can always be solved, even when coupled with any number of additional components. Famous examples of such linear relationships are the kinetic equations of classical mechanics and Kepler's laws of planetary motion, which set off the scientific revolution of modern history. Since their discovery a whole variety of phenomena have been described by linear differential equations, from the flight of a cannonball to the conduction of heat during the oxidation of coal, but in particular the functions of all sorts of machines. In these cases small alterations of causal factors lead to small effects and large effects are induced by equally large causes or by the sum of many small alterations.

However, even before the nineteenth century it was known that there are some physical systems (e.g., unstable processes such as explosions, sudden tears in material or high wind velocities) that cannot be reduced to a linear relationship, regardless of how cleverly they are demonstrated graphically, and that require nonlinear equations for adequate mathematical representation. At that time the mathematical techniques for dealing with such equations were not yet available, and all the way into the 1960s it was assumed that the *qualitative* behavior of nonlinear systems does not differ significantly from that of linear ones. Only after fast processing computers came into existence, with which the dynamics of nonlinear systems could be monitored numerically, did it become evident that there is a fundamental difference between linear and nonlinear systems, namely the phenomenon of "initial value sensitivity" (Eilenberger 1989, 98).

Contrary to a linear relationship between cause and effect, a *nonlinear* one is characterized by the fact that the *tiniest* alteration in initial conditions can result in an unusually *large* effect on the final state of the system. The reason for such a disproportionate relationship may be a series of feedback loops such as those typical for complex physical systems and highly integrated biological ones. Mathematically speaking such systems are characterized by terms that are repeatedly multiplied with one another and thus increase so rapidly that beyond a certain point the descriptive equation can exhibit completely novel behavior (Briggs and Peat 1989, 23, 24).

This leads to very fundamental conclusions regarding the *calculability* of nonlinear systems. Since tiny little inaccuracies in initial values can be augmented in an avalanche fashion, very similar initial states can lead to completely different final ones. Since initial values can in principle not be determined with complete accuracy, a precise prediction about the final state is basically impossible. Systems of this kind, which are unpredictable in the sense of traditional physics, are sometimes called "chaotic systems." In this case the term "chaotic" means simply unpredictable, because even the most chaotic nonlinear systems still completely operate according to deterministic and often also surprisingly simple laws (Bachmann 1990).

The discovery that for fundamental reasons certain parts of nature are neither calculable nor predictable has had considerable repercussions for the scientific worldview that so far have barely been perceived in the general public. In the 1920s, of course, quantum mechanics revealed fundamental limits to predictability by showing that it is basically impossible to accurately measure both the position and the momentum of an elementary particle at one and the same time. But limits to scientific knowledge of this kind seemed only to be relevant in subatomic domains. On the macroscopic level the perspective that continued to prevail was the mechanistic worldview of Pierre Simon de Laplace, a mathematician who in 1776 pursued determinist thought to the ultimate by postulating that knowledge of the position and velocity of all existing particles of matter would allow us to determine the future and the past of the universe with absolute accuracy. According to this view complexity is not a fundamental problem but only a practical one that we will one day be able to reduce to its simple, nicely ordered foundations with the help of continually improved scientific methodology. The discovery of the nonlinearity of complex systems has destroyed this hope, and the mechanistic and reductionistic worldview has finally fallen apart.

This insight is particularly valid for the domain of life processes. As we are more and more beginning to see, many biological systems operate according to nonlinear rules and regulations and now and then behave in an unstable, inhomogeneous, and irregular manner.[13] Even ecological developments that can be described with simple mathematical equations can end up in "deterministic chaos" and thus become unpredictable (Bachmann 1990, 88). In view of these considerations, constant and stable states such as "popular ecology" sometimes tends to associate with a vision of nature untouched by humans must be regarded as an exception (Bachmann 1990, 92; Worster 1993).

At the moment, of course, we cannot yet tell how great the contribution of the still-young theory of chaos to our understanding of ecosystems will be. Some biologists regard the ecological significance of this research field with a great deal of skepticism. They feel that the potential danger inherent in this formal approach is that "phenomena caused by very different things may be subsumed . . . under one and the same mathematical description" and that subsequently the significance of individual biological mechanisms may be overlooked (Remmert, cited in Bachmann 1990, 96). In spite of the fact that the *explanatory* potential of chaos theory is still contested in ecology, the observation that possibly most of the systems it studies operate according to nonlinear laws should be reason enough to reconsider *the way we deal with nature*. Since small causes can have big effects in "nonlinear nature," the consequences of manipulations are most likely not sufficiently predictable. With respect to very complex ecosystems in particular, the widespread idea that they will eventually be exactly calculable and that interventions will therefore be predictable is untenable. Therefore the potential of technology assessment, which critics of technology often tend to demand, should not be overrated and it should not be propagated as a cure-all for the ecological crisis, as important as it might nonetheless be in view of the constantly increasing risks generated by new technologies. Interventions in nature will always be tainted by the fact that we don't really know exactly what we're doing and that there are therefore very definite limits to our ability to predict the results of our actions.[14] Later on I shall deal in greater depth with the question of the consequences that practical reason might tell us to draw from "knowing about not knowing."

4.c. Boundaries

The next problem that stands in the way of completely grasping ecological systems by means of mathematics has to do with the difficulties encountered when trying to determine unequivocally the systemic boundaries of structural units. By referring metaphorically to relationships between "building blocks" at different systemic levels in my discussion of the phenomenon of complexity (Chapter 4.a), I conveyed the impression that to a certain extent these entities are separate units that can in principle be isolated from superordinate levels of the system and that they can always be clearly defined in time and space.

It becomes obvious that this assumption is really an abstraction when ecologists begin to investigate "parts" of higher-order systems. Biomes (plant formations), for example, are not isolated entities as far as species composition is concerned; nor are they self-sufficient with respect to energy and matter. In a well-defined geographic area, abiotic factors (such as soil and climate) can of course generate landscape structures that severely limit the dispersal of many organisms. But when larger areas are examined, more or less *continuous* changes in species composition can usually be observed. Sometimes the overlap area between two ecological units is so great that the overlap area itself could count as an ecological unit, as in the case of forest tundra, which is found between tundra and taiga landscapes. Still, distinctions of this kind should not trick us into forgetting that the particular boundaries we draw will always be more or less arbitrary (Tischler 1976, 102–104).

The observation of nonlinearity in complex systems shows that the "building block model" is also inadequate at lower order levels of a system (such as the level of individuals and species) since it is based on the idea of clearly defined parts. Because all real systems are open with respect to the universe and because very small fluctuations in the environment of a system can evoke disproportionately large effects on the system itself, it is basically impossible to "de-fine" a subsystem or to isolate a part from the whole. In view of these considerations, the idea that we could seal two building blocks in a box and then observe their interactions apart from all the rest of the universe, an idea that is almost self-evident for scientists, appears to be an illusion. *Exact* control of such supposedly isolated nonlinear interactions is destined to fail since with open systems the "surroundings"—that is, influences from the entire system—will always somehow make their way into the box. Briggs and Peat (1989, 148, 149) call these imponderabilities that result from unpredictable contingent conditions and unavoidable inaccuracies of measurement "missing information." Because complex systems cannot be reduced to clearly defined parts, they are not subject to completely exact analysis.

4.d. *Disturbance and Measurement Distortion*

In addition to the theoretical problem of boundaries there is a practical one that is also of a fundamental nature. Any analysis of a living system necessarily involves intervening in the system to a certain extent, and this intervention can distort the results of the analysis. Usually the degree of disturbance varies depending upon how exact and complete the analysis is intended to be. If, for example, one wanted to exactly reconstruct the food web of a simple, Middle European hedge with the thousand species it encompasses, the obligatory task of assessing all the individuals in the hedge would pose an almost insurmountable methodological problem. Moreover, the original community would have to be torn apart beyond recognition. The vision of an "ecological web" is thus misleading since it suggests that it is possible to examine it piece by piece—counting stitches, refastening knots and studying threads—without injuring the entire web (Dahl 1989a, 67).

Theoretically it would be possible to determine the number of *animals* in an ecosystem quite reliably using the capture and release method, which involves catching as many individuals of a species as possible, marking them and letting them go again. When a large sample of animals is captured in a second round, investigators can estimate the total number of members of a species on the basis of the relative number of marked (recaptured) and unmarked (new) animals without having to capture every single individual. In practice, however, this method presents several problems whose effects on the results of extrapolation can barely be calculated. First of all, not all animals can be marked in a manner that exposes them to no risk whatsoever (e.g., amphibians or insects). Second, not all animals can be recaptured that easily. In the case of vertebrates that are sensitive to disturbances, for example, it is quite probable that they will change their behavior in the course of the investigation or even migrate away. Finally, inaccuracies are inevitable when animals are territorial, occur in clusters or are randomly distributed, occurrences that are the rule rather than the exception in nature.

Therefore, in spite of all efforts, there seems to be no method in sight that can be executed practically in a manner that would permit land animals to be assessed with reasonable certainty. "The number of errors that can occur in this instance is enormous," says Remmert (1984, 224). However, according to this ecologist we have to know the exact number of animals per surface area if we want to make quantitative statements about ecosystems. In addition we also have to know how many progeny these animals produce in the system per unit of time as well as their death rate. But this requirement has not been met anywhere so far. "Even in the case of large animals that are relatively easy to see it is hardly possible to determine the number of adult animals in a limited population" (Remmert 1984, 224).

Attempts to determine the number of individual *plants* in an ecosystem are hardly any better, especially since it is not always clear what exactly an *individual* plant is. Even with respect to humans there is also by no means unanimous agreement about the concept of what constitutes an individual (for example with respect to identity of the self). But with plants we are confronted with the even more serious problem of temporal and spatial boundaries. Phenomena such as new trees growing out of an old stump or reproduction by cuttings and runners indicate that this problem cannot be resolved in a satisfactory manner. Plants seem to occupy a borderline position with respect to individuality. It follows, of course, that caution is in order when dealing with many calculations made by plant ecologists that are based on the idea that individuals exist naturally among plants. It is possible that they refer to completely inappropriate units in their data.

In addition to this problem, it is very difficult to determine the mass of the root system of plants, a measurement that is necessary for many quantitative studies. Since the size of the root system of a particular kind of plant may vary considerably under different conditions compared to the mass of photosynthetic tissue above ground, estimates are often unreliable. In order to eliminate error all together the only resort would be to measure the root mass of each and every individual. But neither the root system, the animals in the soil, nor the hedge itself would survive such an investigation.

As the discussion about the problem of boundaries (Chapter 4.c) demonstrated, when it comes to understanding an entire system, not much is gained by assessing the individuals ("building blocks") in such a system. In order to totally analyze the system "hedge," it would also be necessary to determine all the causal chains that exist between the building blocks and represent them numerically. This would require identifying all causes and effects and correctly describing their relationships to one another.[15] Besides the fact that a venture of this kind would be destined to fail due to the mega-astronomical number of relationships involved (see Chapter 4.a), it doesn't take much imagination to realize that such an endeavor could not be carried out without disturbing the very functions that are to be investigated or maybe even destroying the entire system.

As Kornwachs and von Lucadou (1984, 132) have shown, the basic problem posed by the fact that complex systems have to be disturbed in the course of investigating them not only applies to the ambitious goal of totally analyzing a "real system." It must also be taken into account even when operating with a more pragmatic and operationally defined concept of a system that includes only those properties that are relevant for a particular aspect of the topic being studied. This means that because of disturbances caused by investigative procedures, some information may not be accessible that would be required to solve a certain ecological problem. Due to the basic problem of disturbance it is not possible "to completely eliminate errors due to disturbances caused by investigative procedures since the corrections that have to be introduced generally lie within the error tolerance interval." However, as both authors point out, to *control* a system we have to understand its "mechanisms" as completely as possible. "But if the very attempt to control and govern a system causes the system to change its behavior in a manner that cannot be grasped, then the means we have for understanding the system, which are based on "the mechanism" of the system, have become obsolete" (Kornwachs and von Lucadou 1984, 113). Thus the more completely and exactly a researcher attempts to describe a system in order to be able to control it with certainty, the more she seems to have to count on it eluding her in the process.

Thus the dilemma is that we must either accept measurement distortion or disturbance of the system as the price for aiming at exact and complete measurement or put up with inaccuracy and less-complete measurement in order to avoid distortion, a dilemma that resembles Heisenberg's uncertainty principle, according to which it is impossible to completely accurately determine both the position and momentum of a particle at one particular moment (Heisenberg 1969, 144). When a measurement procedure fundamentally affects the results of measurement, the concept of subject/object dualism, which forms the very foundation of the scientific method, ceases to be valid. Thus very basic limits to potential knowledge exist in both sciences, physics and ecology.

4.e. Uniqueness and Generalization

In connection with the problems of complexity and measurement distortion, the following objection to the conclusions reached above is conceivable: Even if ecosystems are really too complex and too vulnerable to be grasped in their entirety by determining all their parts and all the functional relationships in detail, a perfectionist approach of this kind may not be necessary for all practical purposes. After all, we don't have to take a particular watch completely apart and analyze all its mechanisms anew from A to Z in order to repair it when it stops running. Since the defective watch was manufactured according to a general construction plan, a watchmaker can analyze the damage by drawing on all the experience that has ever been made with this type of construction and the defects associated with it. Then he can use this knowledge to repair the watch. If this strategy is applied to our understanding of ecosystems, it would mean that instead of a complete analysis, in many cases it might suffice to know the *types* of subsystems that make up a larger one. It would then no longer be necessary to re-examine those aspects of the subsystems for which knowledge about *general* regularities has already been attained.

As the example of the defective watch was intended to show, the objection discussed above is based on arguments drawn from everyday experience, and nowadays these are made primarily with technical objects. Since mass-produced technical objects are composed of standard and almost identical parts, people often assume that like physical, chemical, or technical systems, ecosystems are also made up of clearly classifiable parts and that the parts that belong to a particular class are *practically identical*. According to this view, the structure, the functional relationships, and the interactions of an oyster bed, for example, could be described by general rules that would also apply to all the oyster beds within the same ecosystem, sometimes even to oyster beds located in a different ecosystem.

Of course we know that rather uniform morphological and functional "construction plans" can be found among many biological systems from cells to organs and to individuals of different species and groups of species. If it were not possible to classify and systematize the diverse manifestations of life to a certain extent, there would be no science of biology nor any successful applications of this science as, for example, in medicine. And yet examples from medicine in particular show that in spite of a common morphological and functional organization, subsystems of superordinate ones still cannot always be regarded as identical, nor can they be treated therapeutically in the same manner. Take the systemic level of organs, for example. These concrete manifestations of biological blueprints differ from one another much more than what we experience with technical objects. The variability of "building blocks of the same type" usually increases with increased levels of systemic order or rather, with increased complexity. An especially striking example of variability among biological systems is the variability found between individual humans (and the structure of their brains, which has been molded by innumerable contingencies) as well as among ecological systems.

In ecology the variability of "building blocks of the same type" is essentially due to the fact that the objects this discipline studies (with the exception of individual organisms) consist of communities of organisms, but these are not organisms themselves. Non-organismic aggregates of this kind usually lack the coherence typical of organisms, organs, or cells. As a result they can attain a high degree of diversity and variability so that we can no longer reasonably talk about inherent "construction plans," not even metaphorically. Even if ecosystems that are influenced by similar climatic factors exhibit convergence in some respects and therefore can be categorized as members of a particular type of ecosystem (as, for example, a forest, steppe, or desert), the differences between members of a type can be quite significant when they are examined in detail. Ecosystems are unique specimens to a much greater degree than organisms are, and as such they are irreproducible in time and space.

Because of this it is particularly difficult to establish general rules and simple laws in ecosystem research. "The processes we may have analyzed in great detail in one system can function completely differently in another one, even if it looks quite similar," Remmert (1984, 195) maintains. Even if basic regularities have repeatedly been discovered in other biological disciplines such as genetics or molecular biology in the course of the past decades, comparable endeavors in ecology have usually been unsuccessful, and according to Tischler (1976, 133, 134) they will probably remain so in the future. Attempts to force ecology along the same straight and narrow path as physics or chemistry are misdirected. "Nature doesn't let itself be pressed into a few rules at the upper levels of its manifestation. . . . The greater the units of life we study and the more we try to investigate manifestations of life in all its forms, the more limited are our possibilities for making generally valid statements." According to Tischler (1976, 133), the theoretical foundations of ecology will "never be able to do more than describe certain tendencies and principles. And they should be satisfied with that."

Thus ecologists usually have no other recourse than to analyze *individual* cases in as detailed a manner as possible. In previous sections (see 4.a and 4.d) I have already indicated some of the methodological problems that stand in the way of anything that comes even close to *complete* analysis. Now let me point out another difficulty as well, the *time* factor. Here it is not so much a matter of the enormous amount of time it often takes to produce an approximately accurate "snapshot view" of an ecosystem but rather the fact that such brief glimpses in principle do not suffice for understanding an ecosystem. Contrary to popular ideas about "ecological balance," ecosystems are not static entities whose species composition can be regarded as remaining constant for years on end. Instead, they are occasionally subject to very great fluctuations due to different climatic conditions. Since the weather is never the same from year to year, the flora and fauna of an ecosystem also vary from year to year to a degree that is often underestimated. As a long-term study of a xeric grassland community in southern Germany showed, animal and plant populations can vary by a factor of ten or more simply as a result of different climatic conditions. Remmert (1984, 257) reports the following: "Not only did the composition of the flora of plant societies vary severely. The total production of matter above the ground and the composition of the fauna did as well. Crickets and grasshoppers were reduced to 1/10 the original count, while flies increased significantly. After a particularly 'favorable' summer the numbers returned pretty much to the original level." Comparable fluctuations due to climate changes and altered food supplies have also been observed for the breeding populations of different species of birds (Berndt and Henß 1967; Remmert 1990, 139). These might be associated with long-term spatial changes in areas occupied by these populations (Vauk and Prüter 1987, 182).

Fluctuations of these dimensions confront ecologists with very fundamental methodological problems, because if taken seriously, they contradict a very basic prerequisite for a causal analysis, namely, the constant *identity of the parts* of a whole. "What macroscopically might appear to be a functionally homogenous entity is actually the result of a myriad of continually changing details for which it is impossible to analytically establish constancy at any level of abstraction" (Breckling et al., 1992, 4). Similar to the insight of the Greek philosopher Heraclitus, who maintained that it is not possible to step into the same river twice, the reproducibility of field studies in ecology is limited, as is the possibility of testing their results by repeating the study. "Nature forces the observer to constantly be prepared for the unexpected, for surprises and for singular events" (Breckling et al., 1992, 4). Even though there is now little controversy about the idea that singular events can also be investigated scientifically (Vollmer 1986a, 53f.), it is still clear that the strong historical nature of ecological phenomena sets serious limits to methodological possibilities in ecology. In addition to the standard requirement of generalizability, that of reproducibility must often also be further qualified in ecology (see, for example, Gorke 1990), although in the exact disciplines of physics and chemistry both are often considered to be indispensable criteria of the scientific method.

In view of the historicity and the strong natural fluctuations and systemic oscillations of ecological systems, it should now be clear that an even approximate understanding of their complex relationships can only be achieved by *long-term* investigations, that is, ones that extend at least over decades. For example, in order to demonstrate that the population size of lemmings in the arctic tundra varies in a ten-year cycle, it is necessary to measure at least three population peaks, which means researching them for more than thirty years. Of course, we have to always keep in mind that periods of time that seem to be long in the context of human experience may be no more than a wink of the eye when the entire developmental history of an ecosystem is taken into consideration. According to Remmert (1984, 269), highly labor intensive ecological stock-taking of the kind that politicians tend to call for as an emergency measure after an ecological disaster should really not be conducted or publicized at all. "What's the use of the most accurate measure of the density of a bird population and all the parameters that accompany it, if it only represents the brief period of a year? It only causes confusion, because a year later the population size might be ten times greater than expected."

It is obvious that in view of such *completely natural* fluctuations in population size it is quite difficult to clearly demonstrate the effects of more recently introduced factors induced by *human activity* such as pollution, carbon dioxide, radioactivity, tourism, and agriculture. To expect ecology to provide sufficient evidence before taking any kind of action, as our legal system often still requires, means expecting the impossible. There are two major obstacles to satisfying the need for long-term studies discussed above. First, because of the newness of the ecological problems with which we are confronted, we lack sufficient data from the past for comparison. Second, because of the urgency of our ecological problems we don't have enough time for long-term investigations in the future. For example, it is impossible to determine with scientifically reliable methods and in sufficient time, which means *before* serious climatic changes might possibly occur, whether or not the unusually warm summers in the past years are due to statistical fluctuations or consequences of the ever-more visible greenhouse effect. Because of the difficulties involved with making generalizations about time or space when dealing with higher order systemic levels affected by contingencies, our knowledge is apparently subject to fundamental limitations, and it is therefore not justified to refrain from executing urgent measures by referring to the lack of such knowledge.

4.f. Quality and Quantity

An aspect that is fundamental to modern science since Newton and Galileo is the mathematical description of natural phenomena. Even though approaches based on observing nature by more qualitative means involving our capacity for "gestalt" perception and synthetic thinking have continually cropped up in the history of science (e.g., Goethe), the quantitative, analytical method has prevailed because of its ability to formalize, simplify, and generate testable hypotheses. In particular in physics and chemistry and the applied areas of technology associated with them this method has proved to be so successful that nowadays it is often regarded as the *only legitimate way* of seeing things in science. In view of the technical success of physics and chemistry it is understandable that other scientific disciplines such as biology, medicine, and psychology have developed a strong orientation toward these "exact" disciplines, and that they try to reduce the variability of the phenomena with which they deal wherever possible to things amenable to quantitative, mathematical analysis. Due to growing pressure for practical success in dealing with environmental problems, the still quite young discipline of ecology is also expected to meet the standards of quantitative analysis above and beyond purely theoretical results.

I have already mentioned that the development of the computer has indeed opened up prospects for using a quantitative approach in ecology in a big way and that this has significantly augmented our insights into complex relationships in nature (e.g., concerning the importance of nonlinearity and feedback loops). At this point I am not in a position to appropriately pay tribute to the successes of quantitative ecology in greater detail. In view of the ecological crisis and the debate about the expectations of technical optimism I have outlined above, it seems to me to be more important to emphasize that there are *limits* to quantitative methods and to demonstrate where these lie.

One of these limits was already discussed in connection with the problem of generalization in ecology (see 4.e). Every mathematical description of an ecological system implies that the terms (subsystems) incorporated in a formula can be regarded as practically identical, provided they are assigned to the same variable. However, it is increasingly difficult to fulfill this prerequisite the more complex and comprehensive the system to be studied is, or rather, the more significant the influence of contingent conditions is. For example, in mathematical computations a population is assumed to be uniform, although the individuals within the population usually differ genetically. This criticism can be countered by the argument that genetic variability can be "smoothed out" by calculations based on statistical averages and that these facilitate the final results. But in doing this investigators overlook the fact that by averaging out genetic "exceptions" sometimes the most valuable information is lost. "With statistics we measure average manifestations of living processes. In nature, however, behavior that deviates from the norm and the assertive power of 'loners' is sometimes decisive for the survival of a population" (Tischler 1976, 134). This is particularly important in the case of nonlinear systems, where properties that might be insignificant from a statistical point of view can assume unforeseen importance under the influence of autocatalytic feedback loops. The qualities that cause these properties to exercise such effects can barely be grasped by numbers and are therefore not calculable.

In ecosystem analyses this difficulty is often avoided by only considering aspects that can be quantified and by restricting the investigation to things like energy flow and metabolic cycles. The temptation to limit oneself to these two areas is great since the larger contexts of metabolism and energy transfer "are the only universal, quantitative principles of ecology that are generally valid" (Tischler 1976, 134). However, when a unilateral perspective of this kind is selected in order to attain scientific accuracy in one area, other kinds of inaccuracy occur and can result in fundamentally erroneous estimates. In terrestrial ecology, for example, animals are sometimes mistakenly considered to be of only marginal importance in many ecosystem analyses, because from a *quantitative* perspective they play only a small part in transferring energy and driving metabolic cycles. However, this view disregards their *qualitative* potential for determining the structure of the system in which they live. Think about deer, for example, which can reduce the production of a forest by consuming shoots and buds, or a beaver, which can generate lakes by building dams. According to an image projected by Remmert (1984, 250), the effects of these two species on their ecosystems are comparable to those of switches and amplifiers in a technical system or to the effects of sensory organs and the nervous system on an individual organism. And just as it is difficult to grasp the significance of hearing ability for an individual in terms of numbers, it is also difficult to quantify the effects of animals as pollinators or seed dispersal agents in all respects. Pollinators and agents of seed dispersal can significantly influence the structure of a plant society, while at the same time these functions barely show up in calculations of energy flow or metabolic cycles. The claim that in ecology, as in other sciences, all qualities can be expressed in terms of numbers and equations, is an indication of "inadequate understanding of biological thinking" (Tischler 1976, 4).

Another barrier to attempts to quantify nature all the way up to systems of the highest order is the tremendous complexity at these levels (as discussed in 4.a). Although it might be possible to reveal explicit cause-and-effect relationships in simple lab and field experiments by isolating, knocking out, adding, or exchanging parts of the system, and although these relationships then can be described with relatively simple mathematical equations, it is much more difficult to correctly interpret information on the frequencies of different properties or processes generated by field observations. If statistical analysis reveals significant correlations between these properties or processes, then these relationships generally can also be described by relatively simple mathematical means. But this tells us nothing about whether the *statistical* relationship we have described is also a *causal* one. "Even refined mathematical analysis can at the most confirm what formally exists, but it cannot reveal any biological causes of relationships" (Tischler 1976, 4). For reasons discussed above (complexity and problems associated with generalization, boundaries, and measurement distortion) it is in principle much more difficult to produce convincing evidence for a causal relationship in ecology than in physics or chemistry. Strictly speaking, in ecology one never gets much beyond weighing the strengths and weaknesses of plausibilities.

Since complex relationships in nature prevent us from recognizing many relationships either directly or in field experiments, it is more and more common in modern ecology to rely on *models* designed to improve our understanding of interactions between organisms and parameters of the environment with the help of graphs or mathematical representations that simplify the matter. Models of this kind depict arrangements that are "analogous to certain aspects of a natural system" (Tischler 1976, 12). By deliberately blotting out all other aspects they permit us to "recognize regularities that we would hardly be able to grasp with a purely empirical investigation of the things that really exist in nature." One example is the model of predator/prey systems with which the reciprocal regulation of population density among bobcats and blue hares can be demonstrated. The population dynamics of this relationship can be described mathematically with so-called Lotka-Volterra equations (Osche 1978, 60, 61).[16]

As useful as such functional models might be for helping us to understand the *formal* side of a *particular* ecological aspect, we should still be cautious about overestimating their ability to permit us to calculate and thus also predict greater effects of interventions in nature. Even though highly complex, generalizing models have been developed with which we could occupy large computers for hours on end, according to Remmert (1984) they all fail to succeed (except for a few cases in special areas of autecology[17]) when it comes to what science is always and exclusively supposed to be all about, that is, when applied to conditions in nature outside of the computer laboratory. The difficulties that crop up have to do with problems that have already been discussed at length, namely *complexity* and *generalizability*. As Wissel (1992) has shown, highly complex models that attempt to approximate reality as closely as possible are the very ones that prove to be inadequate for conveying an understanding of the real relationships in nature. "When the number of details that are taken into consideration becomes too large, it prevents us from recognizing the influence of individual details on the behavior of the model" (Breckling et al. 1992, 4). The complexity of the *model* then impairs causal analysis in the same way that the complexity of *nature* does. A similar dilemma arises when we attempt to accommodate generalizability and applicability with one another. "The models of population ecology and ecosystem research are either so specific that they fit things of the past exactly or so general that they fail to permit any predictions. Just as there is no generalized model of the biologically best investigated organism around, namely human beings, so also—according to many contemporary ecologists—will there never be a generalized model of populations and habitats that would permit us to make predictions" (Remmert 1984, 304).

This quote from Remmert should not be misunderstood. Of course there are numerous models, theories, and generalizations that describe *particular aspects* of ecological systems (or human beings). For dealing with certain scientific problems these approaches can be highly effective. However, the expectations of technical optimists I wish to criticize here are those not restricted to achieving *particular ends* but rather those aimed at the *sweeping* goal of controlling and governing entire ecosystems. A technical optimist is not just interested in forecasting short-term changes in ecosystem processes (comparable perhaps to a short-term weather forecast in meteorology). His goal is long-term system management (comparable to long-term climate manipulation). In view of the specific limits of the science of ecology, these expectations appear to be illusory and in the long run even counterproductive. Only if it were really possible to attain complete (or sufficiently comprehensive) ecological knowledge would the technical optimists' position be legitimate. However, since this knowledge is *in principle* impossible to attain, it is wiser to adjust our attitude toward nature to what we do not know in ecology, rather than to the necessarily fragmentary knowledge we have at our disposal. I will discuss this further in Chapter 16.

5. Limits Set by Epistemology and Theory of Science

Many of the objections I have presented to technical optimism indicate that the *particular* limits inherent to the science of ecology are often related to more *general* limits, namely the limits of human beings' epistemological capacities and limits to the scientific method in general. In the history of philosophy there seems to be no doubt that such limits to everyday and scientific knowledge really exist. Attempts to demonstrate these limits or determine them go back to Plato's famous cave story in his essay *The Polis*, and extend to Hume's *Enquiry Concerning Human Understanding* ([1748] 1999) and Kant's *Kritik der reinen Vernunft* (Critique of Pure Reason) ([1787] 1976) as well as to modern theory of science (Sachsse 1967; Stegmüller 1969a, 1969b; Popper 1972; Kuhn 1970), to mention just a few examples. The common denominator of all these theories is the realization that what humans perceive as reality is not necessarily equal to everything that reality includes.

However, a popular assumption is that this discrepancy is only relevant for immediate sensory experience and the prescientific worldview associated with it, while the scientific method supposedly permits us to recognize the discrepancy and bridge the epistemological gap. And indeed, it cannot be denied that quantum physics and relativity theory have succeeded in demonstrating that innate ways of perceiving the world (such as three-dimensional space or causality) are deceiving when dealing with things beyond the limits of the familiar mesocosmos of humans, but that they can be transcended by scientific reflection. More recently, theories of epistemology based on evolutionary theory have attempted to explain the sources of epistemological limits and errors from a biological perspective on the basis of phylogenetic considerations (Vollmer 1975; Riedl 1980, 1985; Engels 1990). And yet it would be wrong to assume that such studies could be carried out in absence of the very errors they have set out to investigate. The "other side of the mirror," as Lorenz (1973) called humans' cognitive faculties, is itself only accessible by means of these very faculties, that is, as a "mirror image" (C. F. von Weizsäcker 1977, 187f.). This means that scientific knowledge about our cognitive faculties is by necessity limited.

Eddington (1939) described the relationship between scientific knowledge of reality and "real" reality with a vivid parable that compares a scientist to an ichthyologist who wants to examine marine life and therefore tosses out a net. After several rounds of fishing and conscientious examination of the catches, he arrives at the following basic law of ichthyology: "All fish are larger than five centimeters." When a critical observer objects that there might still be smaller fish that slipped through the holes in the net, the ichthyologist replies more or less as follows: "Things that I am not able to capture with my net are in principle beyond the scope of ichthyology. They are not defined objects of ichthyology and therefore for me as an ichthyologist they are not fish."

Even if Eddington's metaphor, which I have rendered only briefly, doesn't apply to the methodological approach of scientists and their relationship to reality in all respects, it still demonstrates some epistemological aspects that are relevant for the present discussion (see also Dürr 1991). It illustrates particularly well the aspects of selection and projection that are characteristic of the scientific method.[18] Just as an ichthyologist is only interested in the fish he can really catch, scientists also filter out those parts of the entire spectrum of nature that can be grasped objectively and reproduced according to certain rules of observation and experimentation. Scientific knowledge is therefore always *limited* knowledge of a metaphysically conceived and more extensive reality.

From the perspective of epistemological positivism the objection could be raised that it is senseless to refer to such a metaphysical and linguistically inaccessible "world as such." But the moment a scientist begins to do science, he seems to practically operate on the premise that there is such a thing as a "real" world independent of the observer ("reality postulate," Wuketits 1983, 2). He assumes in the sense of "hypothetical realism" that the world can be grasped at least partially or approximately (Lorenz 1973; Vollmer 1975, 1985; Riedl 1980). In the words of Eddington's parable, an approximation of reality can be attained by using different nets with varying mesh sizes and by constantly improving them. Each net is developed for a particular purpose through interaction with reality and thus permits us to make increasingly exact statements about special properties of fish relative to the particular purpose of the net.

The difficulties connected with trying to arrange the various statements attained in this manner to form a comprehensive picture indicate that in addition to capturing only a section of reality, each round of fishing of necessity also produces *qualitative changes*. A good example from physics is the electron that sometimes appears to be a particle and other times a wave depending upon the experimental approach used, so that any normal concept of it as an object is impossible. In ecology as well the projective nature of the scientific method and the limits of language permit many phenomena to be described only in such a complementary sense. Thus with respect to complexity (4.a) it might at first seem contradictory to operate with terms such as "building blocks" and "subsystems" when boundary considerations (4.c) eventually force us to conclude that closed units of this kind don't really exist. Similarly, the nonlinearity of biological systems was used in argumentation on the one hand (4.b), while on the other the ability to freely formalize (4.e) and quantify (4.f) ecological systems was questioned. These cases demonstrate that while it is possible to criticize *one* scientific method of projection with *another* and expose its limitations, it is not possible to achieve a kind of synthesis that would accurately reflect reality as it is. There is no way to get around the basic flaws of projection.

As the far-reaching implications of evolutionary thought suggest regarding both biological and cultural aspects of epistemology, nets, or rather pre-scientific and scientific projection methods, are not simply random occurrences. They have proven themselves to be adequate or inadequate for certain purposes in confrontation with (postulated) "real" reality by a feedback process. Thus I must disagree with Schönherr (1989, 28, 34), who, after an otherwise legitimate epistemological critique of such methods arrives at the conclusion that science is "unfounded" and that its experimental evidence is "arbitrary." If it were really true that "our endeavors to attain generally valid, objective knowledge have resulted in a knowledge structure that really has *nothing* in common with nature" (Schönherr 1989, 25; my emphasis), then it would be hard to understand why technical products such as pacemakers for heart patients and planetary probes are able to function. Technical achievements of this kind seem to indicate that such structures "fit" at least certain sections of reality ("partial isomorphism;" Vollmer 1985, 31). In contrast, the ecological crisis and other negative consequences of science and technology suggest that some of the projection methods employed are only of limited value or even useless in *other* areas. Dürr (1991, 46) drew attention to the fact that scientific thinking seems to have been the most successful and thus best suited to reality in cases in which "in first approximation the whole does indeed seem to be equal to the sum of its mentally isolated parts" and "where the functional integration of the different components is weak." These prerequisites are satisfied ideally by technical systems whose isolated parts interact with one another in an readily controllable fashion at only a few nodal points in the system and thus exhibit strictly determined behavior. Strongly integrated and complex systems, on the other hand, like those with which ecology or the social sciences have to deal, pretty much resist such attempts at structuralization characteristic of technical and deterministic models. In these instances ideal abstractions of scientific thinking corresponding to standards of classical mechanics seem to apply only to a very limited extent.

6. Alternative Science?

Recent appeals for a "new," "alternative," or "ecologically oriented" science can best be understood in the context of the faulty correspondence between certain objects and the scientific method described earlier. With a different kind of science it might not only be possible to avoid the destructive consequences of existing science. Perhaps it might also be possible to "regenerate it socially" (Schäfer 1982, 43). "If science is to be able to critically examine the conditions of life it has generated in politics and in nature and successfully control the control of nature it has achieved, it will have to become a different science. Neither the complete destruction nor the complete perfection of the current system will be of help" (Altner 1982, 432). What might a different kind of science look like?

Bossel (1982, 39), who criticizes "established research" as being "particularist" and "ecologically counter productive," envisions a new kind of science whose research approach is "multidisciplinary and primarily holistic" and whose goals are directed at maintaining and developing an "ecological community." With respect to social obligations, however, it is not clear whether they might affect only the *social context* of science or whether they would influence scientific *methods* as well. As Hemminger (1986, 25) rightly points out, the rules of scientific knowledge cannot be altered to match cultural needs, as urgent as these needs might be. "They are determined by very basic qualities of reality, including both the reality of nature and that of knowledgeable human beings." Chargaff (1991, 356), otherwise one of the sharpest critics of the modern scientific enterprise, bases his skepticism about the possibilities of alternative science not so much on the properties of nature as on the properties of current (Western) civilization. "It must be recalled that to turn our backs on the usual methods of natural science would require such effort that it is unimaginable without a previous social, moral and psychological revolution of heretofore unseen dimensions. . . . It would mean giving up the major intellectual tools of our branch, induction and reductionism. To move from small to big, from parts to wholes, is so tightly rooted in our sciences that new ways of thinking would seem like punishment." Furthermore, a multidisciplinary approach such as that called for by Bossel would not be enough, because as cases such as cybernetics, systems theory, and chaos theory have shown, interdisciplinarity does not necessarily lead to a "holistic" view of nature. In essence they are all still based on the fiction of a world composed of parts, except for the fact that they have replaced monocausal by polycausal thinking and linear by network thinking. The ecologist Trepl

(1983, 10, 11) maintains that modern ecosystem research is no exception. In his words it is "just the opposite, a rejection of attempts to understand the unabridged, concrete whole" of nature. Ecology isn't located outside the logics of progress but "is rather the culmination" of this kind of thinking. If ecology can therefore hardly serve as a model for alternative science, it is no wonder that in this context Primas (1992, 6, 7) calls for a "fundamentally new orientation in our thinking," a kind of thinking that "once again regards nature *as a whole* as the object of science." Of course he admits that the "foundations of holistic research have yet to be developed." Other endeavors of this kind such as the idea of "searching for and deciphering the subject side of nature" (Altner 1979, 123) or "searching for science that is free of power structures" have also so far failed to provide methodologically feasible programs or mature solutions. Thus it is probably not very realistic to expect alternative science, if it ever exists at all, to make a timely contribution to winning the battle with our ecological crisis.

Nevertheless, reflections of this kind are valuable in that they question the claims of prevailing methodology to being the only accurate kind of representation and open our eyes to methodological sidetracks that have been neglected so far. In keeping with this line of thought von Gleich and Schramm (1992) do not expressly argue in favor of alternative science, but they do advocate shifting the methodological focus in ecology. Based on the view also outlined in the previous chapter that a mathematical approach rooted in Galilean and Cartesian thought is only of limited value for evaluating phenomena in ecology, these authors suggest that we give more thought to a line of science such as that favored by Aristotle that is more strongly oriented toward natural history and founded on concepts embedded in experience. They do not consider it a step backward to deviate from the modern mathematical orientation of science but rather as something "that is more in keeping with theory of science and which from the standpoint of everyday research would also be more valuable for the further development of ecology as a science." While I agree with this opinion in principle, after having raised mathematical methods to a level of absolute authority we must be careful not to make the same mistake and abandon all other approaches for an "apparently more appropriate theory." The "best theory at the moment" should always be regarded as only a temporary one in the whole process of scientific progress. "This process develops the best when varying and contrasting approaches are allowed, cultivated and set in relationship to one another at different levels" (Breckling et al. 1992, 8).

7. Science and Worldviews

If we look at criticisms of existing science more closely, we will see that most of the arguments do not so much focus on epistemological *methods* but rather on certain scientific *paradigms* as well as on the *worldviews* derived from them and the consequences that ensue. Since these three levels are not always clearly distinguished from one another, misunderstandings and serious misinterpretations often occur among both critics and representatives of science.

As far as the *scientific method* is concerned, the epistemological analysis presented above clearly indicates its biases and limitations. But for two reasons this cannot be regarded as a legitimate argument against the method. First of all, no form of empirical knowledge is free of biases and limitations. Even pure sensory perception represents only a small part of a greater spectrum (e.g., of electromagnetic waves) and "colors" it (e.g., as subjectively perceived heat or light stimuli). On the other hand it is probably the very fact that the scientific method is one-sided, or, more positively speaking, that it operates strictly within certain self-imposed limits (e.g., intersubjectivity, reproducibility, refutability, etc.), which results in the incontestable succinctness of its statements. "A narrow perspective keeps us from looking to the right or left and allows us to steer toward our goals full speed ahead" (Primas 1992, 7). Considering all the different qualities that exist in nature, this shouldn't, of course, be misunderstood as justification for a kind of methodological monism that is only interested in lumping everything together for analytical and mathematical treatment. The problems associated with complexity and nonlinearity outlined by theory of science have shown that in ecology there are definite limits to mechanistic and reductionistic methods. Even if this manner of thinking should prove to be of heuristic value in some cases, it should not be equated with "the right way to think" (Primas 1992, 7). In other words, *methodological* reductionism should not crystallize into *ontological* reductionism.

It is even more important to separate method and *paradigm* when it becomes obvious in various different areas of science that mechanistic thinking, determinism, and reductionism no longer represent adequate paradigmatic elements of scientific reality (Prigogine and Stengers 1984). "Thus the best fundamental theories of matter available nowadays reveal that the material world is a unit *that is not composed of parts* but can be described instead as consisting of fictitious parts that interact in a very special context." This quote from Primas (1992, 7), which primarily refers to quantum mechanics, can also be applied to the situation in ecology. The phenomena of *emergence* and *downward causation,* which are typical of ecological systems, "make the complete success of any reductionist programme at least problematic" (Popper and Eccles 1977, 20). Emergence refers to the appearance of new properties of a *whole* that can neither be deduced from nor predicted by knowledge of its *parts.* Another way of saying this, which is frequently quoted, is that the whole is more than the sum of its parts. Thus "emergence is a descriptive notion which, particularly in more complex systems, seems to resist analysis" (Mayr 1982, 63). One of its most characteristic traits is downward causation (Campbell 1974, 180), also referred to as structural causation. According to the reductionistic point of view, things that happen at a lower level of the system determine those that occur at higher levels, and causation thus operates from the bottom upward. But phenomena such as gravitational pressure in the stars or the regulation of population density in a colony of gulls show that through negative feedback the reverse is also possible, that is, that the macrostructure of the whole can influence the properties of parts at lower levels in the system.

In view of the fact that more recent theories of science have revealed the shortcomings of ontological reductionism with more and more clarity (Wuketits 1983, 129), the question that arises is why this approach still has such a strong, formative influence on many biologists' understanding of nature as well as that of a large part of the general public. One important reason for this seems to be that the scientific method and the scientific paradigm of reductionism have often assumed the role of an insufficiently reflected *worldview.* However, because of their quasi-religious nature, worldviews are basically in danger of detaching themselves from "external reality" when confrontations with this reality lead to internal contradictions. The worldview that various authors refer to as "scientism,"[19] a view associated with very basic "faith" in science, exhibits several signs of such fixation. Although it officially claims that science is its foundation, it doesn't seem to have taken notice of recent paradigmatic shifts in science or the epistemological problems connected with it. Therefore it would probably be better to talk about "science superstition," because the worldview of scientism doesn't seem to have anything to do with reasonable faith in the (temporary and biased but empirically tested) results of scientific inquiry.

The manner in which scientism raises the scientific method to a position of absolute authority can be summarized in the form of two dogmas. The *positivistic* dogma maintains that only things that can be scientifically investigated, measured, and predicted can be regarded as real, while the *technocratic* dogma proposes that science is basically in a position to solve all the problems of the world. With the help of epistemological considerations and a discussion of objections to technical optimism I have attempted to demonstrate that these dogmas are not very convincing. Like Jaspers (1968),[20] Garaudy (1991, 370) refers to them as "superstition" and "totalitarian fundamentalism," since they lead to the "exclusion and edging out" of all the "more profound dimensions of life" (such as art, religion, creativity, and love) as well as the most important of all contemporary problems, the question of goals and values. Because if one reduces the role of reason to investigating the relationships between different phenomena and finding ways of technically utilizing them, as scientism does, then reason is only important for evaluating *means*. According to this point of view *goals and values* are generated "automatically" as a function of what is technologically possible and what appears to be politically and economically profitable. Garaudy (1991, 372) compares reason that is reduced in this manner to "a sleepwalker, who neither knows nor cares which way he's going." The products of reductionist reason are not only grotesque inventions of modern warfare (such as the neutron bomb) but also the consumer behavior of modern society, which continues to spiral upward in spite of disastrous effects on the environment, to mention only two. In this context scientism appears to be not only a case of "post enlightenment mania" but a "social evil" as well (Vossenkuhl 1992a, 98).

If it is indeed this reduced form of reason that has led to our present ecological crisis, then the question is why people still rely on it in their search for a way out of the dilemma. Why is it that scientism and its applied form, technical optimism, are not only still alive and kicking but still believed to be able to show us the way out of the ecological crisis? It appears that the success and unbroken attractiveness of these ideologies can be attributed to at least three things: their function as guidelines, the promise of salvation they hold, and their power structure.

The tenacity with which technical optimism holds onto ontological reductionism suggests that it is the promise of greater *power* over nature that is of central importance here. The idea that every system is calculable, made up of parts that can be isolated from one another, and therefore capable of being taken apart and put back together again is a platform for visions of perfect control of nature by humans. This vision seems to remain unperturbed by arguments from theory of science or signs of the obviously disastrous consequences of the ecological crisis. On the contrary, the idea prevails that anything that has been spoiled by reductionistic science can be "repaired" by it as well. In keeping with this kind of thinking, for example, as a remedy for increasing depletion of the ozone layer it has been recommended "to slingshot frozen ozone into the atmosphere to repair the damage" (Briggs and Peat 1989, 201). As a means of combating the greenhouse effect the economist Schelling (University of Maryland) suggested adding sulfur to kerosene in airplanes in order to generate aerosols that produce a cooling effect. This kind of "geo-engineering" would be cheaper than all the economic ruptures and trade conflicts that would result from trying to reduce carbon dioxide emissions (Schuh 1994, 49). According to the sociopsychologist Richter (1988, 29) examples like these where the power of science and technology is highly overestimated clearly reveal pathological tendencies. They are manifestations of the "infantile delusions of grandeur" characteristic of our modern scientific civilization, which after escaping medieval impotence and religious subordination now lay claim to egocentric, god-like omnipotence. Richter (1988, 5) speculates that the reason for this yearning for technical omnipotence, which he refers to as a "God-complex," lies in the "fear of an unbearable sense of loneliness and impotence in the world."

To counterbalance such feelings of existential insecurity and pointlessness, the technical optimist elicits the vision of unlimited technical progress that continues to generate new values in the course of its everlasting rise to power. While socialism casts utopian visions of progress eventually leading to a world free of all burdens, the liberal and capitalistic versions of this vision sees no end to progress and believes instead in constant innovation. To a certain extent both seem to incorporate the religious idea of *redemption* and the path of *salvation*, with the exception that the distant paradise of religion is brought closer to home in the form of human-made, redesigned life on earth.

The need for *guidance* that emanates from these visions is reflected by the respect that scientific reports and expert opinions are capable of engendering. On the basis of exclusive claims to truth characteristic of scientism, "true believers" of science will grant expert opinion a kind of authority the likes of which were reserved for religious agents and holy scripts in former times. As the biochemist Chargaff (1991, 366) maintains, "an empty space is left behind by the waning of religion, and the human conscience is sub-sequently filled by natural science. If every era needs a religion, then it is science which has become the religion of our century." However, in times like the present in which not only pseudo-religious belief in power of sci-ence flourishes but animosity toward science, irrationalism, and funda-mentalism as well, it seems appropriate to elucidate Chargaff's comments more precisely. In addition to art and music, science is certainly one of the greatest achievements of the human mind. By striving for truth and objec-tivity science can teach people to recognize facts and thus encourage greater modesty in their thinking, not to mention the pleasure and awe that insights into nature's laws bestow upon true scientists. Thus while the ori-entation that science provides in the area of testable matters of fact is un-doubtedly laudable, its expansion into the area of norms and values represents an inadmissible transgression. In this case it is not difficult to recognize the dangers connected with scientism. Since, strictly speaking, values and norms have no place in the positivistic world of scientific think-ing, the inexorable desire for these things that nevertheless exist can only be satisfied by blurring the difference between facts and norms, that is, be-tween what is and what ought to be. Scientism does this by raising what is to the status of a norm (Reichelt 1979, 4). The normative effect of statisti-cal studies in the social sciences and psychology demonstrates nicely just how well scientism is supported by people's needs for (supposedly) scien-tifically based guidelines. Thus scientific evidence that a certain kind of be-havior is "frequent," "normal," or "in keeping with the trend" is often regarded as proof that this kind of behavior is something that everyone should exercise (see Postman 1992, 88).

However, when ecological problems are at stake, the border between what is and what should be is also frequently blurred. But in this case the philosophical background is not so much a scientistic and technocratic one but rather romantic and naturalistic. This will be examined in greater depth in the following chapters.

II. The Science of Ecology as a Normative Authority?

8. The Naturalistic Fallacy

Since the days of David Hume (1711–1776) a generally recognized principle of practical philosophy that is still valid today is that *"what is"* and *"what ought to be"* are two separate categories that cannot be logically bridged. In his book titled *A Treatise of Human Nature* Hume ([1740] 1968) demonstrated that it is impossible to deduce statements about what *should* be from statements about what *is*. This logical impossibility is referred to as "Hume's Law." According to this law one cannot reach an evaluative (normative) or morally binding (deontic) statement from a premise that does not itself contain at least one normative or deontic statement (Ricken 1989, 44).

The basis for Hume's Law is the idea that *normative* or *deontic* properties can neither be defined by *descriptive* properties nor equated with them. Attempts to reach definitions of this kind have nevertheless been made time and again in the course of the history of ethics (e.g., by equating the term "good" with terms such as "happiness," "general welfare," or "species survival"). But G. E. Moore ([1903] 1994) was able to show that in terms of conceptual logics such definitions are *synthetic* and therefore not real ones. Moore rejected the idea that definitions of "good" are *analytical* as an example of *naturalistic fallacy*. However, in contemporary discourse the concept of naturalistic fallacy is usually employed in a more comprehensive sense and usually means a violation of Hume's Law. According to this interpretation of the concept, we are dealing with a naturalistic fallacy when practical claims to validity and moral principles are reached *exclusively* on the basis of natural facts (e.g., scientific knowledge from research on evolution, animal behavior, psychology, or ecology) (Vossenkuhl 1983; Birnbacher 1991). The term "naturalism" refers to epistemological and ethical positions that are based on such argumentation (see Mittelstraß 1984, 964; Wimmer 1984, 965, 966).

Although the term "law" is usually employed very sparingly in philosophy and seems to indicate that the principle underlying Hume's Law is universally accepted, there have been repeated attempts well into the present to contest or bridge the logical gap between facts and values or at least to demonstrate exceptions to Hume's Law.[21] A very well-known example is Jonas' (1982, 284) attempt to justify his ethics of responsibility with a "principle discoverable in the nature of things" and thus to refute the "dogma that no path leads from is to ought" (Jonas 1984, 93). At this point I can pursue neither Jonas' particular argumentation nor the controversial discussions about other attempts to bridge the gap. Instead, allow me to refer the reader to summaries of the is/ought problem by Schurz (1991), Vossenkuhl (1993a), and Engels (1993). According to the literature cited in these reviews, it can be assumed that the basic thesis of Moore's argument, the observation that naturalism "offers no reason at all, far less any valid reason, for any ethical principle" (Moore [1903] 1994, 71), has never been refuted. As Vossenkuhl (1993a, 137) has shown, however, the fallacy of naturalism is not so much that normative statements are explicitly derived from descriptive ones but rather that the normative significance of an obligation is silently regarded as generally accepted or in need of no further explicit justification.

It is not surprising that in particular in discourse about the ecological crisis the self-evidence of certain normative premises is often taken for granted. This is because there is hardly another case in which the gap between is and ought is more profound and provoking than in this one. On the one hand we have "ecological tragedies" such as species extinction, climate changes, and loss of forests that have been well researched and described by science, and on the other there is our inability to logically and conclusively deduce any moral principles, let alone concrete directives for action, from all the facts available. It is against this initially seemingly depressing background that we must consider the fact that if the is/ought problem is recognized at all outside of academic philosophy, it is almost completely ignored in discussions about the environment and practical nature conservancy.

Even more remarkable is the observation that in the more theoretical field of environmental ethics naturalistic fallacies are repeatedly presented without explicitly questioning Hume's Law. This is revealed, for example, by the originally purely *descriptive* concept of "ecological balance," which literature on environmental ethics almost always describes as an ecologically ideal state *to which we should aspire*. I can envision two reasons for such clearly unintended violations of Hume's Law. First, with relationships between humans and *nature* the naturalistic fallacy seems to be more difficult to recognize than in the case of a relationship between one *human* and another. To consult nature as an advisor when dealing with nature is apparently more plausible than when dealing with another human. Second, in public discourse the terms "ecology" and "ecological" are employed so often in an ideological, political, or even moral sense that it is no wonder that their use results in misunderstandings and erroneous interpretations.

But it would be a mistake to think that naturalistic fallacies are a purely academic problem and of no significance for discussions about the environment or the practical problems of nature and species protection. Uncritically blurring the discrepancies between is and ought, facts and values, continually leads to considerable confusion in many different areas of society and also results in ecologically veiled ideologies and illusions.

9. Consequences of Naturalism

9.a. "Ascertaining" Environmental Standards?

In politics and administration it has long been standard practice to present political and thus also normative decisions as logical consequences of undeniable scientific facts. When surrounded by such an aura of scientific proof, decisions that are really relatively arbitrary are thus withdrawn from critical examination. A strategy of this kind isn't necessarily always due to the attempts of people in power politics to get rid of undesired opposition. Just as often it reflects the dilemma of those who must reach decisions and who turn to "objective science" for assistance when faced with barely comprehensible information and multiple interests. That an act of this kind involves shifting the burden of responsibility is supported by Erz's (1986, 11) observation of an increasing tendency in environmental administration agencies "to expect researchers and experts to perform the risk assessment and political and administrative evaluation procedures required for balancing conflicts." This tendency is particularly evident in cases involving the establishment of environmental standards and threshold levels. In these cases there is a widespread naturalistic view that such standards are hidden natural phenomena that *must only be discovered* and whose validity can be directly deduced from statistical analysis or graphical representations. It is based on the idealized thought that "a desired environmental standard can be found in the form of a morphological peculiarity (threshold, discontinuity) of a curve" (Gethmann and Mittelstraß 1992, 16). For example, the naturalist expects the ecologist to be able to tell her the maximal amount of nutrients a river can bear without losing a certain species of fish. But even if such a *threshold dose* is demonstrated, this in no way means that it must automatically be established as a valid threshold level. (Why attach so much importance to this particular, extremely sensitive species of fish?) Furthermore, in ecology we more frequently have to deal with stochastic (random) effects for which there is no threshold dose. For example, according to current evidence it is impossible to determine a tolerance level of ionizing irradiation, below which (statistically speaking) no biological damage can be expected at a later date. If one plots the irradiation dosage (in rem) against the incidence of sickness, the curve transects the origin. Thus strictly speaking, the term "greatest tolerable dose" in the German Ordinance for Irradiation Protection means no more than that a certain number of lethal victims of subsequent effects of irradiation are the price we are willing to pay for achieving some greater end. The very fact that there are different ordinances for threshold levels at the workplace and elsewhere makes it quite clear that such environmental standards are "not natural phenomena but special rules, whose justification depends upon the purpose they are to serve" (Gethmann and Mittelstraß 1992, 18). Thus when these authors call for a more "cultural" understanding of environmental standards, they do not intend to deny the significance of scientific research and factual evidence for establishing them. Their aim instead is "to interpret the particular contribution of scientific research in a non-naturalistic manner" (Gethmann and Mittelstraß 1992, 18).

A highly sophisticated view of this kind is not only of interest from an epistemological standpoint. When representatives of a naturalistic perspective deny that environmental standards are ultimately conventions, they eliminate both the possibility and necessity of justifying these conventions in public discourse. But in view of the fact that such conventions may cause other individuals, the environment, or future generations to accept an often incalculable and perhaps even existential risk, continuous critical examination and public justification of norms can even be considered a moral obligation. In light of this discussion, an expression that suggests that a threshold value was *ascertained* not only invites misunderstanding. It is also a sign of lack of responsibility unless what the *person affected* is willing to accept is explicitly addressed. In the words of Beck (1988, 144, 145) an *acceptable* risk is ultimately always an *accepted* risk. "Neither can we 'prove' by experiments or simulations what people have to accept nor can any risk calculations be established by technical and bureaucratic dictatorship. Risk calculations require what they are actually supposed to generate, cultural acceptance."[22]

9.b. "Ascertaining" What Should Be Protected?

In view of the problematic consequences of naturalism in politics and administration, you would think that their "critical opponents," that is, representatives of nature and species protection, would criticize naturalistic tendencies or at least avoid them because of the impediments they create. But that doesn't always seem to be the case. As the publications of people in nature conservation agencies and scientists active in species and nature protection reveal, the "nature lobby" also tends to use naturalistic arguments to promote their interests. Take, for example, "species diversity," a key concept that is certainly ambivalent as far as its normative content is concerned. This term is often employed in regional zoning and ecological auditing processes as a scientific criterion when evaluating whether or not an area should be placed under protection (e.g., Gerstberger 1991). But those who use this criterion rarely seem to think that it requires any further justification.[23] However, in order to arrive at "statements from ecological science of an appropriately evaluative and judgemental nature," it is not enough to demand "that the means by which data and facts have been attained, edited and interpreted be made transparent and reproducible," as Erz (1984, 2) has suggested, in hopes that "this criterion" might help us to "differentiate between the science of ecology and an ideology based on fragments of ecological thought." It is also necessary to admit that as a *descriptive* science ecology is not capable of making such value judgments all by itself. It requires the "assistance" of a normative discipline (as, for example, ethics). In light of Hume's Law, which maintains that a normative statement cannot be derived from a premise that does not itself contain at least one normative statement, there is no way to get around this (see Lehnes 1994). Thus it is surprising that Erz (1986, 13, Figure 1) explicitly excludes the humanities and philosophical disciplines in his representation of "the difference between the science of ecology, nature conservancy and conservation research." This could be interpreted to mean that he feels that the normative statements contained in the premises of these fields require no further discussion or justification. Support for this speculation is found in his concept of an analogy between nature protection and medical practice, which implies that it is just as easy to describe the "natural functioning" of an ecosystem as it is the "health" of a human being, and that the normative implications are also just as indisputable. As I will demonstrate in the following chapters, this idea is not tenable, and the apparently self-evident normative value of certain "key ecological concepts" raises both logical and factual doubts.

Thus naturalism presents nature conservation and species protection with the same problems as those described for politics and administration. When the norms contained in premises are regarded as scientifically founded or as natural phenomena whose validity is self-evident, they are removed from both the possibility and necessity of critical examination and discussion. This then enhances the risk that under the guise of objective knowledge subjective value judgments creep in unnoticed, which may sometimes even contradict one another. For example, *one* conservationist might call for grazing on an xeric meadow because species diversity can be maximized by these means while *another* might favor natural succession (with less species diversity), which would temporarily provide the bushes that the capercaillie (an endangered grouse species) needs for breeding. The problem here is not the conflict itself. The field of nature conservation is full of them. It is the naturalistic perspective and its failure to understand the source of the conflict, namely the question of values. If this question is not addressed, because nature or science has apparently already answered it to the advantage of the position one happens to favor, the possibility for a rational solution is just about null. Research results and lobbying then remain an opaque conglomerate, and the naturalist ultimately has no other resort than to defend his position by contesting the scientific knowledge of his opponents. This leads to unproductive polemics, which not only discredit the endeavors of nature conservation but also damage the reputation of scientific ecology. Ecology is then forced to deal with the "frequently expressed reproach" that it has been imbued with ideology or that "science and ideology can no longer be distinguished in this area" (Erz 1984, 2).

10. What Do We Mean by "Ecological"?

The criticism of ideological contamination, however, also stems from semantic misunderstandings due to the increasingly imprecise use of the terms "ecology" and "ecological" in public discourse, associated with an almost complete alteration of their meaning. Both terms are currently employed with a multitude of meanings extending from the political to the moral and even ideological, which have little in common with the descriptive science defined by Ernst Haeckel.[24] If you try to make sense of combinations of words such as ecological economy, ecological laundry soap, or ecological convictions, it becomes obvious that the term "ecological" always implies certain values or claims above and beyond simply referring to ecological relationships.

These normative implications are particularly apparent when you consider the opposites "ecological" and "nonecological," terms that make no sense at all on a purely descriptive level. According to its original meaning the term "ecological" encompasses *all* kinds of relationships between organisms and their environment, regardless of whether these relationships are natural and thus desired or whether they were altered by human intervention. From this perspective the relationships in an eutrophic village pond are no less ecological than those in a crystal clear mountain brook. In both cases the science of ecology is unable to decide which of the possible ecological states should be maintained, promoted, or reconstituted.[25]

This may become more plausible by drawing on an analogy to a different science, physics. We would hardly call it "nonphysical," in analogy to the term "nonecological," if someone were to throw his suitcase out the hotel window instead of taking it down in the elevator. In this case it seems obvious that it is not the task of *physics* to evaluate the way the suitcase is transported. All physics can do is to calculate the physical parameters involved.

However, if ecology is continually expected to provide value judgments, this reflects the popular idea that there is something like an *ideal ecological state* of nature that can be determined objectively. According to this view "ecological" almost means something like "heavenly" or at least implies a natural order in which equilibrium, harmony, and general well-being predominate. That this vision of ecological reality is an illusion is illustrated quite graphically by two examples that Dahl (1989a, 57) has outlined: "Let us assume that a common housefly is somehow able to form an opinion about its environment, and who would be willing to swear that it can't? It would probably consider the absence of rotten meat in the room to be an existentially unreasonable demand, and it would not talk about proper ecological conditions until the cat under the couch regurgitates, thus providing plenty of nutritional resources. For a fly the concept of what a favorable world might look like is quite different than for the human inhabitants of the house, . . . and for cholera bacteria it is quite different than for a person suffering from cholera. While one is departing from the world, hordes of bacteria exult over the onset of good times, unless, of course, they have been poisoned by medicine and, fading away, lament the insolent attack on their otherwise so intact ecology." It seems therefore that whether or not a habitat is ecologically intact and whether or not this is thought to be ecologically good or acceptable is apparently a matter of perspective and depends upon the specific needs and interests of the creature who wants to inhabit this "ecology." What organisms *should* live where—this is a question that ecology alone cannot answer.

If it is not possible to determine what is "ecologically good" in a manner that takes *all species* into consideration, we still, of course, have the option of attempting this in an *anthropocentric* context. Within this context we could not define an ideal ecological state that would be equally desirable for *all* species, but at least we would be able to establish this with respect to *human* well-being by objective, scientific means. Retreating to an anthropocentric perspective would be plausible since it would correspond to both prevailing legal practice and the widespread idea that ecology is basically just another word for environmental protection.

Apart from the dubiousness of this approach from an *ethical* standpoint since it restricts consideration to human interests, anthropocentric naturalism is no less illusionary than naturalism that encompasses all species. Just as different *species* have different interests in an ecosystem, we must also expect various ideas from different *people* concerning the ecological conditions in which they would prefer to live. Is it "for humans" most favorable to live in a wild primal forest, a landscape characterized by small farms, streamlined agricultural areas or (as it is so euphemistically phrased) a "flourishing industrial landscape"? It seems obvious that this question can neither be answered by referring exclusively to ecological relationships nor can an answer be found in a universally accepted manner.

For practical purposes it would also not be very productive to try to "ascertain" *minimal ecological conditions* for the survival of humankind. First, the extreme technical optimist might point to the innovative potential of humans and their ability to adapt and question whether or not minimal conditions even exist. Second, no human being would voluntarily choose to live under such conditions. Fictitious minimal conditions might be enough to secure collective survival of humankind, but they would certainly not suffice to guarantee a humane and morally good life. How the latter might be defined more exactly from an ecological perspective is a matter I cannot pursue any further at this point. For the present discussion it is enough to note that this problem itself is not simply an ecological one.

It seems strange to conclude that determining what is commonly regarded as "ecological" is primarily a matter of interests and views about humanity rather than ecological relationships. Even if the term "ecological" in this sense hardly has anything more to do with science, it apparently still almost automatically bears the connotation of science. The problem here is not so much that normative and scientific categories are mixed and interchanged, but rather that people do not seem to be aware of this. *Hidden naturalism* is more dangerous than open naturalism. When normative dimensions are unconsciously and inadvertently superimposed on ecological ones, the impression arises that we are dealing with claims that are exclusively rooted in scientific knowledge and must simply be carried over into the realm of practical execution. The *factual information* involved may have been correctly determined by scientific means but then coupled with *normative premises* in an inappropriate manner. If these are not recognized as such and "dissected" from the factual content, *all in all* they will lead to erroneous conclusions, even if the facts are correct.

Just how important it is to differentiate between ecological facts and normative premises attached to them is shown by the sometimes careless use of ecological slogans in discussions within nature conservation and environmental ethics. Claims about the importance of things like species diversity, ecological stability, closed cycles, and equilibrium are by no means as ecologically "self-evident" as some might think. Contrary to the naturalistic idea that these claims can be *directly* deduced from ecological evidence, their normative content always originates in nonecological premises, to which, of course, reference is seldom made. Since it is quite probable that many so-called "guiding principles" of ecology are often unconsciously and unwittingly loaded with normative content, I wish to subject them to closer scrutiny in the following sections and check in particular for unjustified generalities. In this connection I regard the relationship between normative reflection and the science of ecology as reciprocal. Even though it is not possible to directly *deduce* norms from facts, it is still necessary to *refer to* facts when formulating norms. Therefore, after having rejected *naturalistic* fallacy I also wish to point out the dangers of what might be called *normativistic* fallacy, which consists in the erroneous assumption that specific or concrete obligations can be reached *solely* on the basis of normative considerations (see Chapter 14).

11. A Critique of Guiding Principles of Ecology

11.a. Ecological Equilibrium

Probably the most well-known of all ecological concepts, the term that is regarded as the epitome of ecological thought, is that of ecological equilibrium. In this concept hopes for harmony and peace between humankind and nature, similar to those attached to the term "ecological" when it is applied normatively, are condensed particularly strongly. That this popular but nonetheless illusionary interpretation of ecology extends all the way into contemporary philosophical literature is demonstrated by Maurer's (1982, 28) view, according to which ecology is "the science of equilibrium, of a harmonious relationship between human and nonhuman nature." Schönherr (1985, 22) even assigns a metaphysical dimension to equilibrium when he refers to "ecological equilibrium based on pristine nature" as "the meaning of earth." He equates a "violation of ecological equilibrium" with a "violation of nature" or rather as "lost nature" (Schönherr 1985, 133), whereby visions of banishment from the garden of Eden as described in the myths of the Old Testament are involuntarily elicited. Since it should be clear from the analysis of the term "ecological" that even in pristine nature unsullied by human intervention there is no such thing as an ecological paradise with welfare for all, the question is what the term ecological equilibrium can and should mean otherwise.

A striking thing is its similarity with the concept of equilibrium in *mechanics*. In mechanics one refers to a state of equilibrium when the resulting sum of two forces acting on the same point is zero. In analogy to such physical forces one could imagine species or groups of species in an ecosystem whose populations keep each other in a state of "stalemate." If, for example, in the context of a biological community both barn owls and mice are able to survive over longer periods of time, then (physically speaking) the result of these two antagonistic forces seems to be zero, and it appears that we are dealing with a case of so-called biological equilibrium. The average degree of equilibrium between these forces or rather species is a function of the relationship between them. We may therefore refer to this state as an example of self-regulation (Osche 1978, 57).

When the dynamics of populations are examined more closely, however, a very basic difference between the concept of equilibrium in mechanics and ecology is revealed. In mechanics a system remains in equilibrium as long as no other external forces set this system in motion (static equilibrium). But a characteristic property of a biological community is that it is in a state of constant motion and change as a result of internal and external influences (Stugren 1978, 128). Thus the population densities of barn owls and mice do not eventually reach some constant value but instead constantly oscillate around some statistically defined mean as a result of internal feedback mechanisms (Osche 1978, 60). It only makes sense to speak of an equilibrium if we refer to this mean value and exclude irregular and extreme influences from without. In the metaphoric terms of Dahl (1989a, 59), there is no ecological equilibrium independent of these conditions, "just as there is no car that drives forward in a straight line. At the most there is constant oscillation around some midline, whereby the scale quivers from side to side and usually not only quivers but tips abruptly in one or the other direction and rarely pauses for a moment of deceptive attenuation."

What is true for a simple predator/prey system idealized as a closed system[26] is all the more true when ecological systems are viewed as what they really are, namely as *open systems* that are continually exposed to external influences and therefore also to *disturbances*. But it would be mistaken to think that such disturbances always are due to *human* intervention in the established harmony of once untouched nature. On the contrary, it is often *nature itself* that eventually destabilizes its supposed states of equilibrium. The reason why this is usually not mentioned in popular representations of the ecological equilibrium is probably that the extreme temporal and spatial dimensions of such disturbances compared to the dimensions of human perception are so great that they are rarely perceived first hand.

Thus you have to look very carefully in order to recognize natural processes such as a falling tree or the loss of a section of a riverbank as smaller disturbances of the ecological equilibrium. For some of the organisms and smaller systems affected, events of this kind might even be a kind of ecological catastrophe. Other species, on the other hand, might profit from such sudden disturbances and rapidly reproduce for a short time afterward. These r-strategists (or opportunists) are not well adapted to their surroundings but capable of rapid reproduction and fast replacement of one generation by another. Some of them may even require regular and repeated disturbances of smaller dimensions.

Humans more readily tend to notice disturbances with spatial dimensions that are within the scope of their own unaided perception, for example, forest fires, mass insect or algal proliferation, periods of drought, snowstorms or floods. Since events of this kind are relatively rare by human standards of time and also occur irregularly, they are often perceived as a stroke of destructive fate acting *upon* nature rather than as normal processes *within* nature.[27] And their fundamental significance for the structure and development of ecosystems is thus often underestimated. Fire, for example, is an important natural parameter for grassland savannahs, because it regularly reduces the mass and number of species of woody plants and induces the germination of seeds in the grassy layer. Thus in savannahs fire also often prevents *succession* (i.e., a regular sequence of plant and animal communities in the course of time leading up to a stable "climax community"). Other disturbances, however, such as snowstorms or lesions in beaver dams can create mosaic-like "successions within a system" (Remmert 1984, 201), which are critical for the survival of many currently endangered species of plants and animals (e.g., the black grouse). One of many examples of disturbance due to *natural* mass proliferation is that of the red tides on the coasts of North America and the Gulf of Mexico, described at least as early as 1844. From time to time a certain combination of environmental conditions triggers the explosive growth of a species of red-colored plankton (from the dinoflagellate group), whose paralyzing poison causes millions of fish and other marine animals to perish in one fell swoop (Farb 1976, 158, 159). According to Reichholf (1993, 221) these and other cases of mass proliferation are not a sign of a disturbed equilibrium but simply reflect a particularly favorable constellation of conditions for a particular species of plant or animal.

If you look at ecosystems from the perspective of evolutionary history and include large climatic disturbances in your considerations, the everyday concept of ecological equilibrium begins to teeter even more precariously. If there really were such an ecological equilibrium in a rigorous sense, it would cause the process of evolution to come to a standstill (Kreeb 1979, 93). In a completely balanced biosphere species would probably not suffer extinction the way 98 percent of all the species that ever lived most likely have (Ehrlich and Ehrlich 1981, 28), nor would new ones arise that might then compete with those already existing. Remmert (1984, 1) indicated this misunderstanding when he ironically referred to green plants (eucaryotes) as "the first great environmental polluters," the reason being that "when they created the earth's present oxygen atmosphere through photosynthesis and thus subjected the surface of the earth to oxidation, all those living things were doomed to die which had previously adapted themselves to life without oxygen." For the blue-green algae (procaryotes) that predominated at the time, oxygen was a metabolic poison! According to Remmert (1984, 194), other evolutionary novelties which knocked older, prevailing, and quite well-balanced ecosystems out of equilibrium in a relatively short period of time were the retreat of bony fish from fresh water back to the sea, the rise of warm blooded animals, mammals' migration to the sea and the development of seed plants. These incidences all show that from an evolutionary perspective disturbances are not only inevitable. They can even be regarded as productive since they enhance the adaptability of organisms and open up new pathways of evolution.

Paradoxically, the most catastrophic of all disturbances in the course of evolution seem to have been the most productive in the sense of opening up new pathways, namely the *geological* and *extraterrestrial* ones. It is possible that they contributed to some of the so-called mass extinctions (extinction of many species within a short period of time) and successive evolutionary breakthroughs through "adaptive radiation." Thus according to a theory that is gaining more and more support nowadays is that the extinction of the dinosaurs resulted from a meteorite hitting the earth 65 million years ago. The fiery explosion that ensued released enormous amounts of dust and fine particles into the atmosphere, which blocked out the sunlight for months afterward (Alvarez et al. 1980; Hsü et al.1982; Keller 1992, 108). The resulting changes in climate led to the death of many species of dinosaurs in the millennia that followed. Only after this interlude did an opening occur for mammals, which until then had played an insignificant part in life on earth. A similar effect on the extinction and development of species as that of such "cosmic hits" might have come about through volcanic explosions.

Other events that were not as abrupt but have had just as serious consequences for the ecosystems involved were *climatic* disturbances such as the possibly regular occurrence of warm periods and glacial periods (Stanley 1987, 209f.). If you recall that during the early Tertiary Period (about 60 million years ago) mixed leafy green deciduous and coniferous forests flourished around Spitsbergen, Norway, and that later on in the Pleistocene Period (about 2 million years ago) only periglacial steppes and dwarf shrub heaths were able to exist in Middle Europe (Ehrendorfer 1978, 950), you begin to realize how enormous the alterations in flora and fauna have been. Samples taken by drilling through the ice in Greenland, which permit us to reconstruct changes in temperature during the last 250,000 years, have shown that such extreme alterations in temperature not only took place in periods of millions of years. The results indicate that periods with a constantly warm climate usually lasted for three thousand years at the most, and that catastrophic drops in average climatic temperature occurred regularly in between. Relatively stable climatic periods such as we have experienced in the past few thousand years seem to be the exception rather than the rule (Gerdes 1993c).

In light of these examples the inevitable conclusion seems to be that in the *rigorous sense* of the word there is no such thing as the highly touted ecological equilibrium. In view of the dynamics of evolutionary history and the "normalness" of disturbances, the term must be regarded as at the most "a legitimate simplification which is only valid for certain periods of time and certain situations" (Kreeb 1979, 91; see also Chesson and Case 1986). It reflects the idea that the structure of an ecological system remains constant for longer periods of time *by human standards of time measurement*, even if the species composition and population densities may vary somewhat. A forest, for example, is subject to continual quantitative and qualitative changes, but it can still remain a forest for hundreds of years. Relative stability of this kind is achieved by feedback loops that cause the species composition and number of individuals to return to *approximately* the same state after a disturbance has occurred. If the disturbance is too great, of course, so that the information contained in it can no longer be adequately stored, the system will not be able to absorb the disturbance.

In this case, since it is not possible to return to the original state of the system, people often speak of a "collapse of the ecological equilibrium." This expression is misleading because it implies that for a particular region there is only *one* possible state of equilibrium, which then gives way to a state of nonequilibrium. But in reality, after each and every disturbance, either the original state of the system is reinstated, or else a new state of equilibrium is established. It is hard to avoid the conclusion that from this perspective the term ecological equilibrium is rather meaningless. If a new state of equilibrium *automatically* succeeds an old one, all the term tells us is whether or not an alteration has occurred.

The same holds true for the very popular image of an *ecological web*, which has already been criticized elsewhere in this book (Chapter 4.d). Like the term equilibrium it is based on a static concept of a system, which tends to view the currently existing state as the valid one. Thus the picture of a web undoubtedly nurtures the idea that tearing apart the ties that exist within an ecosystem harms its ability to "function" in the same way as tears in the meshing and knots render a fisherman's net useless. In reality, however, ecological webs are much more flexible than human-made ones. "When a knot in an ecological web tears apart, the rest of the web is readily resealed to form a complete web, sometimes at the expense of a few other knots. Sometimes new threads from outside for which there was no room in the web before are woven into it. The web may even become tighter than it ever was before. At any rate, it remains a 'web' as long as any life at all is around" (Dahl 1989a, 60). Since every web in principle remains a web even after the most serious disturbance, *from a purely ecological standpoint* it must be regarded as just as meaningless to refer to a "destroyed web" as it is to refer to a "destroyed equilibrium." In the end, both terms simply express the rather trivial observation that something has happened to some ecological relationships.

But why then are these two terms used so frequently when they mean so little from an ecological perspective? The reason seems to be that they are not supposed to transport ecological meaning but rather a *normative* message. This can be expressed as follows: "The ecological state we currently experience as being natural is preferable to any other one. Stability is better than change!" Having shown that the concept of equilibrium in the sense of long-term stability is only a simplified borderline case and that it "is not defined in ecology in an unequivocal and generally acceptable manner" (Stugren 1978, 128), it should be clear that a claim of the kind noted above can hardly be justified ecologically. Ecology can provide no evidence for why of all things the state that just happens to exist should remain the way it is. Of course this means that the reverse, which is also frequently claimed, must also be rejected as a case of naturalistic fallacy, namely the claim that simply because disturbances have been shown to occur *naturally*, any intervention whatsoever *by humans* in nature can be vindicated by ecology. Thus current interventions by humans in the climatic conditions of our planet can neither be *justified* ecologically by pointing out *similarities* with natural climate fluctuations in the past nor can they be *criticized* ecologically by referring to some *unusual characteristics* that distinguish current fluctuations from former ones.[28] Demands such as those for stability and preservation of the present ecological state must seek justification outside of ecology, by considering the interests involved and, as I shall demonstrate in Part B, on the basis of ethical considerations. Of course it must be kept in mind that these considerations should not be formulated

without taking ecological evidence into account. If maintaining and protecting species is regarded as something that ethics requires, then ecology
must provide information about whether or not ecological stability is necessary to achieve this end and what exactly the term means.

11.b. Ecological Stability

If we put the normative and sometimes even magical connotations of the
term "ecológical stability" aside and first attempt to grasp its scientific and
descriptive content, then we experience similar problems with analyzing
this concept as we already did when dealing with the term "ecological
equilibrium." Symptomatic of these difficulties is the fact that the term "stability" is not defined uniformly in ecological literature either[29] and therefore is often used in a contradictory manner. Since this can lead to
misunderstandings, particularly in discussions about nature conservation,
a few ecologists have suggested that we distinguish between different kinds
of stability. Depending upon the source of certain disturbances (inside of or
outside of the ecosystem) and the response of the system, three different
types of stability are usually distinguished: (1) resistance stability, (2) resilience stability, and (3) constancy (Pimm 1984, 322).

Remmert (1984, 260) recommended that only those systems be termed
stable that exhibit "no significant change" in response to external influences and that immediately absorb external disturbances. As Zwölfer
(1978, 15) has emphasized, however, this doesn't mean that the number of
individuals remains stable but rather that the "basic stock of species" remains unaltered. According to Remmert, systems that are not stable can be
assigned to two different classes depending upon the way they react to external influences. If they are *resilient*, then contrary to stable systems they
exhibit change but manage to return to the original state relatively quickly.
But if they are *sensitive*, they are unable to compensate for the effects of external disturbances. Once they have been pushed beyond the limits of resistance, systems of this kind must then attain a new state of "equilibrium."

Resistance to *external* factors can be distinguished from continuity with
respect to the *internal* structure and state of an ecological system. In the latter case, ecologists currently refer more and more often to "constancy" or
"variability." "Accordingly, a *constant* system is one in which only relatively
minor alterations can be observed under the prevailing climatic conditions"
(Remmert 1984, 260), in other words, the species composition and the

stock of individuals of each species varies from year to year only within certain defined limits. The importance of distinguishing between constancy and stability is demonstrated by the recurring observation that ecosystems that appear to be stable with respect to population dynamics (i.e., *constant* ecosystems such as a tropical rain forest) may be existentially threatened by disturbances, while ecosystems with pronounced fluctuations in the number of organisms of certain species (e.g., tundra or taiga) usually prove to be stable or resilient. I shall return to this point in the discussion about species diversity and stability.

Even if the distinctions in the concept of stability outlined above seem to have helped to clarify the most serious misunderstandings, one problem still remains to be solved, the problem of *objective* measures of stability in time and space. The difficulties involved have already been indicated by the use of fuzzy expressions such as "*relatively* minor changes" in connection with the definition of constant systems or "a *relatively* rapid return to the former state of the system" with respect to the definition of resilient systems. Just as it would be meaningless to characterize an astronomical object as "relatively small" (small with respect to meteorites, planets, stars, or galaxies?), so also is the term "ecologically stable" meaningless without determining the spatial and temporal frame of reference involved. Wiens and his colleagues (1986, 145) therefore place a great deal of significance on the scaling system used in ecological research. "Some of the most vociferous disagreements among ecologists arise from differences in their choice of scale."

As far as the choice of a *measure of time* is concerned, it appears that the average lifetime of a human being is usually applied as a standard measure (Hoekstra et al. 1991, 154). For example, a forest seems to be stable simply because trees live longer than humans do (Reichholf 1993, 38). Even a lake that cannot be seen to be drying up will appear to be "relatively stable" to someone with a life expectancy of about eighty years, even if the lake is no more than a brief interlude in the history of a landscape when viewed from a greater temporal perspective. And the other way around, a puddle with a lifespan of a few weeks appears to be inconstant by human standards, while from the perspective of a mosquito larva it is undoubtedly constant enough to ensure its development to a full-fledged adult insect. Some organisms of plant plankton can even pass through several generations during the lifetime of the puddle. This shows that the term stability in the sense that it is usually employed can hardly be considered to be scientifically objective. Rather than using the lifespan of *the investigating subject* (a human) as a measure, it would be necessary to define the term from case to case on the basis of the various different lifespans of the members of *the object under investigation* (the ecosystem). It cannot be denied, however, that there may be good *practical reasons* for employing the lifespan of a human as a standard measure of stability. But in this case it would be important to indicate that a defining criterion for the concept of stability has been chosen for methodological reasons (or even reasons of personal interest), not for "inviolable ecological" ones (see Hoekstra et al. 1991, 154).

Similar fuzziness with respect to the term "stability" can be observed when determining the *spatial* frame of reference. By their very nature small systemic units such as a patch of woods exhibit more fluctuations and change than the entire forest zone of middle Europe. This phenomenon is illustrated by the *mosaic-cycle theory*, which is currently in the process of replacing or at least modifying the climax theory (Remmert 1991).[30] According to the mosaic-cycle theory, an ecosystem is usually not a homogeneous entity that eventually attains a stable end state (climax state) after a linear sequence of developmental stages (succession), but rather consists of a large mosaic of subsystems, each in a different state of development and each capable of starting anew with a cycle of succession after experiencing a period of disintegration or local disturbance. In an extensive beech forest, for example, many acres of meadows, birch and meadow areas, cherry and maple sectors, and areas with old beech tree stands probably alternate with and replace one another (Remmert 1984, 201). This example of intra-systemic succession illustrates nicely how important the size of the investigated system is for the problem of stability. No system exhibits uniform stability in each and every one of its parts. A smaller area might appear to be unstable (in the sense of cyclic *succession*) but might also be viewed as part of a larger system that appears to be stable (when regarded as a *system* of cyclic successions). "Thus in the long run, ecological stability is not the stability of stationary systems but the stability of processes" (Zwölfer 1978, 22).

This concept of ecological stability, which is dynamic and thus quite different from everyday thought, has only gradually gained a foothold in ecology, most likely due at least in part to the widespread prevalence of the static terminology of the climax concept. Even in textbooks from the year 1978 you can still find references to a concept of stabilization (climax), which maintain that "in the final stage of succession no further changes in the species composition of the community occur" (Osche 1978, 32), or that the "community continues to exist permanently without any significant alterations" (Stugren 1978, 228). Ecosystems that occur at the usual location for such communities (i.e., those in which the local climatic conditions correspond to macroclimatic ones) are referred to as "final communities" and they are called "permanent communities" when they occur at other locations (Stugren 1978, 228). No doubt expressions of this kind have contributed to the idea that ecosystems are by nature or rather in absence of human intervention permanently stable. Williamson (1987, 368), on the other hand, has contradicted this idea in his essay titled *Are Communities Ever Stable?* in which he maintains that "in the long run no community has been stable to evolutionary change." Having shown that the concept of permanent stability at the end of a supposedly linear process of succession is a misunderstanding, another one that often goes along with it becomes apparent, the idea that an ecological system of its own accord "strives" to attain a state of greater stability (see Osche 1978, 37). As Zwölfer (1978, 23) has remarked, in nature final stability cannot be "sought after." "An increase in stability simply results from the fact that in the course of time any gains in stability enhance the chances that the system will continue to survive as such while reduced stability favors change in the system."

If an ecosystem cannot "strive for" its own stability, and if neither the combination of species woven into the ecosystem nor the length of their persistence can be guaranteed, one wonders what the basis is for the *normative* claims connected with the term "stability" in discourse on nature and species protection. Why and in what respect is stability supposed to be "better" than ecological change? Having already expanded on Hume's Law to demonstrate that it is *logically* impossible to give a straight ecological answer to this normative question (see Chapter 8), the conceptual analysis of the term "stability" should have made it clear that this is also impossible for *factual* reasons.

The main problem is still the temporal and spatial dimensions of the selected frame of reference, without which the term "stability" remains meaningless and any demand for stability as well. This problem is nicely illustrated by small protected areas in which people attempt to keep a certain basic stock of usually rare species or groups of species constant at a *local level* by means of biotope management. This can often only be achieved through strong and persistent interventions by humans (e.g., through grazing, bank reinforcement, or controlling the population size of competing species). If you recall the natural dynamics of small systemic units described by the mosaic-cycle theory, there is no way to avoid the conclusion that "holding onto" certain phases of succession in such an artificial manner can hardly be justified by ecological and naturalistic arguments. In open systems in nature, "animal stocks constantly vary in time and space; everything is in flux" (Reichholf 1993, 220). In view of the limited size of many protected areas nowadays it follows that it is difficult to "correctly" interpret fluctuations in the population size and species composition with respect to stability. A decrease in population size that might appear to be an alarming sign of instability at the regional level could look like natural fluctuation or shifting in species count when viewed from a state or nationwide perspective.

Which of these apparently incompatible frames of reference should be used to define the term "ecological stability," if this is to be a guiding principle for our actions? The short-term local one, the intermediate statewide one, or perhaps even the long-term global perspective? Anyone who favors the *global, long-term* frame of reference in view of the critique of the local, short-term one must be aware that at this level ecological stability can only mean stability of life, not the stability of particular ecosystems or stocks of species. In view of the fact that the evolutionary process tends to "maintain life but destroy species" (Markl 1981, 26), it is impossible to devise any specific instructions about how to act with respect to protecting species on the basis of such a comprehensive concept of stability.

The highly ambivalent consequences that can result from a naturalistic interpretation of the *global* concept of stability are demonstrated by the very different conclusions that have been drawn from Lovelock's (1988) so-called *Gaia Theory*. According to this theory,[31] which gained interest in geophysics, marine research, and climate research, the planet Earth is not simply an "environment" for life but something like a living organism itself, a system capable of self-maintenance that modifies its surroundings to ensure its survival. "The atmosphere, the oceans, the climate and the crust of the earth are regulated at a state comfortable for life because of the behavior of living organisms" (Lovelock quoted in McKibben 1990, 143). As a result of many interconnected feedback mechanisms the system is able to absorb even the most serious damage such as that which has repeatedly arisen due to showers of meteorites in the course of earth's history and eventually return to a dynamic state of equilibrium. According to Lovelock the Gaia System would easily survive the consequences of an atomic war or total destruction of the ozone layer. Massive disturbances of this kind could, of course, lead to a state of the biosphere in which *humans* (and other higher organisms) would no longer be able to live, but life itself would quite probably still continue to exist (at least in the form of unicellular biosystems). On the one hand this can be regarded as a warning to humans to remember that they are dependent upon nature but not vice versa. On the other, however, this has often led to the conclusion that we don't have to worry too much about the fate of our planet. Regardless of what we do or don't do, life on earth will continue anyway. In fact, the subtitle of the German edition of Lovelock's book (1995), "*Gaia, an Optimistic Ecology*," points exactly in this direction. The fact that this title nevertheless fails to elicit undivided optimism but a certain degree of uneasiness instead suggests to me two reasons why the naturalism on which it is based is not very convincing. First, it is apparently not self-evident to regard life as defined by the dimensions of planetary history as a measure of the concept of ecological stability, and second, a decision in favor of such a measure can in no way be deduced from the science of ecology. Once a frame of reference for stability has been established, ecology can certainly show us *what* should be done or not done *in order to* achieve more or less stability of the system involved. But the decision as to *whether* a system should be kept stable and on the basis of *which frame of reference* must be reached outside of ecology.

This must also be kept in mind for the most frequently selected frame of reference that is usually quietly assumed for the term "ecological stability," namely the *intermediate* one defined by the lifespan of human beings. Here too, as already indicated, we are not dealing with an *ecological* criterion for stability but rather with a "naïve, egoistic" one (Dahl 1989a, 62). It reflects the human desire to live in a world of security, continuity, and predictability (Reichholf 1993, 217). It seems that humans would be most comfortable if nature would call its dynamics to a halt and if the same conditions would prevail ten years from now that we have at the present. This is understandable, of course, since nature characterized by constancy would be easier to control and thus also easier to exploit than a dynamic nature, but such straightforward self-interest should not be sold as ecology. Just as ecological stability is not a characteristic of pristine nature, so also is the demand for it not in keeping with "the interests of nature." If there is any such thing in nature as interests other than those of humans,[32] these are far too heterogeneous to all be able to be satisfied by a concept of stability based only on human standards. And if someone expresses the desire to stabilize "the equilibrium of nature," he should admit that he basically has himself in mind as well as a particular vision of nature directed toward his own interests.

11.c. Species Diversity

In ecology and in the environmental movement the conflict between human interests and those of other species indicated earlier is often circumvented by pointing out that in the long run human self-interest must automatically take the interests of other species into account since the ecological stability to which humans aspire can only be attained via species diversity. The more species an ecosystem has, and the more highly integrated it therefore is, the more resistant it is to undesired disturbances of the ecological equilibrium. By referring to the image of the food web discussed earlier, the very plausible idea is transported that every species of organism is a kind of "knot" in a web and therefore contributes to the stability of this web by virtue of its relationships to other species. As evidence for the relationship between the number of "knots" in a web and its stability the example of monocultures in modern agriculture is often presented, which have proven to be more susceptible to mass attack by pests than comparable cultures consisting of several species (e.g., Tahvanainen and Root 1972). Experiences of this kind from biological pest control as well as simply structured scientific experiments have reinforced the general idea that species diversity *generates* ecological stability. And anyone who wants stability must promote species diversity.

If you recall the basic difficulties with which ecology must deal in trying to arrive at general rules and simple laws (Chapter 4.e), it is hardly surprising that results from numerous investigations have been presented in the past twenty years that contradict the simple equation "species diversity = stability." The evidence and conclusions published in this connection are so complicated and heterogeneous that according to current views the *stability-diversity hypothesis* appears to be an illegitimate generalization.

One of the reasons for the increasing skepticism with which the relationship between stability and species diversity is regarded has to do with the ambiguity of the term *stability*. This can be deduced quite simply from the fact that even more conceptual distinctions for the term have been proposed (Robinson and Valentine 1979; King and Pimm 1983) than the categories of resistance stability, resilience stability, and constancy described earlier (see Grimm and Wissel 1997 for a critical review). While some authors consider stability to be the ability of an ecosystem keep its *biomass* constant (biomass stability), others are primarily concerned with its ability to maintain a stable *composition of species* (species deletion stability). In this respect, model studies of plant-herbivore systems revealed that both types of stability apparently are reciprocally related to one another (Pimm 1984, 321). Plant societies with a large number of species do exhibit greater biomass stability, but they are subject to a greater risk of losing one of the competing species of which they are composed when the herbivore is removed.

The competitive behavior of species among one another suggests a second reason for the controversial response to the stability-diversity problem, namely the conceptual ambiguity of the term *species diversity*.[33] It seems plausible that the relationship between stability and diversity is not just a matter of the *number of species* but is also dependent upon the number and kind of functional *relationships* between them, in other words really a matter of *complexity*. For *theoretical* reasons, you would expect greater stability from a complex food web than from a few simple food chains independent of one another. However, it is still a moot point whether or not there even is anything like a web *in reality*. While Stugren (1978, 136) believes, for example, that hemi-zoophagous animals (nonspecialized carnivores that also require a significant percentage of plant food) form interconnections that attach different food chains to one another and thus enhance the stability of the entire structure, Remmert (1984, 228) is skeptical about the significance of such interconnections. If you express all the relationships of a hypothetical web as a number and disregard all pathways that account for less than 1 permille (1/1000) of all the transported nutrients, a few major pathways or chains usually remain. From this perspective the web-like interconnections between different chains appear insignificant. Remmert therefore basically rules out the widespread idea that a web contributes to stability unless quantitative verification is provided. "However, real quantification has so far not been carried out for any ecosystem on earth, and therefore we as yet have no proof of a web" (Remmert 1984, 228).

The third aspect that raises strong doubts about the validity and explanatory potential of the stability-diversity hypothesis is its inadequacy for making *comparisons* between ecosystems with different local conditions. If, for example, one compares dry heath areas with many species and a bog with relatively few species, the bog is often the more stable of the two systems, in spite of its limited diversity. The arctic tundra, northern coniferous forests, and the marine sand-dune communities are also remarkably stable, even though they are composed of relatively few species because of their extreme locations. If you compare these stable systems containing only a few species with tropical rain forests, coral reefs, or middle European alluvial forests that are rich in species but highly sensitive to disturbances, not much will be left of the "favorite doctrine of community ecology." Once again ecology has been shown to be unable to establish a simple rule for ecosystems that is generally valid.

One response to objections of this kind that has been proposed several times is that the "rule of stability due to species diversity" was not ever meant to be used for comparative purposes but rather is "only valid for estimates involving the same type of biotope and for a series of communities that succeed one another naturally at a particular location" (Heydemann 1981, 27). However, in connection with the phenomenon of succession mentioned here it has been found that the most stable stage, the climax stage, is often *not* the most diverse one (Remmert 1984, 198, 199). Moreover, the uniqueness of ecosystems and their contingencies (see Chapter 4.e) make it rather difficult to determine when two biotopes belong to the same type. At the same time the manner in which the stability-diversity hypothesis has been qualified by the author cited above indicates quite clearly that diversity, if at all, is only *one* of a number of factors that contribute to stability and that local parameters might even play a much more important part.

As a matter of fact, numerous theoretical and empirical analyses in the past few years have led to the new and still somewhat unusual idea that "stability is really not a matter of ecology or biology but rather a question of whether or not resources are available when an intervention is undertaken that puts a strain on stability or resilience" (Remmert 1984, 265). Ecosystems are not necessarily stabilized by species diversity but rather *vice versa*: species diversity can flourish in areas where environmental conditions, particularly climate and soil conditions, are stable and where in addition the habitat is characterized by spatial and temporal heterogeneity. In other words, diversity is often not the *cause* but rather the *result* of stability, or more precisely, "a consequence of processes that generate an impression of stability" (Reichholf 1993, 24).

A good example of this thesis is the Amazon rain forest, with its more than forty species of trees per acre (compared to two per acre in moderate zones), which is thought to be one of the most diverse ecosystems on earth. It is characterized by a combination of four things that Tischler (1976, 114) describes as prerequisites for species diversity: (1) constant and favorable climatic conditions that preclude the necessity of adapting to seasonal changes; (2) a very advanced age (more than 100 million years) due to the constant climatic conditions, various different transformations notwithstanding; (3) a highly diversified habitat, and (4) a lack of nutrients in the soil, which encourages the formation of "specialists" that maintain highly diverse relationships between one another as well as with abiotic components of the system. With respect to the fourth point Reichholf (1993, 40) regards the development of diversity as "nature's response to deficiencies." Since almost 100 percent of all the nutrients in the Amazon rain forest are incorporated in plants and hardly anything can be extracted from the leached soil, only very little can be turned over. "But when only very little turnover occurs due to deficiencies, very little will change either. This generates the impression of stability" (Reichholf 1993, 40).

According to Remmert's definition (1984, 260, 261; see also 11.b), we are really dealing here with a case of constancy rather than stability. The delusive nature of the impression of rain forest stability is revealed by the rapidly advancing and irreversible destruction of this complex ecosystem by humans.[34] If the rain forest is extensively cleared and if its biologically active substances are burned or removed, this is equivalent to losing all the resources contained in it. Since the soil has no resources of its own, regeneration is impossible. At the most poor secondary growth could occur, which is a far cry from the original richness of the rain forest. As this sad example demonstrates, the enormous species diversity of an ecosystem may be of no use at all when its *vital resources* are the focal point of a certain kind of intervention.

This leads to a fourth weakness of the general claims inherent in the stability-diversity hypothesis, namely their failure to take into consideration possible *kinds* of intervention and the *time* of their implementation. Thus it makes a big difference from ecosystem to ecosystem whether or not intervention occurs in the form of chemical damage (via air or soil pollution), a physical attack (such as fire or chain saws), or biological disturbance (e.g., through mass proliferation or the introduction of exotic organisms). For example, although desert plants are usually quite resistant to physical stress as, for example, sandstorms, they are extremely sensitive to chemical pollution. The marginal conditions under which they exist give them hardly any energetic leeway for possible detoxification processes. And the other way around, tropical rain forests can hold up relatively well under chemical pollution, while physical stress such as fire, chopping, and clearing damages their "spinal cord," so to speak. The example of the European beech forest demonstrates that the location of such a "spinal cord" may be very different from ecosystem to ecosystem and may basically have nothing to do with reduced species diversity. This forest, which is composed of relatively few species, is able to withstand both regular clearcutting as well as lumber removal and usually regenerates time and time again. In modified analogy to the image of a food web Remmert (1984, 263) describes the paucity of species in such a system as a "uniform wall" and species diversity as "loosely constructed scaffolding" rather than a stable web (as is usually the case). These images make it easier to understand why a highly differentiated "filigree structure" such as a tropical rain forest can be destroyed much more readily than the "fortress wall" of a beech forest, even though the latter has far fewer species.

In light of this new evidence the old stability-diversity hypothesis is not only criticized but turned upside down. Instead of guaranteeing ecological *stability*, species diversity all of a sudden appears to be a highly fragile evolutionary strategy, which might lend a system biological *constancy*, but at the same time also renders it highly *sensitive* to external influences. If a system once succeeds in attaining high species diversity in response to the long-term constancy of local parameters and climatic conditions, then in the future it will continue to depend upon such lack of fluctuation and disturbance. The older and more complex it becomes, the less it appears to be able to survive the loss of certain key species. If a single species of plant becomes extinct, the system also loses all the "specialists" (e.g., pollinators, symbionts, parasites, etc.) that have become primarily or even exclusively adapted to this species in the course of a long process of coevolution. This "domino" or "run" effect is more and more significant the closer the plant species that disappeared originally was to the basis of the food web or various food chains (Heydemann 1985, 594).

From this still somewhat unusual but nonetheless more and more widely accepted idea that constancy and sensitivity are related, it seems to follow that it is not the *constant* and species rich systems that are best suited for satisfying utilitarian human interests in stability and resilience but rather *inconstant* systems, which usually have relatively few species. Whereas *constant* systems, which are accustomed to conditions that are usually more or less free of externally induced fluctuations, require particularly cautious and guarded care, *inconstant* systems, which might be well adapted to fluctuations such as vegetation loss due to fire or mass insect proliferation, would probably quickly recover from human attack with bulldozers. According to May (1976, 162), "this inverts the naïve, if well-intentioned, view that 'complexity begets stability,' and its accompanying moral that we should preserve, or even create, complex systems as buffers against man's importunities." Complex systems are probably far less suited than simple ones for such a buffer function.

Contrary to the concept of stability that was regarded as valid until well into the 1980s (e.g., Kurt 1982, 130; Amery 1982, 128; Wehnert 1988, 140), this new perspective represents a radical change. To determine whether or not it is correct, testing with long-term investigations is necessary. In the meantime "the new concept of stability and resilience" certainly "raises more questions than it provides answers" (Remmert 1984, 265). Therefore, to be on the safe side the reader must be cautioned not to simply turn the stability-diversity hypothesis around and by reversing the slogan originally attached to it claim that anyone who wants ecological stability should promote systems with few species. The same objections that have been levied against generalizations in ecology in general (see 4.e) and against the stability-diversity hypothesis in particular would also basically apply to a "reverse version" of the hypothesis. In view of the fact that the varying effects of different kinds of intervention and the role of resources are not covered by the concept of stability, any attempt to formulate a generally valid statement about the significance of species diversity for stability seems futile.

It is, for example, impossible to unequivocally predict the effect of a simple disturbance such as fire on one and the same ecosystem. Fires that pass through an area against the wind have more serious consequences than those that run with the wind. Fires in spring have different effects than fires in the summer or fall. As we shall see more clearly in Chapter 22.b, the consequences of species loss for a system are just as difficult to calculate since the significance of different species for the ecosystem varies. Thus the loss of oaks in a middle European oak forest would certainly have different consequences than the loss of the pygmy owl, which is rare anyway. In view of the many relevant factors that exist and the uniqueness of each incidence of disturbance Remmert (1984, 260, 261) maintains that a strictly scientific discussion of the effects of sudden interventions in a system is hardly possible. "It is basically impossible to predict whether adding an animal to a system is more significant than taking one away. It is impossible to say whether human interventions by means of fire, bulldozers, insecticides, or herbicides are more consequential than the biological factors mentioned before. And the significance of these various different human interventions cannot be predicted ahead of time." The crux of the matter in all of these discussions about stability and species diversity is that interventions and systems are compared with one another that qualitatively and quantitatively are really not comparable at all. But if comparison is impossible, there is hardly any chance of testing generalizing hypotheses in the field, an absolutely indispensable operation for establishing the validity of any ecological theory.

In view of these methodological difficulties, the heterogeneity of empirical evidence and the paradigmatic changes that have occurred in theoretical ecology, it should be obvious that sweeping demands for species diversity are no longer supported by the science of ecology. The broader form of the diversity-stability theory that was once thought to be valid is now regarded as having been refuted at a theoretical level as well (May 1976, 158f.). Therefore Auhagen and Sukopp's (1983) argument that species diversity is a general prerequisite for "the efficiency of the economy of nature" and "our ability to use nature" can also be regarded as obsolete. On the contrary, "Anyone who wants to generate stability by promoting species diversity is almost always wrong" (Reichholf 1993, 24). Thus *purely ecological* arguments for supporting species diversity are destined to fail for three reasons. The first of these is the *logical* impossibility of deriving any *normative* statements about species diversity from purely *descriptive* ones (naturalistic fallacy); the second has to do with the impossibility of *ecologically* justifying the frame of reference selected for measuring stability; and the third involves the inability to find empirical support for the stability-diversity hypothesis (even when agreement has finally been reached on an intermediate frame of reference).

However, all of this seems to have had little effect since the term "species diversity" is still considered to be a generally valid guiding principle of ecology within the ecology movement and even among professional conservationists, one that can help us to "scientifically evaluate" communities and habitats and "ascertain" whether or not they should be protected (Gerstberger 1991, 318–320).[35] Such well-established terms in nature conservation may be based on a diffuse form of naturalism. At any rate, they undoubtedly encourage the mistaken naturalistic notion that determining and evaluating whether or not an area should be protected is a *purely empirical* matter that consists, for example, of measuring various different ecological parameters such as the diversity index. The impression that determining what should be protected is a purely empirical matter and that maintaining or raising species diversity requires no further justification is reinforced by the results of many investigations that involve a *particular* biotope or area but fail to provide *specific* justification for species protection tailored to the conditions of that area.

However, in numerous cases nature conservation has already had to suffer the consequences of operating with normative generalizations that are either based on or come quite close to naturalistic fallacies. Thus broadly idealizing species diversity has often led to conclusions that obviously contradict the original intentions and intuitions of nature conservation. For example, after the *Waddensee* in northern Germany was partially enclosed by dikes in the area of what is now known as the Hauke-Haien-Koog, a paradoxical situation arose. The artificial flood pool generated in this manner was found to contain more animal and plant species than the open Waddensee area that existed before (Schmidt-Moser 1982, 110). When plans arose at a later date to enclose even more such areas of this unique ecosystem, a system incapable of being enlarged, conservationists suddenly experienced that their very own slogans praising species diversity and ecological stability were turned against them. If the idea is to generate ecological stability by increasing species diversity, shouldn't we encourage even more dikes like the one enclosing the Hauke-Haien-Koog?

The diversity of species in *urban areas* seems to be just as precarious as that of the Hauke-Haien-Koog. In the western part of the city of Berlin, for example, there are more than 120 different species of birds and a quarter million breeding pairs. This corresponds to a breeding density of five hundred pairs per square kilometer, a value that according to Reichholf (1993, 184) is surpassed only by first-rate bird breeding areas in "real nature" in middle Europe. "Cultivated landscapes [outside of the city] have nothing comparable to offer, neither regarding the number of breeding bird species nor with respect to their population density." As Reichholf points out, neither birds as one of many groups of animals nor the city of Berlin is an exception. "The city has long become an important habitat for many species. The only thing is that this has barely been registered or else its significance has been played down!" (Reichholf 1993, 185). Of course, the reason why the greater species diversity of cities has been played down is that in true naturalistic style species diversity is often still regarded as a scientifically objective and thus also universal criterion for evaluating the protective status of an area. If the greater species diversity of cities were not de-emphasized, we might have no other recourse than to place all urban areas under nature protection!

Similarly absurd consequences would result if species diversity were strictly applied as a criterion for judging various different stages of *succession*. If, for example, you cut down a naturally grown middle European beech forest, which contains only a few species, a fallow with many species of plants and animals grows up in its place. In the course of succession, diversity is low in the beginning, becomes greater and greater, and eventually returns to a relatively low level during the climax phase. It is obvious that it would be foolish for nature conservation to try to deduce immediate rules of action from changes in diversity during the course of succession. To do this would mean that we would have to clearcut all our beech forests and maintain the stage of succession that exhibits the greatest degree of species diversity.

By issuing a warning against making normative generalizations I don't wish to deny that species diversity might still be a good measure and a good supportive argument in matters of nature conservation *under certain circumstances*. In many cases, systems that have been modified by humans can be distinguished from more natural ones on the basis of species diversity (Bezzel and Reichholf 1974).[36] However, as the examples I have presented indicate, it is not always possible to make a distinction of this kind. Natural systems may have few species, whereas areas subject to strong anthropogenic influence may be highly diverse. Thus, in this case as well, the concept of species diversity as an evaluative criterion should not be overextended to the extent it has been in the past. According to Remmert (1984, 206), by overestimating diversity indices "their reputation has become so bad that no one uses them anymore in recent publications."

Of course, the dangers of overestimation are not restricted to diversity indices. Other criteria for the protective status of an area, such as community diversity, representativity, rarity, state of maintenance and naturalness (von Haaren 1988, 102), can also lead to paradoxical situations if they are employed universally or exclusively. This is illustrated particularly well by the rarity criterion. In this case conservation measures that may involve a very large area are sometimes established exclusively on the basis of the occurrence of a *single* species (sometimes even on the basis of only a few individuals of such a species).

If you try to avoid the problems of one-sidedness and overextension by referring to the entire set of criteria available, another (and no less serious) problem arises, the problem of weighing the importance of different criteria. As experience in nature protection has taught us, it is seldom possible to optimize all these criteria simultaneously. Sometimes they are even mutually exclusive.[37] Take the example of the capercaillie mentioned earlier (Chapter 9.b) and the xeric meadow on the edge of a beech forest. (a) Should grazing be allowed on this meadow to keep its *species diversity* up? (b) Should *natural* succession be allowed to take place, leading to an old forest stand with very few species? (c) Or should succession be halted at an intermediate stage with shrubs and many clearings in order to secure the survival of *rare* capercaillies? I have already shown that as "ecological" as this problem may appear to be, it cannot be solved solely on the basis of the science of ecology. Anyone who is satisfied neither by feigned naturalistic solutions nor by arbitrary decisions must be prepared to delve through the norms outside of ecology upon which the criteria mentioned above are often implicitly based or at least should be based. Starting with these general norms and perhaps also with hierarchies of norms and working backward, it would then be important to determine from case to case which criteria for protection are the most compatible with the selected norms and therefore to be given preference.

At this point it is not possible to launch a treatise on justifiable *guiding principles* and reproducible *evaluation procedures* for the practice of nature conservation as the above discussion suggests we should. Suffice it to say, so far these problems have usually been either completely ignored or only inadequately treated, the result being that endeavors in nature conservation and species protection are practically "up in the air" when it comes to formulating generally valid arguments or specific aims. This problem has been expressed recently in many publications in relevant journals in which methodological deficits in conservation research have been pointed out and the need for developing reproducible evaluation procedures has been urgently brought to attention.[38] I completely agree with the criticisms that have been presented, but I feel that there are more profound reasons for the deficits in conservation research. Is it possible that naturalistically toned scientism has obscured our perspective? Is it possible that in nature conservation it was simply not opportune to deal with philosophical and thus "nonscientific" problems of evaluation and that the choice was made instead to derive norms from facts without "beating around the philosophical bush"? Even if this meant violating Hume's Law, by doing so it was possible to convey to the pro-science public (and perhaps oneself as well) the impression that the claims and measures of nature conservation are based *exclusively* on the noble laws of modern science.[39] This strategy is, of course, understandable considering the fact that nature conservation usually finds itself backed up against the wall by the supposedly superior rationality of economic interests. But it must be rejected as counterproductive, because in the long run it will not only damage the reputation of the science of ecology but undermine the credibility of nature conservation as well.

11.d. Closed Cycles

A final key concept of ecology that I would like to examine more closely is the term *closed cycle*. This term too is often associated with the idea that we are dealing with a "natural necessity" that ecology has clearly identified. Demanding closed cycles is thought to be "natural" because it is assumed that there are *no exclusively linear* metabolic processes in nature but rather only cyclical ones ("recycling"). This is, for example, the concept upon which Himmelheber (1974a, 66) bases his argumentation, when he describes the principle of closed cycles as a fundamental difference between economic and ecological systems. "Our technology and the production and consumer economics that go along with it operate *linearly* from raw materials to the industrial product and the dump. . . . In nature, on the other hand, we find only *closed cycles*. Apart from the solar energy that our planet constantly receives and the heat that it returns to outer space, the ecosystem earth is in general a closed system consisting of many interconnected cycles. *There is no waste in nature.* This fact, which we usually take for granted without questioning, is quite remarkable." In view of all the foiled attempts to establish universal laws without exceptions in ecology, it would certainly be even more remarkable if Himmelheber's claim held true. Closer examination does indeed reveal that this claim is not only an illegitimate generalization. In those cases in which it seems to apply it also represents an extreme simplification of the facts.

I do not wish to deny that the principle of closed cycles in the sense of constant turnover of matter does occur very frequently in nature. Cyclic turnover of water, carbon dioxide, oxygen, and nitrogen is well known. All of these cycles are for the most part closed and belong to the *gaseous class* of cycles (Tischler 1976, 109; Heinrich and Hergt 1990, 61). This class can be distinguished from the *deposit class* of cycles involving minerals including phosphor, sulfur, potassium, sodium, calcium, magnesium, silicon, iron, and various other trace elements.

As the term deposit class indicates, the flow of these minerals through the system is not always continuous. It can come to a halt for varying periods of time when sediments are formed. Interruptions of this kind often last only a few years (as in the case of calcium in the plant litter on the floor of mixed woodlands), but sometimes they encompass intervals of geological significance. Take, for example, the phosphor cycle. Some phosphates flow with rivers from the soil to the sea, where one part of them passes through the plankton back into the food chain while another part is deposited at the bottom of the sea. Phosphates are deposited as *phosphorite* on the ocean floor, and it may take millions of years before it becomes available to plants again so that the phosphor in these deposits is practically no longer part of a biogenic cycle. Another example of matter that normally circulates but sometimes persists in the form of deposits for long periods of time is carbon found in such organic deposits as coal, oil, and gas. These deposits were formed from the remains of plants millions of years ago under extreme physical conditions (high pressure, exclusion of oxygen), and because of their extremely inaccessible location under the surface of the earth, they have been disconnected from the carbon cycle ever since. But we don't have to refer to such extreme geological constellations in order to refute the claim that there is no waste in nature. Every drying-up lake with its layers of mud that bacteria are unable to decompose, every coral reef, every guano cliff, and every layer of peat at the bottom of a bog can be regarded as a natural waste pile. What is decisive for the term "waste" is the fact that the substances that have accumulated within the system can no longer be decomposed and that they can reenter metabolic cycles once again only after a fundamental change in the system has occurred.[40]

Reichholf (1993, 165) describes a striking example of an ecosystem in which a particular organism continually produces waste (in the sense described above) and as a result not only suppresses the growth of almost all other organisms but eventually also destroys the conditions it requires for its own reproduction. The organism in question is an iron-metabolizing bacteria that gets its metabolic energy from oxidizing soluble iron(II)oxide (FeO) to insoluble di-iron(III)tri-oxide (Fe_2O_3). This chemical reaction, which is similar to the process of rusting iron, requires large amounts of oxygen as the bacteria multiply. Therefore, as oxygen is gradually depleted, conditions are generated in which hardly any organism other than the bacteria can exist. Another consequence of reduced oxygen levels is that proteinaceous materials are no longer able to be completely oxidized by oxygen but instead are converted to the metabolic poison hydrogen sulfide (H_2S). The final product of this chemical process is ocherous mud mixed with black iron sulfide (FeS), which is barely able to support any kind of life so that its components never find their way back into a biological cycle. At the most, iron ore deposits can be generated from it sometime in the distant future, in the same manner as occurred at different places on earth millions of years ago. Typical of the linear metabolic ecology of iron-metabolizing bacteria is that the bacteria destroy the conditions that all oxygen-producing organisms (as, for example, algae and green plants) need to survive, even though the bacteria depend upon oxygen for oxidizing iron dissolved in water in the course of their own proliferation. In view of obvious analogies to ecological exploitation by humans, it is not surprising that the mass proliferation of iron-metabolizing bacteria such as has been observed on the lower Inn River (Reichholf 1981) is automatically considered to be an "environmental catastrophe" and not a natural process. In light of the widespread vision of harmonious cyclic relationships between producers, consumers, and decomposers, the march of iron-metabolizing bacteria straight into the dead-end road leading to iron deposits that is so fatal for other species must appear to be extremely "unecological."

But it would be a mistake to dismiss the ecology of iron-metabolizing bacteria described above as a freak occurrence. For the greater part of the 3.5 billion years in which life has existed, nature has functioned solely in this linear manner. Substances were used and metabolized and end products accumulated. There was no recycling. The oldest and simplest ecosystems on earth maintained themselves primarily through constant proliferation of their constituents, that is, through undeterred growth of blue-green algae and bacteria. As Reichholf (1993, 178) emphasizes, recycling is an "invention" that has come about in the last half billion years. "It has really only functioned properly during the past tenth of the period in which life has existed. Before that frothing production led to the accumulation of huge masses of plant material due to the success of photosynthesis.

The waste product of this process, oxygen, oxidized and poisoned the earth until respiration based on oxygen opened up a new dimension of life." As this "ecological crisis" in the early history of the biosphere shows, environmental pollution and waste production were not first invented by humans. Nature consists not only of cleverly organized cycles that give the impression of being harmonious. It also encompasses primitive and apparently reckless processes leading to the accumulation of waste.

What conclusions regarding the *normative* use of the term "cycle" can be drawn from these descriptive thoughts on evolutionary history? First of all, it should be clear by now that just because waste production and environmental pollution occur in natural systems, this does not mean that it is legitimate for humans to dump waste products on nature. Not only would it be extremely unwise to refer to iron-metabolizing bacteria in this connection. It would also be logically untenable to directly derive rules of behavior from facts of nature in this manner. We would also be dealing with a naturalistic fallacy if the demand for an economy based on recycling were to be mainly or even exclusively justified by maintaining that natural systems are also characterized by cycles and recycling processes. Even if it is probably difficult to demonstrate from case to case that a naturalistic fallacy is being invoked without knowing exactly what the basic assumptions of the person presenting such an argument are (Engels 1993, 120), it seems that at least the risk of a naturalistic fallacy is greater when the grounds presented for an argument are the supposed universal validity of the principle of recycling (as in the case of the quote by Himmelheber cited earlier). Explicit reference to the universality of natural processes and relationships in normative discussions can hardly be interpreted otherwise than that a *normative* statement is being articulated primarily or exclusively on the basis of such a *descriptive* statement. Instead of first clarifying in detail what we *want* (interests) or *ought* to do (ethics) and then examining how this can be achieved in the context of the natural laws and processes described by ecology, goals are derived directly from ecology. Justification for a norm is thus reduced (at least to a certain extent) to the simple motto that "things in nature *always* function this way and *therefore* we should do the same, or recycling can be found *everywhere* in nature and *that's why* human society should operate on the basis of the principle of recycling too." As the preceding discussion has shown, this kind of argumentation is not only weak due to the logical error involved in deducing what ought to

be directly from what is, it is also based on the false premise that the principle of recycling is universally valid in nature.

Of course, one could maintain that it is relatively unimportant whether
demands for closed cycles are justified in a watertight manner or not. In
view of continually growing garbage piles and diminishing resources the
necessity for recycling seems to be obvious. It must be conceded that there
really is no alternative to recycling if our system is to continue to exist in
the long run. Since the publications of the Club of Rome (Meadows et al.
1972; Mesarovic and Pestel 1974) and *The Global 2000 Report* (1981) there
can be no doubt about the necessity of reclaiming and recycling secondary
raw materials. Nevertheless, it can lead to serious misunderstandings "if a
process of this kind based on simple economic logics is presented as being
'ecological,' and if recycling is glorified as a principle of life without saying
that dumps also occur in natural ecology" (Dahl 1989a, 64). Idealizing the
principle of recycling in sweeping statements can sometimes lead to paradoxical conclusions such as those already described in connection with
idealizing species diversity. If recycling were ecologically valuable in and of
itself, there would be no reason to criticize the uninhibited consumption of
fossil fuels and the greenhouse effect associated with it. By burning coal
and oil, one could argue, the carbon cycle, which has been interrupted by
these deposits, can be brought back to full swing operation again. As this
example demonstrates, strictly adhering to the guiding principle of recycling doesn't necessarily lead us down the road to ecological paradise. It
can also lead to climatic catastrophe.

Idealizing the principle of recycling can lead to consequences that are
even more serious than such misunderstandings, namely to overestimating
the power of such processes. The more recycling is conceived to be a kind
of magic formula of ecology because of its supposed universality and normative evidence, one that permits us to continually convert old things to
new ones without any limits, the more we are in danger of losing sight of
real relationships and the limits to this principle that exist as well. If you
take a closer look at cycles, particularly from the perspective of thermodynamics, three unpleasant truths become apparent that are often forgotten
in the euphoria associated with recycling: (1) A cycle is not a "perpetuum
mobile," that is, a kind of hypothetical machine that functions without energy input. It operates only when energy is pumped into it. Even if matter
flows through ecosystems in cycles, energy does not. Energy flow is *non-
cyclic*. (2) Since all ecological systems are *open* systems and coupled to lin-

ear energy flow, strictly speaking there is no such thing as a *closed* cycle of matter. At the most we are dealing with "nearly cyclic processes." "Perfect cycles are only possible in a reversible, theoretical model, not in nature where irreversible processes occur" (Kreeb 1979, 75). (3) In economic processes as well, matter doesn't really flow in closed cycles since a certain percentage of material is *inevitably* lost with each recycling stage. The more we try to reduce such losses, the greater the amount of energy that must be invested. According to the third law of thermodynamics, Nernst's theorem, *perfect* recycling is impossible since totally reclaiming matter from a state of greater disorder by concentrating it in one of less disorder would require infinite amounts of energy (Schütze 1989, 33). Instead of "recycling," it would be more appropriate to refer to "downcycling."

This leads to the following dilemma with respect to *artificial* cycles. If we conduct recycling with very little energy input in order to keep the release of heat and carbon dioxide at a relatively low level, the amount of matter that can be reclaimed in this manner and thus also the number of potential rounds of recycling is rather low while the loss of matter (entropy of matter) is high. If instead we invest a great deal of energy in the recycling process in order to increase the amount of material that can be reclaimed as well as the number of potential rounds of recycling, the price we have to pay is the release of large amounts of heat and carbon dioxide (when fossil fuels are the source of energy). "Thermal pollution" (entropy of energy) is high.

For the *natural* system of our biosphere the inevitable increase in entropy, otherwise known as the "law of entropic doom" (Schütze 1989), also holds true, although its effects are much less pronounced in nature than in artificial systems. There are two reasons for this. First, natural systems operate primarily with solar energy, the only form of energy that affects neither the heat nor carbon dioxide level in the biosphere. Thus these systems can even reduce entropy within certain limits of space and time, and can construct "islands of negative entropy." Second, in nature systems all are adjusted to low energy input. As Reichholf (1993, 179) has pointed out, from the standpoint of evolution their cyclic systems are "products of limited resources" and therefore fare all the better when less energy is converted to entropy that has to be diverted. "The economic processes of industrial society with its mass turnover of free energy and highly concentrated raw materials, with its mass production of entropy in the form of heat or waste, is exactly the opposite of natural systems with their high potential for survival" (Schütze 1989, 96). This fundamental difference between natural systems geared to limited resources and those of industrial societies based on wastefulness is often overlooked when recycling is propagated as a way out of the waste and resource crisis. Recycling is reasonable and necessary, no doubt about it. But it would be a mistake to regard it as a common cure for reconciling ecology and economy with each other by *technological* means. If our artificial systems are to function just as well as natural ones in spite of an inevitable increase in entropy, they too must be geared to minimal input instead of maximal output. As Reichholf (1993, 180) puts it, "Aim for 'low input.' Low input means a simpler lifestyle, less energy consumption, less mobility, less production, just enough to live well." In Chapter 32 we shall return to the question of how this is all related to the death of our planet's species and what consequences individuals and society as a whole should draw from this.

12. Ecological Health?

At the beginning of the book I described the hopes of technical optimists that the science of ecology might some day play the part of a kind of "environmental medicine," which not only diagnoses and treats ecosystem diseases but also calculates stress limits and thereby helps to prevent such diseases as well. I first criticized these expectations from the perspective of epistemology and theory of science. After having analyzed some of the "guiding principles" of ecology in the preceding sections, it is now time to also consider the suitability of ecology for serving as an agent of "environmental medicine" in light of normative considerations. Is the analogy between the normative discipline of medicine and that of ecology, the descriptive nature of which has been repeatedly emphasized, really legitimate in any way at all? What do people mean when they talk about a "sick ecosystem" or an "intact environment" or a "healthy river"? Are we dealing here with scientific terms or simply with metaphors that nicely illustrate a point?

After all, it is not uncommon to find medical terms being applied to ecological topics, not only in popular and philosophical literature about the environment (e.g., Meyer-Abich 1991, 164) but also in textbooks and publications of professional ecologists, as, for example in Chapman (1974, 385) and Clapham (1973, 229).[41] For Remmert (1990, 195) an ecologist is clearly "comparable to a doctor. He diagnoses diseases of the ecosystem earth. He knows that these diseases are dangerous for the life of human beings. As in the case of many sicknesses of a single individual, there is no general and immediately effective cure. Thus the ecologist recommends small therapeutic measures. . . . " Remmert (1990, 199) is quite clear about the fact that he uses medical terms in the sense of a real analogy (not just a metaphor) when he concludes his thoughts on the subject as follows: "The decisions which a trained and knowledgeable ecologist with additional certification in conservation has to make are *scientific* decisions and should not be political ones" (my emphasis). However, they can only be regarded as scientific decisions when their normative premises, namely ecological health, are just as self-evident and self-explanatory as the concept of health is in medicine. But is this truly the case?

In order to answer this question the concept of health in *medicine* must be examined more closely. However, when you do this, you discover that in medicine, in spite of strong intuitive clarity, there are no generally accepted definitions for the terms "health" and "disease" but at the most various different explanations that complement one another. Apart from highly problematic references to the concept of what constitutes "normality" (see Canguilhem 1974), there are primarily two approaches that compete with one another in theoretical medical literature, one with a subjective and social focus and another more attuned to scientific and objective thought. Thus according to the controversial definition of the World Health Organization (WHO) health is a "state of complete bodily, mental, spiritual and social well-being" (Pschyrembel 1986, 587), while a more scientifically oriented doctor would probably be more inclined to associate health with the *proper functioning* of an organism and its organs. Practically speaking this concept seems to be based on a technical and cybernetic model of the body as a machine with interconnected feedback loops that serve to maintain certain "target values" (e.g., a particular blood pressure level or pulse rate). According to this model disease can be regarded as a reaction to a disturbance of intermediate intensity (i.e., not a lethal one), which results in the organism temporarily establishing the "wrong" target value by means of a secondary but as it were "parasitic regulatory system" (C. F. von Weizsäcker 1979, 328). If the outcome of the disease is positive, the system eventually returns to the right target value. A similar concept is expressed in a modern textbook on natural healing methods, in which disease and health are also viewed from the perspective of cybernetics and thus regarded as complementary aspects of a uniform process of auto-regulation. "Disease is a disturbance of the equilibrium or an attempt to establish a new equilibrium at a different level" (Melchart and Wagner 1993, 33).

The use of the terms disturbance and equilibrium seems to make an analogy to the concept of equilibrium in *ecology* almost inevitable. According to cybernetic models of ecology based on systems theory, ecosystems are also regarded as auto-regulatory systems capable of reacting to disturbances of intermediate dimensions and thus maintaining a certain systemic state of stability. In light of these similarities it is not surprising that the term ecological health is usually associated with the concepts of ecosystem stability or equilibrium.[42] According to a definition of Clapham (1973, 229) a healthy ecosystem is one that is either very close to the state of equilibrium typical for the area in which it occurs (stability in a narrow sense of the term) or at least capable of returning to this state following a disturbance (resiliency). Chapman (1974, 385) and DeSanto (1978, 8) express similar views.

It can easily be seen that a definition of ecological health of this kind raises a series of problems. The first problem is that contrary to medicine there is no other complementary and subjective definition of health in ecology. An ecosystem can neither tell us whether it feels like it is healthy nor give us any explicit information about the nature of any troubles. In medicine, on the other hand, the fact that (at least) two definitions of the term health exist side by side seems to indicate that in spite of all the success of a more objective model of health, a subjective perspective is still considered to be indispensable. Whether or not the auto-regulatory system that constitutes a human being is in equilibrium, whether or not it is oscillating around the "right" target value or the "wrong" one, apparently cannot be determined completely from outside of the system, in other words solely on the basis of scientific methods. In light of these considerations the use of the term health in ecology, which is restricted to an external perspective, seems to be dubious, at least as far as claims to direct analogy to the use of the term in medicine are concerned.

One possible rejoinder to this criticism could be, of course, that veterinary medicine also has to get along without subjective expressions of suffering by its patients. Veterinary medicine proves that for *nonhuman* biological systems (for which we can never know for certain whether any kind of subjective perspective exists) it is at least of practical value to define disease as "a functional disturbance capable of being determined objectively." This argument would be convincing if ecosystems were really similar to animals or other organisms with respect to their structure and function, as the common superordinate term "biological system" seems to indicate. But is this really true?

Before answering this question it should be recalled that what science means by the term health (just as what biology and medicine mean by the term "function") is based on a conceptual model that regards biological systems as in principle similar to machines. This observation, which is closely connected with the fundamental role of the experiment in modern science and Vico's "verum-factum principle" (Vico [1709] 1963; Vossenkuhl 1974; Hösle 1991, 58, 59),[43] is not fundamentally altered by newer and supposedly "alternative" approaches in the context of systems theory and cybernetics (see Cramer and van den Daele 1985). Cybernetics originated in engineering science as a discipline designed to study control and regulation processes in *technical* systems. It was applied in biology and sociology afterward.

As shown in Chapter 5, however, there are fundamental limits to the mechanistic projections so commonly used in science including the application of cybernetic machine models to *nontechnical* systems. The more complex, highly integrated, and thus less deterministic a system is, the less appropriate the methodological model of a piece of apparatus made up of isolated parts is for understanding it. As far as the term health goes, this means that restricting its definition to an objective, scientific perspective becomes more and more problematic the more complex and less deterministic the system is to which it is applied. While a definition based solely on science may be sufficient at the physiological level of organs and for simply organized plants and animals, problems are inevitable when higher and more complex levels of systemic organization are brought into play (e.g., in ethology, psychology, or sociology). This is nicely exemplified by highly controversial topics such as the well-being of animals in zoos or objective criteria for the mental or psychological health of human beings. It seems that both problems cannot be dealt with satisfactorily solely on the basis of a cybernetic approach.[44] Since the objects studied in ecology exhibit a degree of complexity that is more similar to those of ethology or psychology than those of veterinary medicine, it is faced with the same problems as these disciplines. Not only is a subjective perspective lacking; *complexity* is a second major barrier to applying the medical concept of health to ecological problems.

The third objection that arises and marks a fundamental difference between animals and ecosystems has to do with the term "functionality." At a physiological level what this teleological concept means is relatively clear and unambiguous. Thus a kidney is regarded as "insufficient" when it is unable to fulfill the function of osmotic regulation, one of the defining factors of its role in the entire system. In analogy to the concept of function at the level of organs it also seems quite plausible to refer to a disturbed function of the *entire organism* when certain species-specific aspects of an organism's life are impaired by organ damage or when its very survival is threatened. It makes sense to talk about functionality in these two cases since the purpose is obvious. With respect to organs it is a matter of the function of one such part in the whole system, and in the case of an entire organism the matter at stake is its survival or species specific development.

However, when the analogy is carried further, it becomes less clear what the concept of *ecological* functionality might mean, in spite of the frequently heard reference to "intact ecosystems." If the functionality of an ecosystem is equated with self-maintenance, that is, with stability, the same difficulties for the concept of health crop up as were already discussed in Chapter 11.b in connection with the guiding principle of ecological stability. A major aspect of the problem was found to be the question of finding objective measures of space and time for defining stability in view of the basic dynamics of all natural processes. Objections raised against the idea of a generally valid concept of stability therefore also apply to the concept of ecological health. Whether or not an ecosystem can be described as functioning properly or as being stable is an ambiguous matter and varies depending upon the temporal and spatial reference frame that is chosen. The selection of this frame of reference, on the other hand, as already emphasized, is not a matter of scientific objectivity. It reflects instead the subjective interests of the person who chooses such measures or rather who wants to live in the ecosystem in question.

In view of the highly divergent interests that exist in an ecosystem, it seems to be obvious that such a relative concept of ecological health makes little sense. If you take the interests of all species into consideration, the term is of no practical value at all, since the many different vital interests are often mutually exclusive. If, however, priority were to be given to certain individual interests over others, this would not be scientifically objective. Wouldn't one obvious solution to this dilemma be to postulate some sort of *general interest* of the ecosystem as a whole, however this might be determined?

This idea takes us back to a previous question that has only been answered in part, namely whether ecosystems are comparable to animals or other organisms with respect to their structure and function. From the standpoint of modern ecological science this question must be negated and thus provides a fourth objection to the concept of ecological health. Ecosystems are not "super-organisms" to which interests can be attributed in a meaningful manner the way this is done for normal organisms. Even though there was a movement in ecology in the first half of the twentieth century that attempted to interpret communities of organisms as "quasi-organisms" with individual organisms in place of organs (e.g., Clements 1936), this "super-organism theory" has "practically disappeared from ecological discourse nowadays" (Trepl 1988, 177). Of course a few elements of the old organismic concept have made their way into the *holistic version of ecosystem theory* (process-functional approach; e.g., Odum 1971), which has influenced the widespread idea of nature as a kind of "cybernetic super-machine" in the current ecological movement. Advocates of this view regard nature as a whole that functions according to rules of emergence, whereby every part contributes to the maintenance of the others and thus also to maintenance of the whole. This approach is just the opposite of the so-called *individualistic approach* in modern ecology (population-community-approach), which attempts to interpret natural processes primarily on the basis of the antagonistic behavior of species toward one another, where these processes are located and historical factors as well (e.g., Krebs 1985). For advocates of the population approach, only the organisms, species, and populations really exist; ecosystems, on the other hand, are merely scientific abstractions. For Müller-Herold (1992, 29), for example, the concept of an ecosystem is simply a "heuristic concept" that permits us to apply ideas of systems theory to processes in living nature. "Systems simply do not exist as such. They are conceived for practical purposes in order to reduce the complexity of the real world and thus to provide general insights and to create new possibilities for action. . . . Forming systems is therefore not a self-evident matter, and the systems that are generated have to prove their worth."

Even if the controversy about both fundamental concepts continues to persist and can only be touched upon here,[45] there does seem to be general consensus that ecosystems differ from organisms in the following points, which also undermine the concept of ecological health: (1) As unique entities they exhibit neither a universal organizational structure nor a definite developmental program, nor do they have any inherent goals that might or might not be attained. (2) Thus the identity of an ecosystem is ambiguous, since its temporal and spatial boundaries consist of relatively arbitrary limits established on the basis of certain methods. (3) The so-called equilibrium of an ecosystem is not something it ultimately aims for in the same sense as an organism strives to live or maintain itself. It is rather the "result of the activities of numerous individuals of very many species" (Erbrich 1990, 7), which, as opposed to the organs of a body, are all pursuing their own goals. (4) Contrary to organisms, ecosystems can assume various different states of equilibrium (both in succession and in the course of oscillation; Remmert 1984, 251). If one state of equilibrium is irreversibly destroyed, another is bound to arise automatically. (5) Ecosystems do not suffer individual death; instead they merely experience transformations.[46]

Having shown the idea to be erroneous that an ecosystem is a kind of organism for which therapy can be prescribed *in its own* interest when its equilibrium is disturbed, the question that now arises is whether anything can be gained from employing the concept of ecological health, or whether it might not be better to avoid using this term in the future. Bayertz (1988) suggested that if we talk about ecological health it would be better to refer to the interests of the *subject* who exercises therapy (human beings) than the controversial interests of the *object* at the receiving end of therapy (nature). Since human beings are by nature "dependent upon controlling nature and exploiting its resources," it would seem reasonable "to define the 'health' of an ecosystem as that state which provides humans with the greatest number of opportunities for using it. The relevant criterion for the health of a river would then no longer be pristineness or the number of species it contains but rather the extent and variety of human needs it is capable of satisfying. For example, the Rhine River is currently used primarily for waste disposal and transportation purposes; however, the predominance of these forms of utilization impairs or prevents the satisfaction of other kinds of needs. According to this definition a 'healthy' river would also be capable of being used for fishing, swimming, and as a drinking water reservoir" (Bayertz 1988, 97).

Even if it cannot be denied that this definition of ecological health reflects a popular way of using the term, four objections to it can be raised. First, the criterion of providing "the greatest number of opportunities for utilization" is probably not of very much practical value since it is not at all clear how "the extent and variety of human needs" should be determined and how they should be balanced against one another in cases of conflicting needs. Second, this definition leads to consequences that are both ecologically untenable and counterintuitive. This is illustrated by Bayertz's suggestion (1988, 97, 98) that our "model of a healthy ecosystem" should "no longer be untouched nature . . . but rather a garden landscape cultivated by humans." Ultimately this concept of ecological health would mean that national parks, which are consciously exempted from utilization, would thus represent a state of reduced ecological health. The suggestion is further unrealistic because it requires that humans be able to alter nature at will to meet their needs and desires with the help of ecology. In Part A.I this quite scientistic and optimistic vision was criticized from the standpoint of theory of science as well as from an ecological point of view. Third, this definition is ethically unacceptable because it fails to take the interests of other species into consideration. Even if ecosystems as *wholes* have no interests, this does not mean that we can assume that their parts, that is, plants and animals, also have no interests whatsoever.[47] Fourth, Bayertz's definition runs the risk of nurturing ideology because of the way it confuses interests with objective science. Since it is probably impossible to reach a consensus about the "best" way to use nature, we must count on being confronted with as many different "ecological diagnoses" about an ecosystem as there are interests in using it. By referring to the term equilibrium Reichholf (1993, 214) has vividly shown how such an endeavor might wind up. "A lynx will quickly be seen as a disturbance to equilibrium because it preys on deer; cormorants suffer the same fate because of catching fish; peregrine hawks or chicken hawks do too because of hunting pigeons, among which a few carrier pigeons might be found, etc., etc. Each and every person defines 'equilibrium' in his own way depending upon the relationship between species which he happens to prefer."

In this respect the term ecological health, like the term equilibrium, is just as vulnerable to being usurped and employed by interest groups for instrumental purposes. People who use this term in their argumentation automatically profit from its aura of scientific authority and from intuitive evidence by way of analogy to the corresponding concept in medicine. However, as I have attempted to demonstrate, the normative claims connected with the term ecological health are not redeemable, and since it tends to blur conflicts of interest that exist among human beings as well as between different species, it seems wise to forget about this term all together. If it is occasionally used to illustrate a point, it should be made clear that in using this term we are dealing with a popular metaphor of nature conservancy, not with a theoretical concept derived from ecological science. Ecology is not a medical discipline for dealing with the environment.

13. Ecologism

The dangers of converting ecology to an ideology that I have outlined in the preceding chapters and also discussed in connection with the term ecological health must be examined more carefully with respect to their fundamental meaning for ecological discourse. After all, according to the landscape ecologist Haber (1993, 102) "about 80 percent of all the things that are discussed as if they belonged to 'ecology,' even in scientific circles, are really instances of 'ecologism.'" According to Haber this includes "most of what environmental and nature conservancy organizations or green parties propagate," their significance for criticizing developments in society or their political importance notwithstanding. However, in a discussion about the role of ecology, the often uncritical way ecological science is used by these groups should not be ignored. "We should not forget the political consequences that in the past resulted from practicing biologism in an uncritical manner!" (Haber 1993, 103).

What are the specific characteristics of ecologism and how can it be distinguished from the more comprehensive pursuits of biologism? What the two have in common is both unreflected naturalism and a superficial or rather a scientifically obsolete understanding of nature. While biologism often tends to reduce nature to a ruthless battlefield of constant strife (at least in its most notorious form, that of social Darwinism), ecologism seems to tend to the other extreme, to a harmonizing transfiguration of natural processes. If, for example, one analyzes the most common forms of argumentation employed in the ecology movement in support of the "guiding principles" of ecology discussed above, regardless of the legitimacy of their claims one often finds a romantic but rather narrow vision of nature. According to this not completely erroneous but nonetheless one-sided view, pristine nature unsullied by humans is in and of itself an infallible cybernetic system capable of maintaining its subsystems in constantly self-regenerating steady-state equilibrium. The stability of the systems is guaranteed by a wide diversity of organisms, which are all woven together in a kind of eco-social web in which every species is obligated as a "paying member" and thus safeguarded against crises. In this connection Capra (1983, 440) refers to the "wisdom of nature" or rather the "intelligence of ecosystems," which manifests itself in "cooperative relationships" and results in "harmonious integration of the components of the system at all levels of organization."

As fundamental and widespread as cooperation and integration in nature truly are, it would nevertheless be dangerous to declare these principles absolutely valid, thereby generating an idyllic vision of mutual well-being in nature. This vision is dangerous in that it courts ecological naturalism. If nature really were the best of all imaginable cybernetic worlds and ecology the discipline in which all its laws are compiled, what would be more reasonable than to derive the norms for how to deal with nature directly from this collection of laws? But nature isn't "wise" all the time and in every respect, and ecology is highly ambivalent. "For every soft pathway that can be identified [in nature], you can find a hard one somewhere else" (Dahl 1989a, 65). On the one hand ecology does indeed describe remarkably subtle interplay among species, but on the other hand it also points out the ruthless competition and mutual pressure that seems to exist among them as well. As harmonious as ecology's world might sometimes appear to be, it just as frequently describes natural disturbances and catastrophes (see Chapter 11.a). As Markl (1981, 29) pointed out, by its very nature evolution is prone to crises and "not an instrument of nature for maintaining species. On the contrary, the mechanisms of natural evolution lead almost automatically to species loss." Hundreds of millions of species have arisen in the course of earth's history and then become extinct, either because of a crisis due to rapid changes in environmental conditions, climate fluctuations, or an evolutionary cycle to which they were unable to adjust. From this perspective it is hardly possible to talk about "harmonious equilibrium" and "systems that function free of any disturbance," attributes often assigned to pristine nature.

Another concept that seems all the more dubious in light of this discussion, one that has shaped ecological discourse far beyond the boundaries of the ecological movement and is sometimes even employed as a definition of ecological science, is that of *nature's economy*.[48] It is a classic example of a slogan of ecologism that is neither theoretically well-founded nor practically useful. As a result of the analogy to economics this term generates the impression that the biosphere is something like the national economy with individual businesses, the ecosystems, whose efficiency can be calculated in the same manner as balance sheets. The *descriptive* problem of this approach is that such auditing claims are not restricted to *individual* processes that readily lend themselves to quantification (e.g., water balance or irradiation balance). Instead efficiency auditing is explicitly extended to include *all* of nature and its systems without clarifying which systemic levels, which boundaries, and which ecological "currency" are to be taken into account.[49] In connection with the term ecological health, I have already shown why a universal audit procedure that reflects the *functionality* of all systems at all levels is out of the question. The arguments listed there against projecting the concept of functionality from organs and organisms to ecosystems and populations also applies to the concept of the economy of nature. As Honnefelder (1993, 257) rightly criticizes, an "economy" requires a subject, a housekeeper, or manager, geared toward attaining certain goals. "But it is this very kind of teleology which is foreign to modern interpretations of nature on the basis of evolutionary theory."

Having thus shown that the concept of economy seems to be inappropriate at a descriptive level, the *normative* claims associated with it can also no longer be upheld. Thus like the term ecological health it conveys the impression that there is hypothetically something like intact ecology at all levels of the system of nature, a kind of "ecological welfare for all" provided there are no interventions by human beings. However, the vision of complete harmony in nature that this expression implies obscures the view of its other side, the chaotic, fundamentally contradictory and often also destructive effects on species that characterize it as well. It is by no means always possible to coordinate the various balance sheets of different individuals, systems, and systemic levels. Calculations that might seem to be "optimal" for a particular system and its individuals are not necessarily favorable for its subsystems or the entire system of which it is a part. Instead contradictions and conflicting aims often seem to be the rule. Remmert (1984, 303) voices this phenomenon in anthropomorphic terms when he emphasizes that the interests of an individual are not necessarily congruent with those of the population to which it belongs, and the interests of a population do not necessarily coincide with those of the entire system. "When it comes right down to it, evolution and co-evolution have predisposed one and the same individual to a kind of 'schizophrenia' which you have to recognize and take into consideration if you want to do ecology."

However, according to Remmert (1984, 303) this "schizophrenia" disappears "when the whole system is evaluated on a long-term basis." Does this mean that when the whole system is taken into consideration it might be reasonable after all to refer to the "economy of nature?" In keeping with this idea, Kreeb (1979, 72), for example, referring to Odum (1975), defines ecology as the "science of the economy of all of nature, not only that of subsystems." However, it seems to me that not much can be gained from such a broad understanding of the term "economy of nature." Since strictly speaking "the whole system" can only mean the entire biosphere, interpreting the term economy in this manner would result in the same conceptual vacuum as that associated with the global version of the term stability (see 11.b). From the evolutionary perspective of a global context, the term "nature's economy" can only refer to auditing aimed at maintaining *life itself*, without further defining the exact form that manifestations of life, that is, organisms, animals, plants, and communities, might assume. If "economy of nature" is defined in this sense, it cannot be *further qualified* to permit us to decide whether this economy is best served by a diversity of complex communities of organisms or by a few species of bacteria such as there were three billion years ago.

In view of this *theoretical* ambiguity it is not surprising that the term "economy of nature" also fails to be useful for *practical* purposes (see Eckschmitt et al. 1994). If the "economy of nature" is something we are supposed to protect—and indeed this term can be found in German laws on nature conservation as well as in laws on chemical pollution and waste—then we ought to be clear about how we are to protect this economy and against what. With respect to this problem Haber (1993, 103) maintains that "neither among ecologists nor outside of their circles can an answer be found to which everyone agrees." It follows that inconsistencies, inappropriate estimates, and false expectations in judgments of natural processes frequently ensue. Reichholf (1993, 8) criticizes that "'nature's economy' is much too frequently regarded as having been damaged when all that has happened is that the landscape has been altered, a landscape which was produced by humans, that is, modified land, not original, pristine nature. Or nature's economy is brought into play in order to explain why a species is necessary—'It's an important element of nature's economy!'—or why it's justifiable to eliminate one: It's all right to shoot crows in Bavaria if they disturb nature's economy."

The risk of ecology being exploited for instrumental purposes by interest groups (as, for example, hunters, fishermen, farmers, and conservationists as well) immanent in such use has already been elucidated elsewhere. But yet another risk inherent in the term "economy of nature" brings us back to the line of thought that this section intended to pursue, to the general problem of ecologism. By conjuring up a vision of purposefully designed (or designable), "idyllic and eco-socially oriented nature" this term not only trivializes nature itself but *human beings' relationship to nature* as well. The tendency to trivialization is obvious in such expressions as "eco-logically compatible" or even "reconciliation with nature," which gloss things over and conceal the fact that as consumers in the biological sense of the word human beings *in principle* live and must live at the expense of other organisms. In so doing we fail to comprehend in all its variety and contradictoriness the question of the right way to deal with nature, which also touches upon the problem of what we can and ought to expect from ourselves and other organisms. Moreover the perspective of ecologism boils the question down to merely a matter of ecological auditing. Instead of facing up to the *ethical* challenge of the problem, it is construed as a purely *technical* one. Remarkably, this trend parallels that of the scientism that is typical of technical optimism since ecologism also relies mainly on science to solve the ecological crisis, thus overlooking the fact that this is not only asking too much from ecology from an *epistemological* standpoint but from a *normative* one as well. Ecology cannot assume the role of a "new science for providing guidelines" as, for example, Amery (1982, 39) pro-poses. It is also not a "science of equilibrium, of a harmonious relationship between humans and nonhuman nature," as Maurer (1982, 32) conceives it to be. These and similar ideas (see, for example, Maren-Grisebach 1982, 32; Bookchin 1971, 26, 27) are dangerous not only because they rely on false promises but also because they divert our attention from the core issue, namely the *ethical* dimension.

After having shown what ecology *is not* and what it *cannot* master, the question that, of course, arises is just what part it might play in connection with the ecological crisis and the loss of our planet's species. What *positive* function can ecological science assume in this respect? These questions will be pursued further in the following sections.

III. What Ecology Has to Offer

14. The Normativistic Fallacy

In the previous chapters I repeatedly emphasized that we cannot derive directives for the "right" way to deal with nature *solely* from ecological knowledge. In the present chapter I would like to point out that such directives also cannot be formulated *without* taking empirical evidence into consideration. Just because ecological knowledge isn't *sufficient* for formulating and justifying norms doesn't mean that it is therefore *irrelevant*. An extreme conclusion of this kind would suffer from an argumentative weakness that, according to Höffe (1981, 16), "has hardly been given any attention in general ethical discourse," namely the so-called *normativistic fallacy*. Höffe uses this term to refer to "the opposite of naturalistic fallacy, the idea that specific or even very definite obligations can be reached solely on the basis of normative considerations." It stems from a concept of ethics as a closed system of rational argumentation for which only internal rules of justification and criticism must be taken into account. A concept of ethics of this kind can be found in particular among deontological schools of thought such as the extreme rigorism of Enlightenment. Because it strictly separates the areas of mind and nature, freedom and necessity, and rationality and experience, this concept of ethics is particularly prone to normativistic fallacy. Thus Kant ([1785] 1965), for whom morality consists exclusively of respect for oneself and moral laws, rejects any empirical supplements to a priori principles of practical reason as superfluous and even as a subversion of the obligatory nature of morality.[50] In his essay *Ethics without Biology* Nagel (1979) assumes a similarly negative position toward scientific evidence in the context of ethics. In this essay he not only proposes the certainly correct thesis that biology cannot serve as a source of moral judgment, but also comes to the even more far-reaching conclusion that biology is *altogether irrelevant* for ethics.

In my opinion the logical flaw of such normativistic "overkill" (Scarre 1981, 243) consists of its failure to differentiate clearly enough between three levels of philosophical ethics. According to a system of classification described by Höffe (1981, 15), the first and most general level is that of a moral principle, which is also the final measure of morality; the second is that of principles associated with facts, and the third that of contemporary and context dependent criteria for judgments. Decisive for the relationship between the different levels is that even though principles associated with facts must be in keeping with fundamental moral principles, a moral principle can never *suffice* as a criterion for reaching a decision about a specific course of action. "Only when relevant facts and the general rules connected with them have been considered can the contents of intermediate principles associated with such facts be clearly determined" (Höffe 1981, 15). Thus normative considerations provide only a general rule of thumb for making judgments. In order to be able to apply such a rule to specific situations, it must be adjusted to comply with specific matters of fact (at the second level) and the particular circumstances of the respective context of action (at the third level). In this connection Vossenkuhl (1993a, 134) refers to the "dependence of normativity on descriptivity in ethics," whereby he maintains that for purely semantic reasons normative meaning is only possible in the context of descriptive meaning. Similarly the descriptive meaning of rules is what defines the "space of moral obligations." Only when this space has been described with sufficient clarity, and only when pertinent facts about premises and consequences of an action are known, can a reasonable and moral decision be reached (Vossenkuhl 1993a, 149).

From these considerations the following conclusions can be drawn regarding the relationship between empirical science and ethics: On the one hand, "ought" statements can never be *derived* solely from factual ones, but on the other, they must always *take relevant facts into consideration*. The methodologically correct way to steer a course between the precipices of naturalism and normativism consists of combining value judgments and facts in an appropriate manner. The more specific an ethical problem is, the more scientific information must be included in the process of normative decision making. This shows quite clearly that any kind of environmental ethics that intends to go beyond very general and abstract considerations with respect to ecological problems cannot be regarded as an exclusive subdivision of philosophy but must be seen instead as an interdisciplinary venture (Vossenkuhl 1993b, 13). In this case philosophy and the natural sciences must work together (and perhaps include economics and the social sciences as well), providing each other with relevant information and engaging in a process of continuous mutual feedback involving viewing the premises of one discipline from the perspective of the other.

Just as people in the natural sciences should avoid reducing the ecological crisis to purely a matter of ecology, so also should representatives of the humanities be cautious about prematurely attaching the label "naturalism" to references to scientific knowledge that are absolutely necessary for concept formation in environmental ethics. Sometimes one can observe that philosophers who are so inclined tend to employ the naturalistic fallacy argument as a kind of bludgeon against any kind of reference to nature. As Birnbacher (1991, 68) emphasizes, however, this argument is often not applicable. When it comes right down to it, the naturalistic fallacy argument is only valid in cases in which nature is regarded as a source of moral values, that is, when rules of morality are derived directly from nature in a *logical sense*. It fails to apply when reference is made to nature as a *criterion* for establishing moral principles, that is, when nature is assigned the (weaker) role of increasing the plausibility of an argument. Of course, it is often difficult to distinguish between these two positions. As Engels (1993, 120) has shown, in order to identify an argument as an example of a naturalistic fallacy it is necessary to carefully analyze the implicit assumptions on which the argument is based and to interpret the normative concepts involved. In individual cases this may often not be possible. But in spite of these practical difficulties, one fundamental distinguishing criterion that remains is that naturalism considers moral rules to be *immanent* to (within) nature while the position favored here is that such rules arise through an autonomous but not arbitrary decisional act, which, contrary to normativism, results from reciprocal reference to facts and values.

One indication that reference to nature is used as a guiding principle in the sense described above rather than as a defining one is when a rule for ecological action exhibits a *hypothetical structure*. A logical structure of this kind exists when the directive can be converted to an "if-then" statement. In order to do this, of course, it is necessary to determine the more or less implicit normative premise upon which the descriptive component of the directive is based. For example, if someone demands that nutrient input be reduced in order to increase the diversity of indigenous species, it should be made clear that a normative premise is involved, namely *the idea that* species diversity should be maintained, for example, for ethical reasons. *If* diversity is to be maintained for ethical reasons, *then* a moral obligation to reduce nutrient input can be derived from empirical proof that nutrient input leads to loss of species.

In the context of such a hypothetical imperative, ecology can assume a number of different functions. First, it can contribute to estimating the external circumstances that are relevant for reaching an ethical judgment about a particular course of action. Second, it can provide information about how a certain goal can be achieved and what consequences might be involved. Finally, the other way around, it can examine whether a certain course of action is in the long run favorable for attaining a desired end (see Knapp 1986, 29). At this point I cannot explore all the practical applications of ecology in the context of nature conservation that result from the *instrumental* role of the discipline described here in very general terms. Instead, the reader is referred to numerous sources in literature on applied ecology.[51]

15. "Ecological Thinking"

A survey of studies of this kind reveals, of course, that the positive, instrumental role of ecology is to a great extent restricted to problems requiring a *retrospective* analysis of the causes and effects of ecological damage that has already occurred in order to formulate recommendations for the future. In contrast, ecology seldom succeeds in calculating threshold levels for human interventions in ecosystems *ahead of time*. In other words, it rarely determines what can or cannot be allowed in the present in order to ensure that we and future generations will be able to live in a manner compatible with human dignity and that other species will be able to survive. The reasons for the limited predictive capacities of ecology were discussed in Part A.I and shown to be *fundamental* limits of human knowledge, that is, limits that no amount of progress in science will be able to master completely. One conclusion from this insight is the illusionary nature of any hopes that with the help of ecosystem analysis, simulation models, and technology assessment ecological systems will one day be able to be "managed" in a manner that guarantees safety. The ecological crisis cannot be eliminated by such technical means. Does this mean that ecology is useless for helping us to make provisions to master the ecological crisis? (Mind you, species extinction can only be stopped by such provisionary measures.) Is it in the long run irrelevant for ethics, not for basic *theoretical* reasons (as normativists suggest), but because its predictive shortcomings prevent it from providing the *practical* information required for making ethical judgments?

A misunderstanding of this kind could arise if dealing with nature in the "right" way simply required that we work out appropriate bans and requirements for each and every ecological case. However, this purely casuistic approach would not only overextend the capacities of environmental ethics. It would also *not suffice* to meet the unusual challenges posed by the ecological crisis. One of the main propositions of this treatise is that it is not enough to work out a catalog of ecological and ethical norms, but rather that if ecologically supportive action is to achieve more that alleviating symptoms, it must be rooted in radically different thinking.

Therefore in order to master the ecological crisis I consider the *direct*, instrumental application of ecology to be less important than its *indirect* role as the forerunner of a *change in attitude* with respect to the relationship between humans and nature.[52] Ecology's general insights give us pause to subject one or the other traditional view of the world and humanity to critical revision along with the basic ideas associated with them about the way humans are to treat nature. As a result of such reflection a new view of the relationship between humans and nature is taking on form that has been referred to as "ecological thinking" (Birnbacher 1989, 394) or as "new thinking about nature" (Zimmerli 1991, 389), the consequences of which will extend far beyond the science of ecology. Among other things these terms characterize the realization that (like all other organisms as well) humans and human civilization are embedded in a multitude of complex, natural interactions of which only a small part can be understood. Thus when humans intervene in nature, the major intended effects of such interventions will always be accompanied by more or less serious side effects. Ecological thinking means taking these side effects into account to as great an extent as possible by (1) extending one's perspective beyond the limits of the system under consideration; (2) counting on delayed effects, exponential developments, and long-term effects; and (3) assuming that not only linear causal relationships exist but also web-like ones interconnected with numerous feedback loops. A further result of ecological research with far-reaching economical and political consequences is the realization that "there are limits to growth and that the freedom of an individual has limits set by the system" (Schulze 1993, 274). Obviously this insight is not entirely new,[53] but through ecology it has been given a rational context of justification capable of being reconstructed in detail.

The same is true for an aspect of "ecological thinking" that in my opinion is the most important one: *knowledge of our lack of knowledge.* Granted, even Socrates was aware of this paradoxical fact and expressed it in the famous statement, "I know that I know nothing." But nowadays for the first time this insight can be justified scientifically (see Chapter 5). What is remarkable is that this justification can be produced by the very institution that was once expected to (and that many contemporaries still expect to) someday be able to (at least potentially) know everything. By recognizing its own limits science itself refutes such superstitions about what it can or cannot do.[54] For a number of reasons ecology more than any other scientific discipline is well suited to expose the basic limits of the scientific method. This has to do with its objects of investigation—their enormous complexity and individuality, the unusually great significance of historical factors and specific conditions associated with them, and the relationships involved that are often stochastic (random) ones and sometimes only capable of being demonstrated qualitatively. Ecological awareness is therefore to a very great extent awareness of limits.

In view of the previous discussion about naturalistic fallacies, it should be evident that when I talk about limits, what I am referring to in the first instance are purely descriptive limits, that is, limits of knowledge. However, only a few normative premises are required to proceed from acknowledging limits of knowledge to prescribing limits of action in a logically consistent manner. For the time being I propose to characterize these premises by temporarily classifying them as "avoiding undesirable ecological side effects." One example of such a premise is the generally accepted desire to "maintain the natural basis for sustaining life." Since the possibility of destroying this basis has become more and more real, the sense of awareness that goes along with "ecological thinking" seems to favor an attitude of general caution. If I do not (and never will be able to) know exactly how nature "functions" and what consequences my interventions in nature might have, I should be particularly careful about massive interventions. Ehrlich and Ehrlich (1981, xi) compare the naïve attitude of modern human beings and their illusions of grandeur regarding species extinction with the ignorance of a rivet popper who sets out to remove rivets from the wings of an airplane before knowing anything about their function. In response to the objections of a terrified observer he replies, "Don't worry. I'm certain the manufacturer made this plane much stronger than it needs to be, so no harm's done. Besides, I've taken lots of rivets from this wing and it hasn't fallen off yet." This example demonstrates that at least in cases where the conditions of life are at stake, in other words, where the possibility exists that "spaceship earth" as an environment for human life might crash, our actions should not be guided by the patchy ecological knowledge we have acquired but rather by insight into our lack of knowledge. As Schönherr (1989, 100) puts it, applied ecology should first and foremost be understood as "negative ecology." "Ecology warns us that no kind of explanation or mastery of nature is sufficient to grasp nature. Thus negative ecology admonishes us to be especially cautious in dealing with nature since even rational actions, as ecological as they might purport to be, cannot exclude [the possibility of] destructive effects. Negative ecology therefore recommends that we intervene as little as possible, that we technically manipulate nature as little as possible."

Of course, Schönherr's somewhat abstract recommendation to "intervene as little as possible" makes it obvious that it is hardly possible or even reasonable to interpret "practical ecology" *solely* as "negative ecology" on the basis of exaggerated epistemological pessimism. Since we *have to* act, that is, since our biological and anthropological heritage offers us no other choice than to technically intervene in nature (see Haverbeck 1978), we must resort to positive ecological knowledge over and over again, in spite of the priority we may have given to the negative side of ecology. Only by drawing on this admittedly modest knowledge can we make a relatively reasonable choice between different options, and only then do we also have a limited chance to minimize the undesired side effects of our actions.

If lack of knowledge is to be relevant for practical dealings, it must be presented as *informed ignorance*. According to Guggenberger (1986, 56), ignorance of this kind is characterized by a tendency to renounce the idea of achieving general social progress by direct intervention, that is, by designing and realizing a plan. Since the knowledge of reality that we as humans are able to acquire is always far less than the knowledge we would need to "manage" the future by planning and construction (see von Ditfurth 1991, 411), direct ecological intervention would be more likely to result in a "design for disaster" rather than a "design for progress." Therefore instead of executing plans with self-satisfied gusto, "informed ignorance" requires us to restrict ourselves to promoting "progress" *indirectly* by *creating conditions that are conducive to non-catastrophic development.*

The example of natural evolution shows us just what these might look like. Although even evolution is not devoid of mistakes, systemic breakdowns, and catastrophes (whereby, of course, the meaning of these terms always depends upon the perspective involved; see Chapter 13), for this very reason it seems worthwhile to think about the mechanisms and principles that are responsible for the fact that *in spite of* these setbacks the "business" of evolution has "not gone bankrupt in four billion years" (Vester 1980, 87). In the context of this book I am not able to expand on all the specific attempts that have been made to learn something from nature and ecology that can be usefully applied in other areas, particularly in economics (see, for example, Ring 1994). Instead I shall restrict myself to pointing out four ecological (but not solely ecological) principles (see Kafka 1989) that could help us to generate conditions favorable for "non-catastrophic" development and might also provide a guideline for sorely needed changes in our perspective of nature and the role of humankind.

16. Principles of a Change in Attitude

The first principle, from which the other three can be derived, is the *Principle of Error Friendliness*. This term refers to the peculiar combination of error proneness and error tolerance that is typical of the evolutionary process (von Weizsäcker and von Weizsäcker 1984, 1986). Although internal imperfections and external disturbances are always a threat to biological systems (think, for example, about disease among organisms or the severe loss of offspring among amphibians), at the same time these systems possess the ability to react to such threats and to a certain extent cope with them (for example, by healing processes in the case of organisms or surplus spawn in the case of amphibians, that is, by redundancy). The reason that this relatively high degree of error tolerance exists and has to exist has to do with the *constitutive* part that error plays in evolution. When it comes right down to it, evolutionary progress is simply what is left over after an infinite number of unsuccessful trials and errors in the course of "groping about in the space defined by evolutionary possibilities." "Fleetingness is to time as errors are to evolution" (Kafka 1989, 72). However, if errors are also a fundamental part of *human nature*—as indicated by both the proverb "To err is human" and the theoretical analysis of science (e.g., Popper 1972)—then human activities should be exercised in a manner that allows us to live with errors and learn from them. "Since complete infallibility is a utopian vision and inhuman as well, a basic principle of construction and application in technology should be 'error friendliness,' a quality that is also a decisive prerequisite for evolution or the capacity for evolution" (E. U. von Weizsäcker 1992, 224). Specifically this means that we should avoid technologies and interventions in nature that *exclude* the possibility of gaining experience through trial and error, because under certain circumstances they might lead to irreversible consequences (e.g., the release of genetically modified microorganisms). Technologies must also be avoided that *prohibit* any kind of error since the dimensions of the consequences would extend beyond the scope of any of kind of ethics (as in the case of nuclear energy).[55]

In addition to refraining from certain technologies the Principle of Error Friendliness is also connected with a second principle, the *Principle of Leisureliness*. This is fundamental for the dynamics of ecological processes because it is a basic prerequisite for their *ability to evolve*. Trial and error only functions if enough time is allotted to reliably test whether a novelty is both compatible with the system and better than its older counterpart. A decisive factor in this connection is that the criteria of natural selection also have to have an opportunity to change in the course of evolution and that this takes time. While *biological* evolution usually requires periods of time extending from several to numerous generations in order to be able to test the adaptive value of certain shifts in gene frequency in a population, in *cultural* evolution the lifespan of an individual is probably the pacemaker for the rate with which change is evaluated. Only if the testing rate is approximately proportional to the lifespan of the tester can we be certain that sufficient feedback occurs in the time available, that is, *before* complexity disintegrates and the entire system becomes destabilized. In light of this observation it is certainly not very error friendly, for example, for chemical industries to have introduced 50,000–60,000 chemicals to the biosphere without previously testing as carefully as possible their long-term effects on the climate, ecosystems, or the health of organisms (Dahl 1989b, 50). When science and technology move on to implementation before completing investigation in this manner, they "cause the barriers between society and the laboratory to be broken down," as the sociologist Beck (1988, 203) maintains. Moreover, by doing this the "logics of scientific discovery" postulated by Popper (1959) are "falsified" in a classical sense, at least as far as the status quo is concerned. When the world is more or less converted to a laboratory, as in the case of fluorochlorocarbons that endanger the ozone layer, science pulls the rug out from beneath the logics of experimental science since these logics basically rest on the concept of containment and experimental control.

This example points out the risk of unintended "global experiments" taking place and leads us to a third ecological principle as well, the *Principle of Diversity*. In order to prevent mistakes from spreading throughout the world in an uncontrollable manner, technologies should be avoided that involve uniform blanket coverage, and preference should be given to smaller, independent developments. In this connection we can learn something from studying the organization of ecological systems (e.g., in the context of the mosaic-cycle theory), which show us that compartmentalization, that is, dividing the system into a number of largely independent subsystems, reduces the susceptibility of systems to disturbances. If a given ecological factor changes in a manner that is unfavorable for the entire system, compartmentalization increases the chances that at least one of the subsystems might develop a solution that could help the entire system to survive. The importance of diversity in ecosystems under human control is illustrated by the increasing loss of plant varieties in agriculture and the greater risk of large-scale crop losses that accompanies this development.[56]

The decentralized nature of diversity is closely related to a fourth principle, the *Principle of Self-Organization* (autopoiesis). This term refers to the emergence of complex structures in thermodynamically open systems. Eigen and Winkler (1975, 197) define it as "the capacity of certain forms of matter to generate self-reproducing structures as a result of specific interactions and combinations while strictly adhering to defined conditions." In the present discussion, however, what is important is the fact that ecosystems and organismic communities never develop through a form of centralized regulation. Their functions are regulated solely by means of interactions between their components, compartments, and other factors. Thus the organization of ecosystems is completely decentralized. In view of this property and the fundamental limits to ecological knowledge that exist, it is not surprising that all attempts to generate centrally regulated ecosystems through so-called "environmental planning" have thus far not produced the desired functional reliability that we are accustomed to with natural ecosystems (see Haber 1986). "Environmental planning and modification are in fundamental opposition to self-organization and chaos theory, a contradiction which at the most can only be partially resolved" (Haber 1993, 105). As a practical rule of thumb for nature conservation Remmert (1990, 114) therefore recommends abstaining from corrective and developmental measures, at least in protected areas for which no overriding protective goal exists (such as maintaining a particular species of bird). "If nature conservation wants to protect the water, soil and air of our surroundings, the best way to do this . . . is to let natural ecological processes take whatever course they wish." Professional conservationists such as Scherzinger (1991), Obermann (1992), Gerdes (1993a), and Thiessen (1988) also advocate granting nature as many opportunities for self-regulated development as possible. Gerdes and Thiessen stress the significance of the *psychological* effect that the idea of "protecting nature through inactivity" expresses and might also promote. It would be a real sign that we have finally left technical optimism behind together with its scientistic claims that we will one day be able to manage nature completely. At the same time it would be an indication of a new or renewed inclination toward self-denial and equanimity in human dealings with nature, which, according to the main thesis of Part A, cannot be *sufficiently* analyzed to permit it to be manipulated and regulated *at will*.

Of course I cannot deny that the warning implicit in the thesis presented here, namely that excessive interventions in nature can under certain circumstances threaten the entirety of life as we know it, offers only very few specific suggestions for how to deal with the ecological crisis. This is because it is impossible to demonstrate just how much ecological knowledge is *sufficient* to be able to intervene in nature without serious side effects, nor is it obvious when an intervention can be considered so serious that the risk of undesired side effects associated with it is unacceptable. Mastering the ecological crisis is not "just" a matter of preventing an "ultimate global catastrophe" as, for example, the extinction of humans (the possibility of which technical optimists would deny). It is also a matter of the many different intermediate stages that we might pass through on the way to a global catastrophe, each of which *in itself* would be worth preventing since it is *ethically unacceptable*. Thus the example outlined above of an airplane in danger of crashing (Ehrlich and Ehrlich 1981, xi) represents a highly problematic simplification of what the ecological crisis is all about, even if the cautionary message it carries is legitimate and important. It is oversimplified because by virtue of its all-or-nothing character (fly or crash) it seems to acknowledge the ethical dimensions of smaller or middle-sized ecological catastrophes only to the extent that they might be preludes to a greater one. However, the main question that arises in connection with the ecological crisis is not just whether or not humans will one day force themselves and a large number of other species into extinction. In addition the problem is *what conditions of life* human beings can and should consider acceptable for themselves, for future generations, and for the natural environment if they do manage to survive. The last question in particular shows quite clearly that "an authority other than the prognostic rationality of science is necessary in order to perceive what must be corrected. An authority of this kind is ethics" (Zimmerli 1991, 403). But what kind of ethics is meant here? Traditional philosophical ethics? Is it really capable of meeting the completely new challenges posed by the ecological crisis?

17. Questions for Ethics Posed by Ecology

Considering the numerous calls for a "new ethics" that have been issued and the many new specialized ethical disciplines that are cropping up all over the place (e.g., ethics of science, business ethics, technology ethics, bioethics, etc.), the inevitable impression that arises is that traditional ethics cannot contribute much to dealing with contemporary problems and that in view of the environmental crisis it must therefore be *replaced* by an alternative and more efficient, specialized discipline, environmental ethics. This impression is misleading because it suggests that it is indeed possible to develop a new concept of morality or new ethical and ecological forms of judgment. However, the highest principle of ethics can always only be the moral responsibility of human beings. In this respect environmental ethics cannot really be a new kind of ethics. What is new is the *extended application* of the concept of morality to another area that traditional ethics so far considered to be ethically irrelevant and to which it therefore failed to grant sufficient attention (see Landmann 1981, 168, 169). Because of the increased power of human actions and the ecological crisis, for the first time in history the way we deal with nature has been recognized as being ethically relevant. "Nature as an object of human responsibility is . . . a *novelty* in ethical theory" (Jonas 1973, 74).[57]

As Vossenkuhl (1993b, 6) has shown, several different innovations have occurred in the history of ethics that were prompted or forced into existence by altered circumstances and that caused the area for which ethical norms are considered valid to be extended. "In the course of all these innovations not only was the scope of ethical validity extended. Its whole structure was changed. The principles of ethically acceptable behavior were not rendered invalid but assigned new levels of priority" (Vossenkuhl 1993b, 6). It is safe to assume that the newly incorporated ecological dimension in ethics will also result in similar alterations in ethical theory. For future developments in ethics the following twofold aspect of ecological problems appears to be particularly noteworthy. On the one hand, as shown in Part A.I, many ecological relationships cannot be comprehended sufficiently to permit human interventions in nature to be judged in an ethically sound manner solely on the basis of expected consequences. For ethics this means that an exclusively consequentialistic approach (as, for example, that of utilitarianism) is inadequate. On the other hand, many actions (such as the release of CO_2 when fossil fuels are burned) can only be

recognized as ethically relevant and thus become subject to ethical judgment when their collective consequences become apparent. Thus it follows
that a purely deontological approach is also inadequate. The "ethics of our
technological era" must therefore be an "ethics which takes consequences
into account (ethics of responsibility) and one which at the same time acknowledges the fact that our reasoning capacities may be too narrow for
this purpose, too limited, and in many cases perhaps even misleading and
to a fatal degree also self-reinforcing" (Zimmerli 1991, 404). In light
of these thoughts Lenk's (1977, 936) estimation that "the unity of morality
. . . nowadays is to a certain extent an empirical and a posteriori matter"
seems plausible. As the discussion on "ecological thinking" should have
made clear, ecology can contribute a sizable part of the scientific knowledge that morality of this kind requires.

Since the process of extending ethics to include an ecological dimension
began only thirty years ago—disregarding a few exceptions—it is not surprising that no solid, generally accepted body of theoretical knowledge has
arisen as yet. So far environmental ethics appears to be quite heterogeneous. Moreover, the problems with which it deals are generally more controversial than those discussed in social ethics. There are two reasons for
this. First, the problems of environmental ethics are usually more closely
connected with certain worldviews (as, for example, an ethicist's particular
view of the world, humans and nature). Second, it is still not clear exactly
how nature's moral claims can be described, neither in principle nor in detail. I feel that ecology has an opportunity to make a significant contribution toward evaluating and clarifying these two problems.

When examining *people's views of humans and nature*, what we are dealing
with is certainly not just the scientific component of our knowledge about
humans and nature. On the other hand, it would be hardly justifiable to
ignore the evidence that evolutionary research and ecology have brought
forth. Thus to me it seems to be inconsistent with the findings of evolutionary biology to *exclusively* emphasize the special part that humans play
in nature without taking into account *at the same time* the phylogenetic
background of humans as one species among millions of others. In the
same sense it seems reasonable for ecology to question the perspective of
humans that prevails in anthropocentric ethics in which humans are still
regarded as a kind of "closed society" (Meyer-Abich 1987, 66) and ask
whether the interconnectedness and mutual interdependence of all living
things have been sufficiently taken into consideration. Of course I am well

aware that very different views of humans and nature can be constructed from the raw material of scientific knowledge depending upon which phenomena are studied and the significance attached to them. *Absolutely conclusive* interpretations cannot be expected in this area. At this point I am also not concerned with *deducing* ethical theory from scientific evidence as controversially discussed in the field of evolutionary ethics (see Vollmer 1986d, 1987; Morscher 1986; Stöckler 1986). Instead I am interested in a more modest attempt to secure *compatibility* between ecology and ethics, even though it is clear that possible "contradictions" between the two fields cannot be shown in a logically exact sense.

Similarly, modest expectations must accompany attempts to determine *nature's moral claims*. As I will show later on, for this project it is basically impossible to circumvent conclusions based on analogies between humans and nonhuman nature. This leads to the necessity to establish limits of validity when making value judgments. In this connection it is the duty of biological science to examine whether or not and to what extent the principle of equality that is so fundamental to ethics can be applied to nonhuman organisms. In my opinion, this critical function on the part of biology with respect to ethical value judgments provides a certain guarantee that in the process of drawing conclusions by analogy we go beyond naïve anthropomorphism and attempt to meet the moral claims of other creatures in as *objective* a manner as possible.

At this point I wish to depart from a discussion about extending traditional ethics to include environmental ethics at an abstract level and pursue it instead on the basis of the main topic of this book, namely species extinction and the protection of species. From the standpoint of ethical theory several reasons can be given for concentrating on this particular topic. First, species protection places the greatest demands on an ethics that includes nature since it requires biotope, landscape, and ecosystem protection as indispensable components and since it depends upon supportive measures involving ecologically oriented technology, production, and consumption in addition. Ethical considerations of species protection extend even beyond the scope of pure "land ethics" (Leopold [1949] 1968) since an ethics of species protection involves not just regarding an ecosystem as a collective but granting an equal amount of consideration to its parts. Second, species protection poses the greatest challenge to traditional moral systems, which are closely connected with human qualities such as personality, individual interests, and accountability. And, finally, protecting

species is a touchstone for testing the seriousness of an ethical attitude toward nature since its economic, political, and personal consequences
would have the greatest impact, as we shall see. If species protection were
exercised in a consistent manner, it would reflect all the levels of action and
goals of environmental ethics that exist (Altner 1984, 43).

B. The Debate about an Ethical Solution

18. A Typology of Positions in Environmental Ethics

As soon as the moral dimensions of the ecological crisis became obvious, philosophers began to develop and justify norms for how to deal with nature in a manner that is not purely instrumental but acceptable *in an ethical sense*. The main problem that arose was whether the "right way to deal with nature" could and should be directed solely toward securing the survival of human beings, their health, and their well-being, or whether any kind of direct moral responsibility for nature and its subsystems exists in addition. In other words the question was, Does anything other than human beings— other animals, plants, maybe even abiotic nature and non-organismic entities such as ecosystems and species—have *moral standing*?

Although a few species of mammals (e.g., dolphins) exhibit behavior that is strikingly *analogous* to moral behavior, consensus generally exists that only humans are capable of acting morally and that only we can be *moral subjects* (moral agents). Only humans possess the capacity for reflection and free will (at least relatively free will) that it takes to act morally. However, philosophers do not agree on whether other organisms and natural entities can be considered *moral objects* (moral patients), which means being objects of *direct* moral responsibility, and which properties qualify them for such moral consideration (Warnock 1971, 148). According to Frankena (1979, 5), this is the controversy by which various models of environmental ethics can be distinguished from one another. It is not the usual distinction between teleological and deontological ethics or between value ethics and duty ethics that is decisive but rather the particular *scope* of direct human responsibility. According to Meyer-Abich (1982, 588) the scope of various positions can be divided into five different classes: "(1) Everybody respects himself. (2) Everybody respects himself and other humans. (3) Everybody respects himself, other humans and all consciously sentient beings. (4) Everybody respects all living things. (5) Everybody respects everything." Since this way of classifying the scope of moral consideration involves progressive expansion whereby each successive level includes all the objects of the previous one, the image it evokes is that of concentric circles surrounding the moral subject, located at the center of moral consideration (see Figure 2.). This *"moral circle"* describes a varying extent of *direct* moral responsibility. On this basis Teutsch (1985, 92) established a system of classification that covers most concepts of environmental ethics and involves four main types: anthropocentric, pathocentric, biocentric, and holistic (physiocentric) concepts of environmental ethics.[58]

According to the *anthropocentric* position in environmental ethics humans have moral obligations only toward other humans. Since humans are the only creatures capable of reasoning and morality, only they have intrinsic value, and only they can be objects of *direct* moral responsibility. Since only humans can be "the locus of intrinsic value, . . . the value of all other objects derives from their contribution to human values" (Norton 1987, 135). The moral relationship to nonhuman nature is thus always an *indirect* one. Whether an intervention in nature can be justified or not depends upon whether adverse effects for humans are involved and the extent of these effects. A classical example of an anthropocentric type of argumentation is Kant's ([1797] 1990, 84) scheme of justification for animal protection. In his treatise *Metaphysische Anfangsgründe der Tugendlehre* (The Metaphysical Principles of Virtue) he does not reproach cruelty to animals because of the consequences for the animals but because it dulls human empathy, which "is a very useful natural quality for morality in relationships to other humans."[59]

This kind of argumentation in matters of animal protection is what distinguishes the anthropocentric position in environmental ethics from a *pathocentric* one, which regards forbidding cruelty to animals as a duty *toward animals*. According to the pathocentric approach, not only human beings have intrinsic value but all creatures *capable of suffering* as well. All creatures that can feel pleasure and pain are subjects of conscious purposes and thus have corresponding interests. According to the principle of equality, which is a mainstay of ethics, their interests are to be considered regardless of whatever other properties and capacities they might have (e.g., the species to which they belong, rationality, etc.) (Singer 1979b, 18).[60] By defining the ability to suffer as a critical property, not only human beings but also all "*higher*" animals (which basically means vertebrates) are granted moral standing while "lower" animals, plants, and inanimate matter are excluded from the sphere of *direct* human responsibility. In pathocentric ethics our treatment of these parts of nature is only considered to be morally relevant to the extent that it might indirectly cause pain and suffering to sensitive creatures.

Representatives of a *biocentric* position in environmental ethics do not accept limiting the scope of human responsibility to organisms capable of suffering and grant moral standing to *all living things*, regardless of their level of organization. This position is usually justified by referring to a more comprehensive concept of interest that includes the *unconscious* striving to live exhibited by plants and lower organisms. The intentionality of these living beings is regarded as proof that organisms without conscious feelings are also subjects of purpose and therefore also exist for their own sake. Among advocates of biocentric ethics, whether or not it is ethically justifiable to assign different values to different species is still a moot point. While representatives of "moderate" biocentrism assume that a hierarchy of value and interests exists among all living things ("scala naturae"), proponents of "radical" biocentrism maintain that all living things are in principle equal.

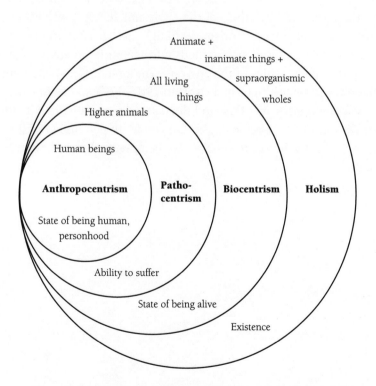

Figure 2. Basic types of environmental ethics and the limits of direct human responsibility they encompass. In the upper part of the drawing the natural objects are listed to which intrinsic value is accorded. The lower section indicates the criteria that are decisive for moral consideration.

The *physiocentric* or *holistic* position in environmental ethics has the most comprehensive standpoint of all four types. It includes not only all living things but also inanimate matter and systemic wholes in the scope of direct human responsibility. By granting intrinsic value to all natural things it abolishes the duality of means and ends characteristic of the other types of ethics. No part of nature exists *solely* as a means for another part of nature. Everything in nature exists for its own sake *in addition* to being means to certain ends and therefore at least potentially has the status of a moral object. Frankena (1979, 11) pointed out that the term "everything" can be interpreted in a more distributive or a more collective sense. In a *distributive* sense it refers to the entirety of all natural *objects* that we conceive of as being separate (e.g., humans, animals, plants, mountains, minerals, and water), while *whole systems* (such as populations, landscapes, and ecosystems) are meant when the term is applied in a *collective* sense. Strictly speaking the term "holistic ethics of the environment" only applies to the collective version of physiocentrism, although the terms "physiocentric" and "holistic" are often used synonymously in literature on environmental ethics. Within holistic ethics of the environment a monistic position and a pluralistic one can be distinguished (Norton 1987, 177). In *monistic* holism only the whole has intrinsic value, and the particular value of its individual parts derives from their relationship to the whole. In *pluralistic* holism, on the other hand, both the entire system and its individual parts have intrinsic value.[61]

19. The Scope of the Discussion

In the ethics of the Western world the anthropocentric concept has been the most prevalent one in both philosophical and theological contexts (MacIntyre 1966; Brumbaugh 1978). For more than three thousand years only moral duties toward humans (or God) were conceivable. Every once in a while pathocentric approaches cropped up, as, for example, among the Pythagoreans (Hughes 1980) or in the writings of Montaigne, Rousseau, Voltaire, and Schopenhauer. But it wasn't until ethical utilitarianism came on the scene (Bentham [1789] 1970; Mill [1871] 2000) together with the criterion of "interest" it advocates, one that extends beyond the boundaries of a single species, that the anthropocentric perspective was perforated to any extent. Nowadays a radically anthropocentric position such as the one evident in the quotation from Kant cited earlier is seldom expressed *explicitly*. The idea that we should not subject animals to unnecessary pain is widely accepted to be a direct duty toward animals. However, this more or less intuitive consensus in the special case of animal protection should not obscure the fact that many new (nonutilitarian) concepts of ethics that claim to be universal are still anthropocentric in nature. These include the concept of "discourse ethics" (Apel 1973; Habermas 1981), the "morality of reciprocal respect" (Tugendhat 1984), and "contract theory" (Rawls 1973). Since they require a reciprocal relationship or rather a symmetrical relationship between rights and obligations, they practically exclude animals, plants, and inanimate nature from the focus of their theory (see Mackie 1990, 194f.). The question of the right kind of behavior toward nature is either not considered at all or left distinctly open (as, for example, in Rawls 1973, 512).

From this systematic perspective and in the context of the history of philosophy it is not surprising that most philosophers who expressly concern themselves with the right way to deal with nature also tend to respond to this problem from a purely anthropocentric standpoint. Many philosophers not only consider conventional anthropocentric ethics to be completely adequate for dealing with ecological problems, they also regard any shift away from anthropocentrism as a dangerous relapse into irrationality and mysticism (e.g., Passmore 1974; Watson 1983; Wolf 1987). The impression that an anthropocentric standpoint continues to dominate within the still young field of *environmental ethics* is augmented by the fact that ethical utilitarianism, which is widely spread in Anglo-Saxon countries, also usually relies on anthropocentric arguments for justifying nature and species protection. As I shall show more specifically in Chapter 24.b, the reason for this is that when it comes to judging *ecological* problems the pathocentric approach of utilitarianism almost coincides with the anthropocentric one. Therefore, in keeping with the clarification of terms proposed by Ricken (1987, 3), when I talk about species protection I shall occasionally refer to the utilitarian position as one of "moderate anthropocentrism."

In strong contrast to the paradigm of anthropocentrism (evident in anthropocentrism and pathocentrism) are the other two nonanthropocentric types of ethics listed earlier. The biocentric position in environmental ethics is also firmly rooted in a long tradition represented by several Eastern religious cultures (e.g., Hinduism, Buddhism, and Jainism) as well as by a few branches of Chinese ethics (May 1979, 162f.). But within Western philosophy it is rather a minority position. Anyone who wants to extend the scope of direct moral responsibility beyond humans and animals capable of suffering must strongly defend his or her position in Western ethics. In this respect holistic ethics evokes even greater contradictions than biocentrism does since it at least partially departs from the concept of interest and claims moral consideration for supraorganismic systemic wholes, a completely new approach compared to the mainstream of ethics. Its premises and consequences have hardly been examined in detail and are therefore often a source of contention.

If you look closely at the controversial discussion that prevails between representatives of different positions in environmental ethics, it becomes apparent that the differences of opinion are basically restricted to the *theoretical domain*, meaning problems of conclusive reasoning, argumentative coherence, and universalizability. With respect to *nature protection itself* and the need for practical implementation, however, there seems to be no serious dissent. "In spite of all the differences in theoretical premises, the same specific matters are regarded as practically imperative" (Birnbacher 1989, 396). Does this mean that the different basic types of environmental ethics can be considered identical with respect to their consequences in practice and that ecopolitics based on anthropocentrism is no different from politics of nature conservation based on nonanthropocentric ethics? Many philosophers as well as economists and politicians do indeed seem to advocate this so-called "convergence hypothesis."[62] If, however, this hypothesis were really correct and the different types of ethics were indistinguishable in the long run, wouldn't that mean that any theoretical debates about the "right" kind of justification are merely the insignificant mental acrobatics of a few philosophers?

I wish to demonstrate that both claims are not valid by addressing the "touchstone" case of species protection. Neither do these different positions lead to the same consequences with respect to their *contents* nor is the way they are justified insignificant for the *implementation* of these consequences. My thesis is that for effective and comprehensive protection of endangered species nonanthropocentric positions are superior to a purely anthropocentric one from both a factual and a psychological standpoint. For *practical* reasons I shall therefore argue in Part B.I in favor of extending the scope of morality to *holism*.

I. A Pragmatic Approach: Is Anthropocentrism Sufficient?

20. Species Protection as an Intuitive Postulate

Before the possibilities of anthropocentric ethics for achieving species protection can be examined, I first have to respond to an objection that might be raised against proceeding in this manner. Isn't it a classic example of a *petitio principii*[63] when I start out by looking at the *practical* problem of which kind of environmental ethics is best suited for achieving species protection before examining the anthropocentric and holistic approaches *in theory*? My first job should be to explain *why* species protection is important. Two arguments against this objection can be presented, a pragmatic one and a meta-ethical one.

First, I could forget about a theoretical analysis of holism or justification of this position if it turns out that the practical consequences of anthropocentric and holistic ethics are the same. This is the conclusion one comes to when the economy principle of Ockham ("Ockham's razor") is applied, which says that one should try to get along with as few explanations as possible. In this case, anthropocentric ethics would win hands down since a first appraisal indicates that it is based on fewer assumptions than holism is.

Second, by proceeding in the manner proposed above, it seems to me that I am following a procedure characteristic of ethical reflection itself. In my opinion the starting point of reflection on good and bad is not an ethical *theory*, from which definite rules for the right kind of behavior are logically deduced, but rather *elementary intuitions*. As Spaemann (1990, 157) remarks in this connection, "We are fortunate that we don't have to wait for philosophy to provide us with knowledge about what is usually good or bad, good or evil. Instead, the job of philosophical reflection is to expose the principles that are inherent in the knowledge that already exists."[64] According to this idea, a reflective act that is performed after the fact has two important practical functions. In the first place it can reveal contradictions and inconsistencies in our normal behavior as well as in our unreflected moral convictions and correct them from a position of greater insight. In the second place an act of this kind can provide "orientation in borderline cases or in problematical new areas, that is, in instances in which traditional rules of behavior and basic intuition are insufficient."

On the one hand, the ecological crisis is undoubtedly just such a problematic new area for which intuitions alone are not enough. As the analysis on the basis of theory of science presented in the first part of the book demonstrated, the limited imaginative capacities of human beings and their misconceptions regarding the complexity and nonlinearity of ecological systems may even lead intuition astray. On the other hand, the example of the *concept of animal protection* (i.e., the conviction that we owe it *to animals themselves* to protect them from suffering) shows that some intuitions regarding our dealings with nonhuman nature are able to withstand repeated critical examination and "purification" through reflection. The discrepancy between such fundamental intuitions and (formerly strictly anthropocentric) ethical theory has caused pathocentric ethicists to adjust theory to intuition. Advocates of a traditional anthropocentric approach, on the other hand, attempt to diminish this contradiction by introducing additional premises (e.g., Tugendhat 1994, 189f.). Both ways of handling the problem seem to support Spaemann's (1990, 165) thesis, according to which "compatibility with our basic intuitions is ultimately the only criterion we really have at our disposal in matters of morality." Of course, it cannot be denied that the question of the role of intuition in ethics is a highly controversial matter among ethicists. Wolf (1988, 223), for example, thinks "that referring to common intuitions doesn't take us very far." I wish to counter this idea with the position proposed in this book according to which intuitions are certainly not a *sufficient* condition for morality, but nonetheless a *necessary* one. "Something is good only when it is both objectively *and* subjectively the right thing" (Spaemann 1986, 82).

I consider *species protection* to be a moral postulate that like animal protection is *rooted in intuition*. There are at least three indications that support this idea. The first of these is that most people in our society *spontaneously* consider species protection to be something good and that numerous individuals and organizations throughout the world are actively involved in trying to keep endangered plant and animal species from becoming extinct. And most people engaged in nature conservation are so certain of the moral value of such activities and their necessity that they seldom bother about justifying their position. It is often only under the pressure of rival interests that they feel themselves compelled to justify their actions, after the fact, so to speak. However, it is doubtful whether the arguments presented by people forced into a defensive position truly reflect their *real* motives. Bierhals (1984, 119), for example, suspects that the scientific, ecological, and economic reasons proffered in such situations are often "not at all the reasons why nature is of such concern to us." That's why I consider it so important to seek out the real, elementary substance of intuitions in nature conservation later on in a different context (Chapter 23.c).

I see a second indication that species protection is rooted in intuition in the increased tendency to establish *corresponding laws* (see Eser 1983, 350f.). Thus according to the Convention on International Trade in Endangered Species of Wild Flora and Fauna (CITES) signed in Washington in 1973 the purchase, ownership, and commercial use of endangered species on the verge of extinction is in principle prohibited. According to the German national ordinance on species protection from 1980 it is also forbidden "to wilfully disturb specimens of explicitly protected animal species, to pursue them, capture them, harm them or kill them" (Lippoldmüller 1982, 5). The fact that "eggs, larvae and pupae . . . are subject to the same kind of protection as adult animals" shows quite clearly that we are not dealing with "animal protection" in a traditional sense, which means protecting *individuals* of a species but rather that what is at stake is the protection of entire *genera* and their populations. Even when the so-called environmental compatibility of certain measures is assessed, as required in almost all cases of road construction and other major alterations of the landscape, it is not the *absolute number* of animals and plants affected by the measure that counts but more often the diversity and rareness of the *species* that occur in the area. In this connection it is interesting that the criterion of rareness has developed into one of the most powerful arguments that nature conservation has. If a nature protection agency wants to prevent the destruction of a landscape as a result of some construction project, its interests will usually be taken all the more seriously by administrative officials if it can demonstrate that a large number of species in the area are on the "red list" of endangered animal and plant species (Blab 1985, 614).

The species with the greatest impact are those that are not only endangered locally but also threatened by extinction beyond regional borders or even worldwide. In the United States, for example, in the state of Tennessee, operation of the Tellico Dam was at first stopped by a court order, because the *snail darter* (*Percina tanasi*), which occurs naturally only in this particular area, was in danger of becoming extinct. The grounds for this decision were provided by the Endangered Species Act from 1973, which strictly prohibits willfully and consciously causing the extinction of a species. Nevertheless, in 1979 the supporters of the dam managed to secure a special court injunction that allowed construction of the dam to be completed, even though there was no guarantee that relocating the fish to another body of water would be successful (Ehrlich and Ehrlich 1981, 182f.).

The controversy about the snail darter demonstrates quite clearly that species protection is still far from being able to assert itself when plotted against massive economic interests, but that it is gaining increasingly more weight in the public eye and in jurisdiction. No way would an 8-centimeter-long (3.2 inches) fish have been able to block a $120 million project for even an instant before 1970, thus claiming the attention of both Congress and the Supreme Court of the United States. It appears that perception of a tangible threat to the natural basis of human life and increased insight into the interdependence of all living things may have increased our sensitivity for the fate of other species as well.

A third indication that it is reasonable to regard species protection as a moral postulate rooted in intuition is finally the "*Convention on Biological Diversity*" passed at the summit meeting in Rio de Janeiro in 1992. No less than 156 nations signed a contract at this meeting obligating themselves to protect biological diversity *in its entirety*. In the preamble of the convention the contract partners express explicit recognition of the "intrinsic value of biological diversity and of [its] ecological, genetic, social, economic, scientific, educational, cultural, recreational, and aesthetic values" as well as its "importance for evolution and for maintaining life sustaining systems of the biosphere" (United Nations 1992).

One could, of course, claim that neither this international convention nor the laws that support species protection in individual countries, neither all the activities of activists in nature conservancy nor broad and spontaneous public support have anything whatsoever to do with moral intuition. It might very well be that all these "indications" merely reflect the subjective preferences of a few overly sensitive and influential friends of nature. However, I tend to share Birch's (1993, 317) view that proof for claims of this kind is in order: " . . . if it walks like a duck, quacks like a duck, etc., then it is a duck unless the skeptic can do *his or her* job of showing that it's not really a duck." The decisive point is that "the burden of proof is on the skeptic."

Having elucidated the idea that the concept of species protection is rooted in intuition, I now wish to specify it further and make more sophisticated distinctions. The example of the inconspicuous snail darter indicates that modern agencies for species protection concern themselves not only with species that are obviously useful for humans but rather with *all* species. "Species protection is . . . an attempt to maintain a species regardless of the properties of individual members of the species. It is the attempt to maintain the diversity of species we currently have on earth as completely as possible" (Rippe 1994, 810). Instead of the *restricted* and *specific* type of species protection that has been exercised for centuries in the form of laws governing hunting and nonhunting periods (Kirk 1991, 6f.), the moral intuitions of many people seem to demand *comprehensive* and *general* protection of species. Moreover, the establishment of lists of species that are endangered in particular areas and the many regional activities designed to protect species that are not yet endangered worldwide (e.g., the peregrine falcon) suggest that the intuitions involved are not only directed toward *global* species protection but also *regional* protection as well. Thus the idea seems to be that species protection should not only prevent the irreversible extirpation of species but secure existing species diversity as well, wherever this is possible and reasonable.

After having approached the topic of species protection by clarifying the semantic meaning of the term, it is now necessary to further specify the question posed at the beginning of this section, namely the question of whether anthropocentric ethics is "sufficient" for species protection. The crux of the matter is whether anthropocentrism is capable of securing not only restricted and global species protection but *comprehensive* and *regional* protection as well.

21. Anthropocentric Justification for Species Protection

In the context of an anthropocentric worldview species have no intrinsic value. They are only valuable if and when they somehow contribute to satisfying human needs and desires. Advocates of an anthropocentric concept of environmental ethics justify the necessity of protecting species exclusively by referring to the *usefulness* of species for contemporary human beings or future generations.

The ways in which species can be useful to humankind are highly varied and multitudinous so that it is impossible to describe them completely and in detail in this book. Instead I wish to refer the reader to the very detailed illustrations in books by Ehrlich and Ehrlich (1981), Daily (1997), and Baskin (1997). In the following section I shall restrict the discussion to a brief survey of the most important utility arguments presented in these books. Graduated differences between these arguments notwithstanding, it is possible to divide these arguments into three categories: (1) direct use of species as a material resource, (2) indirect use in the form of an ecological "service," and (3) immaterial use as a source of mental and psychical enrichment of human existence.

In the first category, that of *direct* use, the first thing that comes to mind is the use of species as food for humans. Even though the food supply of four billion people and their domestic animals currently consists of less than a dozen cultivated plant species, it is still absolutely necessary to maintain as many wild forms of these plants as possible as sources of resistance and vitality factors that can be introduced by crosses to generate new varieties in the course of breeding (E. U. von Weizsäcker 1992, 132). Biological pest control and pollination in orchards are also dependent on the existence of many wild species. As Ehrlich and Ehrlich (1981, 53f.) illustrate with many different examples, another area in which biological species are useful is that of medicine and pharmacy. They maintain that "you could write whole books just about the medicinal use of plants." And it is also possible that many plant and animal species that have not yet been discovered, particularly in tropical rain forests, might provide solutions that would further medical and pharmaceutical progress. In addition, the multitude of species that still exist also represents an almost inexhaustible treasure chest of chemical and physical properties and solutions that could serve as models for use in technical design and engineering (bionics).

As the consequences of the ecological crisis become more and more tangible, the *indirect* instrumental value of species becomes more and more apparent, that is, their importance for maintaining ecological system functions. As ecological research is able to demonstrate more clearly than ever, human life and well-being on earth would be impossible without the "services" of natural ecosystems and their species. Think, for example, about the fundamental significance of marine plankton for global oxygen production, the role of forests in climate regulation, or the importance of microorganisms for decomposing waste and generating humus. According to a statement by the German Advisory Board on Landscape Management (Deutscher Rat für Landespflege 1985, 538) about why we should protect species, both the possibilities for using nature as well as the protective functions of nature all depend directly or indirectly on species. "They govern metabolic cycles and the flow of energy; they construct ecosystems and maintain their stability; and they serve as a natural source of food for humans. Each species has a particular niche in the entire system, and its loss can have serious consequences." Even as they disappear many species provide a service to humans by functioning as bioindicators and "whistle blowers" that direct our attention to dangerous forms of pollution and other forms of damage to the environment (see Arndt et al. 1987).

A form of instrumental use in an *immaterial* sense that is usually underestimated is the significance of species for the psychological and emotional well-being of human beings. Natural landscapes whose character is defined by certain species not only serve an increasingly important function in supporting relaxation during holidays and recreation. In addition, for many people they are part of their concept of "home" and their cultural identity as well. Both the general impression of biological diversity as such and the observation and perception of individual species are a source of aesthetic enrichment for many people, the scope of which extends from basic sensual pleasure to spiritual experience. In addition to such psychological and emotional payoffs there is also the intellectual attraction that dealing with species in a scientific context provides. The more exact and intensive this occupation is, the greater the number of interesting problems and fascinating puzzles with which nature confronts a scientist. And as knowledge increases, so also does the capacity for perceiving the particular beauty of a species, which usually remains incomprehensible when viewed only fleetingly.

This was a brief summary of three major categories of species utility. In view of the enormous potential of species for usefulness, which was only outlined earlier, it seems at first that species protection speaks for itself *for purely anthropocentric reasons*. Well-meaning personal interest, wise foresight, and a feeling of responsibility for future generations ought to be reason enough to stop the ever-increasing course of species extinction as soon as possible. It follows that publications about species death are full of metaphoric imagery aimed at supporting the plausibility of the utility argument. For the ecologist Janzen (cited in E. U. von Weizsäcker 1992, 131), for example, global species extinction is "as if all the nations on earth had decided to burn their libraries without bothering to check out what's in them." Ehrlich and Ehrlich (1981, xi) introduced the image discussed earlier of a rivet popper in the process of prying the rivets from the wings of an airplane. And the biologist Wilson (cited in Mutz 1992, 10) compares his position with that of an "art historian watching the Louvre burn down." Wilson predicts that "of all the madness to which we may subject ourselves and our planet, the last thing future generations will forgive is the destruction of the earth's natural gene pool."[65]

There is no reason whatsoever to doubt Wilson's prediction. It is impossible to overemphasize the importance of biodiversity for future generations. But as striking and disconcerting as these images of the tragedy of species extinction are, two questions still remain to be answered. Do the images really go to the heart of the matter? And is the utilitarian argumentation upon which they are based really sufficient to guarantee that not just *a certain degree of species diversity* is secured but rather, as the intuition of many people demands, that in principle *all endangered species* on earth are granted protection?[66] In order to answer these questions we have to look more carefully at the categories of usefulness described above and the most important lines of argumentation contained in them.

22. The Limits of Utility Argumentation

22.a. Economic Arguments

A characteristic of anthropocentrism is that species are regarded as a *resource*. As complex groups of biological resources, species are in some respects fundamentally different from other natural resources, as we shall see. Nevertheless, they still have two things in common with them. They can be conceived of as instruments for satisfying human needs and they are scarce. Scarcity means they are both valued and limited and that they can be increased only at the cost of forgoing something else that is valued (Randall 1986, 79).

If species are scarce resources and economics is "the science of reasonable, rational ways of dealing with scarcity" (Hampicke 1992, 184), it seems appropriate to consider the loss of species diversity primarily from an *economic* perspective. The ethical and economic aspects of species extirpation are then two sides of the same coin. The idea that species protection really amounts to nothing more than wise, long-term (sustainable) resource allocation and can therefore be most convincingly and effectively supported by arguments derived from welfare economy is indeed quite widespread (e.g., Myers 1976). Advocates regard the strength of the economical approach to be that it permits the instrumental value of species to be determined with a uniform system of calibration (e.g., monetary value) so that their value can be compared with that of other objects by means of cost-benefit analysis. By these means, as many proponents of species protection hope, the priority of species over the satisfaction of other human desires could be demonstrated in nickels and cents, so to speak. Myers (1979, 56) expresses this belief in the power of economic argumentation quite clearly in the following statement: "If species can prove their worth through their contributions to agriculture, technology, and other down-to-earth activities, they can stake a strong claim to survival space in a crowded world."

However, this argument, which is really intended to be a kind of advertising for utility argumentation (see Ehrlich and Ehrlich 1981, 53), seems to draw attention to a structural weakness of the anthropocentric and economic approach. The burden of proof rests basically upon the endangered species. Apart from the dubious *ethical* content of this approach, the burden of proof that it implies is problematic from a *practical* standpoint because *proof* that a species is worthy of being protected is only accepted as valid when three prerequisites have been met. First, its utility must be known at least in first approximation; second, it must be quantifiable; and third, when subjected to a cost-benefit analysis it must be shown to weigh more than potential costs or competing utility values. In the section that follows it should become obvious that satisfying these three prerequisites represents an almost insurmountable hurdle.

As far as *knowledge* of the utility of species is concerned, the problem is that for comprehensive value judgments, what is really needed is *comprehensive knowledge*. However, complete knowledge is impossible, for reasons grounded in theory of science (see Chapter 5). Of course, one could always object that as in other spheres of life we simply have to make do with *temporary* knowledge. When it comes to making value judgments, however, this would mean that irreversible decisions would have to be made on the basis of an inadequate and constantly changing knowledge basis, at least as far as global species extinction is concerned. There is no doubt about it that some species have conspicuous properties whose usefulness for human beings is obvious (and that are therefore worthy of being protected). For most species, however, this is not the case. In these instances the results of value estimates will certainly depend to a great extent on the amount of scientific knowledge that is available. In this connection the knowledge we have about some well researched species should not detract from the fact that "the information basis for valuation of species is very weak" (Randall 1986, 85). Of the 10 million species thought to currently populate the earth, only 1.7 million have been registered scientifically since the introduction of taxonomy in 1753. Of these, in turn, only a fraction has been examined in greater depth with respect to economic utility (Wilson 1985, 702). If *ecological* aspects are included in the valuation, it is quite clear that we are far from an even approximately comprehensive database. Of the innumerable interactions between species and ecosystems that exist, there are only a few whose significance for human beings was able to be demonstrated. However, it would be a mistake to place all bets on prospects of additional knowledge in *ecology* for making value judgments. As I demonstrated in detail in Part A.I, for reasons of principle, ecology is only capable of providing very limited knowledge associated with a great deal of uncertainty.

All attempts to *quantify* the usefulness of species must, of course, be viewed from the perspective of this fundamental lack of knowledge. Since in general only those properties of a species can be quantified that are known *right now* and for which some kind of interest exists *right now*, attempts at quantification always represent a rather arbitrary portion of all the potential possibilities of utilization that exist. Values for which there is yet no demand (so-called "option values," Randall 1986, 84) or that may not be recognized until later on ("quasi-option values"), and that might possibly be the most important ones for future generations (Norton 1987, 36), fail to be taken into consideration.

But the limited time frame is not the only problem of the economic approach. Even in instances in which the possibilities for usefulness are already *well-known* and for which there is a *demand* as well, it is often highly uncertain whether they can be quantified at all, and how this might be carried out. As the ecological and aesthetic values of species clearly indicate, the properties of species are not simply "commodities" for which there is a market and corresponding prices. Economists would have to rely on special procedures to identify the *hypothetical market* to which they belong and to attach a so-called *shadow price* to them.

This can be achieved by means of *utility value assessment*. In this kind of analysis the costs are calculated that would accrue if a certain service had to be implemented by technical means instead of using a species. In his book titled *Der Wert eines Vogels* (The Value of a Bird), for example, Vester (1984) calculates the utility of seed dispersal by birds on the basis of the hourly wage of a planter and arrives at a figure of about $10 per year. The calming effect of birdsongs and the sight of bird flight is estimated to be comparable to three hundred valium tablets or approximately $15. This example shows that assigning a quantitative value to various kinds of utility, if it is possible at all, can only occur under extreme simplification and distortion of complex phenomena. Since these processes cannot be verified by others, as Vester (1984) himself admits, they are unable to provide utility argumentation with the very thing its supporters are looking for, namely a scientifically objective and reliable system of evaluation.

Instead, a closer look at things reveals that the choice of criteria employed in utility value assessment is highly variable and absolutely opportunistic with respect to practical application. Thus according to Ehrenfeld (1988, 213), for example, we must count on a decrease in the instrumental value of plants and animals in the tropical rain forests that remain the more successful pharmaceutical industry is in synthesizing medical products by modern biotechnological means. "Pharmaceutical researchers now believe, rightly or wrongly, that they can get new drugs faster and cheaper by computer modelling of the molecular structures they find promising on theoretical grounds, followed by organic synthesis in the laboratory using a host of new technologies, including genetic engineering. There is no need, they claim, to waste time and money slogging around in the jungle." In Ehrenfeld's (1988, 213) words, this would mean that "in a few short years, this so-called value of the tropical rain forest [would fall] to the level of used computer printout." At this point what matters is not whether or not the estimates of the scientists are right. The example shows instead that arguments in favor of species protection that are based on the material usefulness of a species run the fundamental risk of being countered by *substitution* arguments. There's no denying the fact that this risk is probably on the uprise in view of the growing possibilities of biotechnology and gene technology. Unfortunately utility value assessment seems to enhance this risk since its scientistic belief in our ability to quantify everything lends support to technical optimism, according to which almost every function of a species can be substituted by technology. I shall return to the *psychological* consequences of quantifying utility values in particular in Chapter 23.a.

In order to get around the problems of scientism and dependence on experts that *direct* quantification of utility values raise, many economists prefer to use an *indirect* method of evaluation, namely *willingness to pay* analysis. In this process, instead of trying to determine the supposedly objective value of a species, one attempts to simply determine the subjective economic value society places on some aspect of utility (Hampicke et al. 1991, Schweppe-Kraft 1992). Using a representative survey, for example, people are asked how much they would be willing to pay for well-defined programs of species and biotope protection if these programs were on the market and could be purchased as "commodities." Since it would be practically impossible to evaluate all animal and plant species by these means, let alone particular useful properties of them, this procedure must, of course, be restricted to a few popular species of animals or some rather general criteria such as the "species protection value" of certain areas (however that might be defined) (Schweppe-Kraft 1992, 92). A typical example of this kind of calculation is Auer's (1992, 40) analysis, which estimates the "value of biological diversity" in Germany to be about $2.5 billion per year. The author further observes that this value is four times greater than what it would really cost to maintain biological diversity.

Even though Randall (1986, 95) maintains that surveys of this kind "not

infrequently" exhibit general appreciation of the value of nature conservation, it is obvious that their usefulness for species protection is limited. As
I shall discuss in greater detail, most species probably wouldn't be able to
muster up sufficient willingness to pay in surveys of this kind. Apart from
the ambivalent consequences of willingness to pay analysis, doubts about
the conclusiveness of this method have also been raised (e.g., Müller-Christ
1995, 115f.). First, the results apparently depend to a great extent upon
methodological choices and imponderabilities such as the group of people
surveyed, their distance from the species in question, the knowledge of the
participants, and the reliability of their replies. In view of these contingencies, surveys seem hardly able to achieve what proponents hope to obtain
from them, namely "a reduction in the lack of scientific argumentation"
(Auer 1992, 440). Second, among welfare economists and social choice
theoreticians it is still a moot point whether it is really possible to deduce
morally legitimate and socially acceptable rules of behavior from such expressions of individual preference. As Randall (1986, 89f.) explains, both
the representatives of philosophical individualism and advocates of the theory of public interest find cost-benefit analysis unsatisfactory. A special
problem of this approach is that the interests of future generations are not
taken sufficiently into account or may even be discounted as a result of the
so-called time preference attached such procedures (Birnbacher 1988, 29f.,
87f.),[67] even though strictly speaking *all* future generations will at least potentially be affected by the irreversible extinction of species. And third, it
seems highly doubtful whether it is a boon to species protection when the
variable and opportunistic results of surveys serve as the basis for a system
of values that is supposed to guarantee the existence of biological diversity
for tens or hundreds of years. As Regan (1986, 196) puts it, our obligation
to protect species "demands a more solid foundation than our own change-
able needs and fancies."

Although this criticism of assigning economic value to species indicates
that caution is in order when attempting to *justify* species protection primarily on the basis of economic arguments, this does not mean that it
might not sometimes be reasonable to *view* certain aspects of nature and
species protection from an economic vantage point. Economic analysis can
be politically useful, for example, by drawing our attention to the fact that
"a striking discrepancy often exists between the value judgments of an individual and the actions of society in general" (Schweppe-Kraft 1992, 124).
It can also be employed to refute the widespread prejudice that nature and

species protection are *in principle* activities that run *counter* to economical logics (see, for example, Helbing 1995). In reality the view of the economist Hampicke (1992, 201) is often closer to the truth, namely the idea that it is "economical to exercise nature conservation and uneconomical not to do it."

Nevertheless it is important to realize that the consensus between economics and nature conservation that this statement implies is not absolute and that trying to *justify* nature protection with economic arguments can sometimes backfire. Anyone who argues in favor of protecting particular species by primarily referring to their economic usefulness must be prepared to pay a high price under certain circumstances, for example, when an opponent provides proof of their uselessness or maybe even detrimental effects. This is particularly critical in the case of species protection since contrary to the opinion of many conservationists, it is by no means reasonable to expect that a large number of species would come up with *positive* cost-benefit results subsequent to analysis of their utility or analysis of public willingness to pay.

One can be quite certain that obvious "pests" and traditional "enemies" of human beings (such as rats, cockroaches, and tsetse flies) and simply annoying species (like mosquitoes and houseflies) as well would probably receive a thumping *negative* utility value (see Glasauer 1991, 10).[68] And it would be no surprise if all those species that compete with humans for the use of certain resources (e.g., herons, cormorants, and weeds) would also come out with a negative rating following a calculation of benefits and costs (at least from a regional standpoint). However, many species would simply lose out in a cost-benefit calculation because they have no obvious or at the most only minimal benefits to offer in compensation for the *costs* that maintaining them requires. The term "costs" in this connection doesn't just refer to expenditures required for elaborate biotope management or for looking after and guarding a species (as, for example, a white-tailed eagle). In the language of economics costs also include "lost utility," that is, costs incurred by "refraining from using valuable goods and services in order to attain certain goals that have been set" (Hampicke 1992, 185). With respect to species protection the term "costs" usually means that humans have to refrain from using the areas that certain species inhabit (or at least avoid certain *forms* of usage).

An example of this situation is the case of the endangered North American *spotted owl* (*Strix occidentalis*). At the moment a population of only six to ten thousand individuals of this shy species of owl exist and inhabit the temperate rain forests of the Pacific coast. According to Knauer (1992, 41), in order to prevent its extinction the coastal state of Oregon would have to prevent timber harvesting in an area of almost 50,000 square kilometers of rain forest, an area comparable to about the whole state of Lower Saxony in Germany. This protective measure could lead to the loss of 40,000 jobs in the timber industry. Under these circumstances it is hardly imaginable that it would be possible to use economic arguments to defend protection of the spotted owl or to prove that it is economically indispensable. No utility value assessment or willingness to pay analysis would succeed in assigning an instrumental value to this owl that would outweigh the value of the timber and the equivalent social value of 40,000 jobs that are at stake. Neither the direct economic utility nor the aesthetic or ecological value of this rare and secluded bird would provide ample weight in a cost-benefit analysis.

And yet, we would do injustice to the spotted owl if we were to dismiss it as an exceptional example of a *particularly* useless species. According to estimates by Leopold ([1949] 1968, 210), one of the founders of American environmental ethics, most species in a community are of no economic value in the usual sense. Thus "of the 22,000 higher plants and animals native to Wisconsin, it is doubtful whether more than 5 percent can be sold, fed, eaten, or otherwise put to economic use."[69] As early as the 1940s Leopold complained about conservationists letting themselves be persuaded to devise "circumlocutions" in order to demonstrate that species that are valuable in a noneconomic sense really are of economic value. And indeed, eccentric attempts to determine the value of a species such as the case mentioned above, in which the psychological value of a bird is calculated on the basis of the value of compensatory valium tablets seem to be counterproductive to the goals of species protection, if only for reasons of credibility. In addition they are detrimental even in the context of the value system of which they are a part. The price that has to be paid for attempting to interpret elusive properties of species (such as aesthetic or ecological effects) as economic properties is oversimplification or distortion, and this may sometimes cause important qualities of these properties to be ignored. In view of this "methodological filter" and the fundamental lack of knowledge that prevails in any kind of evaluation, it is not surprising that the instrumental value of many species is often assessed as only minimal. Thus both Randall

(1986, 97) and Norton (1987, 44) maintain that the "true benefits" of species are "systematically understated" in cost-benefit analyses.

In order to circumvent this obvious deficit of quantitative evaluation processes, advocates of an anthropocentric position often present utility arguments in a broader, more "enlightened" form, that of the *insufficient knowledge argument*. Since we can really never know whether a species that appears to be useless might not be useful to humans after all, and since extinction is irreversible, it is important to protect species, "just to be on the safe side."[70]

As conclusive as this argument seems to be in view of known aspects of utility and extrapolation into the future (see Myers 1979, 81), it is nevertheless hardly convincing in the context of an economic approach. The first problem is that for logical reasons it is never possible to pursue all the promising goals that exist at one and the same time. We are usually forced to make a choice, which means that we have to weigh different possibilities against one another. In everyday life this often means that if the utility of something is merely hypothetical and not really tangible, we will usually drop it in favor of something whose utility is real and known. As Gunn (1984, 315) noted in this respect, "one spends one's money on groceries rather than using it to play roulette or buy lottery tickets." The same thing is expressed by the proverb that a bird in the hand is worth two in the bush. This is probably even more true when you're not even sure if there's *anything in the bush at all*, which is often the case in matters of species protection. In order to demonstrate how unconvincing reference to potential but completely unknown utility is, Norton (1987, 124) mocked the "insufficient knowledge argument" as an "Aunt Tillie's drawer argument." Aunt Tillie is an eccentric person who collects string, screws, and jelly jars in her drawers, cupboards, and garage, just in case these things might be needed someday. Obviously, in spite of her passionate collector's tendencies and cautious foresight, sooner or later she will have to weigh the importance of things when all the drawers, cupboards, and garages are full or when she decides she would like to use them for something else. Applying this to the case of anthropocentrically justified species protection this means that when species are regarded as *useful commodities*, they are then forced to compete with other useful commodities for the limited space available in human beings' dresser drawers.

This leads to the second problem. Since the "insufficient knowledge argument" is still a *utility argument*, as soon as weighing importance becomes necessary, it will automatically lead us back to cost-benefit analyses and the necessity of comprehensive knowledge. If, for example, a species such as the spotted owl is to be defended with the "insufficient knowledge argument," this argument is only convincing in the context of utility considerations when it can be shown by cost-benefit analysis that the *potential* value of this apparently useless species of owl is greater than the *real* value of timber and 40,000 jobs. In this case it's not enough to postulate some kind of abstract potential value, because "the potential of everything is infinite!" (Trepl 1991, 428). Instead, at least "a kind of general usefulness" would have to be demonstrated and quantified. Although this sounds like attempting the impossible, a few studies have really been conducted (e.g., the analysis of Fisher and Hanemann 1984) that set the ambitious goal of calculating such potential (only in the future recognizable) values by means of complicated extrapolations (Norton 1988, 202). Since such calculations and the conclusions drawn from them are by necessity highly speculative, it is doubtful whether they will be able to come out on top very often against short-term, clearly defined rival prospects for profit in the context of political contest.

There is good reason to fear instead that such attempts to calculate universal utility functions (including option values) might be misused to justify the calculated *elimination* of species. The problem is that even though they may give the impression of time-spanning universality, in reality they may discount values that might exist far into the future. It seems that for methodological reasons this asymmetry in time frame ("preference for the contemporary," see Birnbacher 1988, 87) may not be completely avoidable. Strictly speaking, the irreversible extinction of species would have to be judged in an infinite time frame. Economic analyses, however, by their very nature only operate within a finite and often relatively narrow time frame. As Randall (1988, 219, 220) emphasizes, conventional methods for evaluating alternative investment possibilities and the discount processes associated with them are usually designed for a time period of a single generation.

In view of this point of departure, it is not surprising when economic analyses occasionally come to the conclusion that under certain biological and economic circumstances it is worth it to eliminate a species rather than try to extract some economic utility from it for time on end. An example of a case of this kind is Clark's (1973) analysis. On the basis of a mathematical model he was able to show that it would be more economical to slaughter all the blue whales left in the ocean as quickly as possible and to invest the profits in a growing industry than to wait until the population recovers enough so that yearly catches can be made. This example shows that in matters of species protection, it can also be dangerous to rely exclusively on economic arguments when so-called useful species are at stake. Economic arguments may be useful for protecting *some* species, but they may increase the risks for others. Regardless of how great this risk might really be, there is no question that it is impossible to guarantee the protection of *all* species with economic arguments. *Comprehensive* species protection of the kind that the intuition of many people demands cannot be achieved on the basis of economically oriented anthropocentric ethics.

22.b. *Ecological Arguments*

Some advocates of anthropocentrism may agree with this conclusion but still fail to see how it might apply to anthropocentrism in general. They maintain that the reason why cost-benefit analysis often operates against species protection is very simply that the most important kind of utility that species have to offer is not economical; it is their role in maintaining the ecological functions of systems. As already discussed, the interactions between species and ecosystems are too multivariate and too complex to ever be able to be described sufficiently, let alone determine a quantitative value for their contribution to human well-being. Nevertheless, as representatives of anthropocentric ethics may argue, ecology has proven quite clearly that *every* species has a certain function within nature (even though it may often not be known), and that therefore *any* reduction in species diversity will disturb proper functioning of nature's economy. Since humans depend upon many services that ecosystems provide, *any* act that causes the death of species will endanger their own well-being as well. Species protection is therefore not just a matter of maintaining possibilities for usage (which could at some time be challenged in the course of a cost-benefit analysis) but rather a necessary prerequisite for a worthwhile environment, perhaps even for the survival of humans beings.

Typical of this concept is the title of a standard book on how to protect birds that came on the market in the 1970s, *Rettet die Vögel—wir brauchen sie* (Save the Birds—We Need Them). Except for a marginal remark about the ethical responsibility of humans for nature, in this book protecting birds is justified with the major thought that "the extinction of birds could also endanger humans" (Schreiber 1978, 9). For the most part this thesis is further explained by referring to the functions of birds as biological indicators and to ecological aspects. The broad selection of bird species presented in the book indicates that the authors regard their focus on "ecological self-interest" to be convincing enough to secure protection not only for a few species but for all of them. But is this assessment correct? Do we really *need all* currently existing species for ecological reasons?

One of the main reasons why conservationists usually respond to this question affirmatively is no doubt the image of the *food web* frequently mentioned in this book. Although this theoretical concept of ecological science really only "describes which kinds of organisms in a community eat which other kinds" (Lawton and Brown 1993, 258; Paine 1980; Pimm 1982), in popular ecological literature it is often regarded as pictorial "proof" of a *positive* relationship between species diversity and stability. If the species of an ecosystem are like the knots in the mesh of a web, this web must be all the more stable the more species are incorporated in its structure. The scientific counterpart of this idea, which was long considered to be self-evident, is the *stability-diversity hypothesis* discussed in Chapter 11.c. As already mentioned, on the basis of this hypothesis conservationists deduced the argument that it is fundamentally in the interest of human beings to protect as many species as possible, since *every* species stabilizes the environment of humans through its interactions with other species. If a species is eliminated from its original ecosystem, this reduces the degree of interweaving and thus also the ability of the ecosystem to resist disturbances of its ecological equilibrium (Commoner 1972, 38).

However, since the stability-diversity hypothesis in its *general* form has since been refuted, this attempt to justify species protection with ecological arguments must also be regarded as having failed. According to the current state of ecological theory, the relationship between stability and diversity is too heterogeneous to provide the basis for a *generally valid* argument in favor of maintaining as great a degree of diversity as possible. "Stability may decrease or increase with reductions in species number in a given system, and the effect may be different in temperate, tropical, and arctic habitats" (Schulze and Mooney 1993, 507). For the game biologist Schröder (1978, 34) the take-home lesson of this heterogeneous evidence is that "ecological theory is not necessarily always on the side of nature conservation." As many publications (also newer ones) show,[71] people in nature conservation and the ecology movement don't seem to have learned this lesson very well yet. In these circles the stability-diversity hypothesis is sometimes treated as an axiom.

However, it is not surprising that people cling to this theory, because if more recent results in ecology were taken seriously, arguments based on ecological utility would have to become much more sophisticated and complicated. While nature conservation used to be able to rely on the stability-diversity hypothesis as a blank check argument for all kinds of situations, it is now forced to formulate specific arguments from case to case. That is, it must differentiate between different types of ecosystem, species, and stability. The suggestive power of reference to an ecological web and the necessity for a maximum number of knots no longer corresponds to the current status of ecological knowledge. In view of the difficulties connected with trying to reach generally valid statements about the consequences of species loss for the stability of a system, it is all the more necessary to consider the structure of the web, the species involved, and the nature and strength of the relationships between these species in each and every case.

In a number of such investigations in the last three decades the not particularly surprising conclusion has been reached that ecological evaluation of species loss not only depends upon the factor "complexity" but to a large extent also upon *which* species or groups of species are involved (Pimm 1980). In the context of an ecosystem different species apparently have different "valances." In some cases removing certain species appears to have no readily recognizable effects on the rest of the species of the ecosystem. In others, however, if only a single species is lost, a whole chain reaction of further species loss occurs, and in the worst case might lead to a total restructuring ("collapse") of the ecosystem. A notable example of such drastic effects was demonstrated by Brown and Heske (1990) with their long-term experiments in Arizona. The two scientists were able to show that if three species of kangaroo rat (*Dipodomys* spp.) are removed from a desert shrub habitat, the whole community undergoes a drastic change. The shrub area is converted to a dry grassland; long grasses flourish instead of shorter ones; the seed-eating birds migrate away; rodents enter the system; and the snow melts more slowly. In this ecosystem kangaroo rats occupy the position of what has been described as *keystone species* in Anglo-Saxon literature. They have "a disproportionate effect on the persistence of all other species" (Bond 1993, 237). The term "keystone" refers to the uppermost stone in a archway made of masonry, without which the arch will collapse. In the same sense a "keystone species" seems to have a strongly supportive role in the structure and dynamics of a particular ecosystem.

It is clear that once Paine (1966, 1969) introduced the term "keystone species" to theoretical ecology, it rapidly aroused the interest of people in nature conservation as well. Had not the discovery of keystone species provided unequivocal and striking evidence of the ecological significance of species for human beings after all? While some people interested in species protection such as Myers (1979, 56) came to the very fundamental conclusion that "some species are 'more important' than others and thus deserve greater efforts to preserve them," others regard the concept of keystone species more as an aid to making decisions about which species are to be given protective *priority* in view of limited political and financial possibilities and time restraints. A ranking list seems to be inevitable for the very obvious reason that it is practically impossible to save *all* the species that exist (Lovejoy 1976; Roberts 1988; Gibbons 1992). Myers' (1979, 51) response to this is as follows: "Since we are clearly going to lose many hundreds of thousands of species before the end of the century, we need to know which ones we can 'best afford' to lose, which ones would certainly leave major ecosystems with critical injury if they disappeared, which ones should be saved because their loss could precipitate ecological breakdown whose dimensions we can hardly start to envisage, and which ones should be preserved virtually at any costs." He admits that applied ecology "so far" has hardly been able to provide any support for answering these questions, but he ventures that nature conservation will one day metamorphize into a discipline capable of making predictions about future developments. Is this assumption justified by the experiences that have so far been made with the concept of keystone species?

After the detailed discussion of the epistemological limits of ecological science presented earlier (Part A), it is not surprising that the answer to this question is "no." In spite of its uncontested heuristic value, critical appraisal of the keystone species concept has increased in both theoretical ecology as well as in practical nature conservancy (Mills et al. 1993). Three problems in particular seem to make its scope and practicability in the context of species protection appear questionable: a theoretical problem, a practical problem, and a philosophical problem.

The *theoretical problem* is that an exact a priori definition of the term "keystone species" is still lacking. In other words, it is not clear what criteria can be used to determine straight off which species is a keystone species and which is not. An a posteriori definition based on the remarkably strong effects of removing such a species from an ecosystem is usually not sufficient for clearly identifying such a species, nor can it be operationalized. Certainly there are cases in which the character of a species as a keystone species becomes obvious *after the fact* (as, for example, the experiment with the kangaroo rats demonstrated), but in many other instances it will probably be difficult to distinguish keystone species from others. This leads to the risk that in the long run subjective estimates may be used to decide whether or not a species is classified as a keystone species.

A definition based solely on the *effect* of some action is also problematic since ecological mechanisms that are qualitatively quite different from one another may thus be lumped together in the same class. Bond (1993, 239) lists no less than eight different functional types of keystone species, which in turn can be subdivided into further subclasses. Considering the extreme diversity of the cases described so far, it is not surprising that almost no properties that are *generally valid for all* keystone species have been identified. They may be common or rare and generalists or specialists for certain kinds of food. Some play a decisive role in metabolic cycles or as energy transformers, others not at all. In view of the problems involved with making generalizations about keystone species, Bond (1993, 249) is very skeptical about the "possibility of a general theory for keystone interactions." In his opinion "too much work on keystone species remains anecdotal."

Furthermore, it seems hardly realistic to assume that the theoretical deficits that result from limited possibilities for generalization can be significantly minimized by greater research efforts. On the contrary, investigations in the last few years seem to have made the matter more complicated by clearly demonstrating that keystone processes vary not only with the species involved but also often from area to area and from time to time. A species that seems to be a "pillar of support" in *one* biotope may play a completely insignificant ecological part in a *different*, only slightly distant one (Brown et al. 1986). And sometimes species develop unusually strong effects on the whole system only in combination with other species. These results indicate that we must abandon the popular but mistaken idea that the quality of being a keystone species is a species-specific characteristic of an organism that, once determined, can be found in all populations of the species. As many investigations have shown,[72] this quality is instead a *unique* property in time and space that results from a narrowly defined combination of local factors, a particular assembly of species and feedback loops with other, defined species.

The *practical* problem for species protection that this raises is obvious. If it is nearly impossible to make generalizations about the effects of keystone species, then Myers' (1979, 56) idea quoted earlier about being able *to predict* the consequences of species loss must also be discarded. Generalizations are the foundation of predictability. Since ecological theory has once again left nature conservation in the lurch, as in the case of the stability-diversity theory, the only recourse for advocates of species protection is to look for evidence of keystone species effects on site. Claims to special protection of species on the basis of their role as keystone species must be supported by experimental evidence for each species individually (and rigorously speaking also for each area in which the species occurs). But this is an almost hopeless venture for both methodological and quantitative reasons. In Chapter 4, I already discussed in detail the *methodological* difficulties that are in principle connected with any ecological experiment. The limits of ecological analysis described in that chapter (complexity, nonlinearity, boundaries, spatial and temporal uniqueness, disturbances, and measurement distortion) should not be construed as arguments against corresponding research efforts in ecology, but they do make it seem hardly realistic to expect empirical evidence to provide a convincing argumentative basis for protecting species. After all the experiences that have been made so far, we must assume that it is only seldom possible to clearly and unequivocally demonstrate all aspects of the real and potential significance of individual species.

In addition to the almost insurmountable burden of proof connected with collecting empirical evidence, another, *quantitative* aspect must be considered, which makes the practicability of the keystone species concept really seem dubious. The number of species is simply too great to be able to examine much more than a fraction of them with respect to their "ecological utility." Just in the area of Middle Europe, for example, there are more than 70,000 species of organisms that inhabit 130 different types of ecosystems. In order to document the functions of all these species within their ecosystems in an archive, it would take hundreds of thousands of time-consuming research projects. Apart from the fact that even in wealthy Middle Europe no one would be able to finance such an undertaking, and that ecologically and taxonomically competent specialists are lacking for approximately 60,000 less-frequent species,[73] a project of this kind would not be very helpful. Since ecosystems and their species composition vary constantly, the search for keystone processes would have to constantly start anew (like Sisyphus' stone-rolling efforts).

The *philosophical* problem connected with the strategy of finding ecological grounds for protecting species has already been touched upon in my critique of an economic approach. If species are protected primarily because of their individual utility, then in practice all those species will be denied protection for which no such utility can be demonstrated. In this connection the keystone species concept is particularly ambivalent for species protection, because it postulates by definition that keystone species, with their unusually strong effects on other species in a community, are an exception. According to Lawton and Brown (1993, 262, 263) we don't really know just how many species in an ecosystem might be keystone species, but it is certainly not the majority. If the keystone species concept is in essence only useful for guaranteeing protection for a minority of species, this predicament is all the greater insofar as most keystone species are probably *not endangered* ones. According to quite a few ecologists, the most frequent mechanisms at the root of what constitutes a keystone species are either trophic links or so-called "engineering" properties, that is, the capacity to make resources either directly or indirectly available to other species (Lawton 1994, 373). However, both the energetically and food-wise highly significant species at the bottom of the food pyramid as well as many "ecosystem engineers" (such as earthworms or forest trees; Jones et al. 1994) are usually *dominant* (frequent) species and thus only rarely a matter of concern with respect to species protection.

The other way around, it seems that many endangered species play only an insignificant ecological role in their communities *for the very reason that they are rare.* Caughley and Lawton (1981, 138) refer to such ecologically inconspicuous species as "non-interactive" species. If, for example, the small-leaved helleborine, the spadefoot toad, or the Apollo butterfly were forced out of their original ecosystems, this would certainly not have any measurable effect on energy flow or metabolic cycles in these systems. It would not be very convincing to demand protection for such endangered species by referring to their importance for the "economy of nature," because their effect on this economy is just about zero. It is not even certain that *other* species would be seriously affected by their loss. As Heydemann (1985, 582) emphasizes, the loss of some species often has "only *unilateral* consequences, that is, consequences for the species itself but not necessarily for others." One currently favored explanation of this observation among a number of ecologists is that links within a community are apparently mostly weak and variable (see Hubbell and Foster 1986, 327; Lawton 1992, 20), with the exception of ecologically highly specialized (stenoecious) species. If a rare species such as the bluethroat is lost and can thus no longer serve as prey for a sparrow hawk, other, more frequent species can usually take over as resources. Another reason is that many relationships between species are asymmetrical, that is, one species may depend upon another and its functions but not necessarily *the other way around,* at least not to the same extent. For example, as members of a lake community, a population of cormorants may be completely dependent upon the fish in the lake, but the fish could easily do without the cormorants. In a natural aquatic system, the size of the fish population is usually regulated by factors other than predator pressure. According to a metaphor proposed by Walker (1992, 20), those species that have a strong influence on the population dynamics of others are the "drivers" of an ecosystem, while other species like cormorants are comparable to "passengers" that are only carried along. Of course both categories must be regarded as extremes on a continuous scale, because most species probably assume an intermediate position between the two (that can vary from time to time and location to location).

In spite of the difficulties involved with trying to locate a species on such a scale, there can be no doubt about it that so-called *top species,* the final consumers in the food pyramid, usually belong to the passenger category. Top species, which currently are among the most highly endangered ones, are certainly the least important for "proper functioning" of an ecosystem. Thus from the standpoint of system ecology, for the North Sea it is quite irrelevant whether it is inhabited by the harbor porpoise or the grey seal.[74] And just as the continued existence of a sand dune landscape does not depend upon a population of natterjack toads, so also can a freshwater ecosystem get along without fish otters, bitterns, and the last pair of ospreys. But these are the species for which protective programs have primarily been developed and which are still of central importance in species protection efforts. Since top species are often relatively big, ostentatious, and well-known, it is usually possible to argue in favor of protecting them on the basis of anthropocentric arguments geared toward their aesthetic appeal or their potential as biological indicators. Arguments based on ecological science, however, which many conservationists and layperson consider to be the most reliable and convincing, are hardly applicable. In the case of the majority of endangered species, ecology is far from being able to demonstrate that we really *need* these species. According to Stanley (1981, 197) it is "simply absurd" to try to depict certain species as "essential cogs in the machinery of their ecosystems."

Certainly, even if many endangered species appear to be ecologically insignificant and unnecessary from the standpoint of contemporary system ecology, this doesn't mean that they really are insignificant in every respect and in the long run. After having demonstrated the epistemological and practical limits of ecological research in general and those of the keystone species concept in particular, it should be obvious that such estimates of the ecological significance of species are only temporarily valid and only under certain conditions. Krebs (1985, 574) expresses the fundamental uncertainty with which we have to reckon when he responds to the question of the frequency of keystone species in the following manner: "Keystone species may be relatively rare in natural communities, or they may be common but not recognized." Thus the problem remains unsolved, although recent data suggest that (at least in terrestrial ecosystems) only a few keystone species can be found. As far as ecological argumentation in support of species protection and the establishment of priority lists are concerned, this means that, strictly speaking, we can never be absolutely certain whether or not we are dealing with an ecologically insignificant species, even when the results of a detailed analysis in a particular instance indicate quite clearly that a species is most likely a "passenger." As Pate's and Hopper's (1993, 320) results show, some species may appear to be insignificant at *one point in time*, but in the *long run* this estimate may prove to be wrong. It sometimes happens that rare species become more frequent as a result in changes in environmental conditions and then assume the systemic functions of other, formerly frequent species that since have disappeared.

This residual uncertainty in determining the value of a species is the basis for an argument commonly found in anthropocentric ethics, which could be called the ecological version of the "insufficient knowledge argument." Since one can really never know for certain whether an apparently ecologically insignificant species might not still play some important but as yet unknown role in an ecosystem now or in the future, and since species extinction (at least the global variety) is irreversible, it is wise to protect *all* species "for security reasons" (van Dersal 1972, 7).

As convincing as the "insufficient ecological knowledge argument" may seem in view of the discussion on "ecological thinking" unfolded in Chapter 15, *in the context of utility reasoning* it proves to be just as dull a knife as its economic counterpart. As I demonstrated in the previous section, in the context of utility reasoning the goals of species protection are never an isolated matter. They must always compete with other utility values for recognition (see Morowitz 1991). Species can only come out on the winning side of a cost-benefit analysis by virtue of their ecological utility if we have at least an approximate indication of the dimensions of such utility. Another way of putting it is that the insufficient ecological knowledge argument is only convincing if it is coupled with *proof* that eliminating a particular species would result in a *real* risk to human well-being. A risk that is only *theoretically conceivable* is not enough, because almost anything is theoretically conceivable. No one will avoid digging in his garden because of the abstract possibility of hitting upon a mine unless there is *some kind of evidence* that war battles once took place on that spot in which mines were employed. Why should anyone protect the spotted owl merely "for security reasons" unless something indicates that this species may sometimes function as a keystone species and that its extinction would cause undesirable ecological consequences? In the context of anthropocentric utility reasoning and the necessity for *positive* knowledge that goes along with it, prophylactic protection merely on the basis of *missing* knowledge would be irrational. Justifying species protection in this manner also seems to be inconsistent, because it relies on the authority of ecological science on the one hand, but at the same time distrusts it when its results contradict the interests of species protection.

You don't have to be a pessimist to predict that if there is a conflict between "positive" and "negative" ecology, in other words between on the one hand knowing that a species is "noninteractive" and on the other hand knowing that such knowledge is always incomplete and uncertain, *positive* knowledge will almost always win. If ecological facts clearly indicate that a species only functions as a "passenger," the abstract notion that this knowledge might be mistaken will not have much of an effect. The risk associated with the erroneous evaluation of a single species is just too small. Thus ecological utility arguments are just as ambivalent as economic utility ones. Instead of fundamentally supporting the protection of endangered species they are sometimes just as effective for justifying their loss. Moreover, the "insufficient ecological knowledge argument" cannot be expected to alter this situation in any way whatsoever.

In view of this highly unsatisfactory balance Norton (1988, 205) proposes that we shift the focus of ecological justification for species protection from the level of the *individual species* to the more holistic level of *biodiversity*.[75] Instead of trying to demonstrate the *specific* utility of individual species, which is often difficult or not possible at all, it would be better to refer to the essential function of biological diversity *as a whole*. According to Norton (1986, 119f.), the individualistic approach is not only inadequate because of chronic lack of information, as discussed earlier. It is also unable to account for a central problem of species death, namely the problem of the so-called *zero-infinity dilemma*, a dilemma that often crops up when dealing with modern environmental risks (e.g., the greenhouse effect or others resulting from the use of mega-technology). According to Page (1978), a dilemma of this kind is characterized by the following properties, among others: (1) extremely high potential for damage, (2) low subjective probability of incidence, (3) ignorance of mechanisms, (4) collective risk, (5) long latency periods between cause and effect, as well as (6) practical irreversibility. In the case of species extinction the zero-infinity dilemma refers to the fact that the extinction of one, ten, or even a thousand species may often have no recognizable effects (zero end of the dilemma), but that increasing species loss a cut above the regional level may sometime or other result in a situation in which universal system functions such as metabolic cycles and energy transfer no longer occur free of interference.[76] In the worst case, major systemic alterations in the entire biosphere are conceivable, which might eventually even destroy the basis of human life on earth (the infinite end of the dilemma). A fundamental error of any *isolated* considerations of the extinction of individual species is clearly that they ignore the long-term increase in risk associated with such events, irrespective of what the exact course of such a development from zero to infinity might look like. Because of the small effects of many individual incidences of species loss, it is mistakenly concluded that further losses will also have little effect, with the same probability and to the same degree as the previous ones. That a conclusion of this kind is at least premature is nicely illustrated by the parable of the rivet popper proposed by Ehrlich and Ehrlich (1981, xi) and discussed in Chapter 15. If you recall, the main figure of this parable justified popping rivets out of the wings of an airplane and selling them by maintaining that he had often removed rivets from the airplane and that it still was able to fly without any difficulty.

Picking up on the main idea of the parable of the rivet popper and expanding on it Norton (1986) developed an argument for species protection that claims to be able to grant instrumental and ecological value to *all* species regardless of their *specific* utility. This argument is built upon two premises. The first of these is based on various investigations in theoretical ecology and evolutionary biology and maintains that diversity is an auto-catalytic process, that is, existing diversity begets further diversity. According to the second premise the reverse is also true, in other words, the loss of diversity is also an autocatalytic process. Extinction of one species results in a kind of downward spiral leading to increasing numbers of further losses. If we then assume that below a certain level of diversity certain ecological system functions that human beings need will no longer be able to operate, then, according to Norton (1986, 121), we have no other choice than to regard each and every instance of diversity loss as an existential threat to humankind, its well-being and even its survival. "Increases in the global rate of extinctions increase the vulnerability of the human species to extinction." In view of this relationship and the fact that the global downward spiral is in full force, Norton is convinced that he has presented an anthropocentric argument that is capable of guaranteeing protection for *all* species. Since each and every case of species *extirpation* increases the risk of global systemic collapse via the downward spiraling effect, each instance in which a species is *maintained* can be regarded as a kind of insurance policy against this greatest of all possible catastrophes. By virtue of its contribution to total diversity and its dynamics ("upward" or "downward") each species can be assigned prima facie value with respect to human well-being, so-called "contributory value." According to Norton (1986, 132) contributory value is significant enough to justify the protection of *all* species, provided, of course, that the social costs are not "unacceptably large" in individual cases.

Even though Norton's argumentative vision of "downward spiraling" and its appeal to the zero-infinity dilemma is basically an "insufficient knowledge argument" (since we don't really know which level of diversity is critical for the onset of a global catastrophe) this more holistic version seems to be more convincing than the individualistic one criticized earlier on. While the potential damage to humans that might occur through the extinction of individual, ecologically important species is too small to halt thoughtless species extinction when these cases are considered *individually*, it is the "navigation system of spaceship earth" and perhaps even the survival of humankind that are at stake in the case of a reduction of *global* biodiversity. In view of these dimensions, any reduction in biodiversity seems to be pure stupidity. When the degree of possible damage is in the range of infinity, then even the slightest possibility of occurrence should be enough to cause this risk to be avoided (Birnbacher 1988, 204). You don't gamble with your own existence or that of humanity (see Jonas 1984, 81). However, a conclusive argument in favor of species protection can only be deduced from a rule of this kind if the postulated causal relationship between individual cases and the pending catastrophe can be demonstrated with some degree of certainty. More specifically, the conclusiveness of Norton's argument depends to a great degree upon whether his second premise concerning downward spiraling as an autocatalytic process is true.

Three points make this rather doubtful. First, an autocatalytic, downward spiraling process requires rather strong coupling between all species so that each incidence of extinction leads to at least one more. This is certainly the case for some species (e.g., keystone species, mutualists, and some host species). But as discussed previously, a domino effect of this kind is least probable among the species currently most in danger of becoming extinct, which often seem to be only "passengers" (see Heydemann 1985, 582). Second, the concept of *global* downward spiraling only holds true if we assume that species in *different* ecosystems also interact closely with one another. This is certainly only true to a limited degree. Thus the only interpretation left is that what we are dealing with is not a single, global spiral but rather many individual and regional downward spirals. But then the downward spiraling argument loses some of its import since regionalization reduces some of its potential for inciting fear. It follows that it would probably be rather difficult to apply the argument of global downward spiraling to justify protecting endemic species in small, isolated habitats (e.g., isolated islands) if the human beings who live there prefer ecosystems transformed by humans (e.g., pastures for sheep) to natural ones. In a case like this a loss in diversity would have no effect on other areas (see Rippe 1994, 813). Third, to me it seems to be highly speculative to interpret the postulated phenomenon of downward spiralling as a *positive feedback* process, as Norton (1987, 59) suggests, that is, as a process that continues automatically, so to speak, once an initial impulse has been given. Wouldn't this mean that once diversity loss has begun it can no longer be stopped? As far as I can see, neither evidence from evolutionary biology nor from ecology supports such a view.[77] Certainly it is a generally accepted fact that positive feedback loops in animate and inanimate systems of nature do indeed occur more frequently than previously assumed (DeAngelis et al. 1986). But it is also clear that these autocatalytic processes are usually kept under control by means of negative feedback and only very rarely lead to "excursions," in other words, to uncontrolled chain reactions (as in the case of a supernova). From an evolutionary standpoint, biological systems with positive feedback loops that are unchecked by counteractive forces are apparently unable to survive.[78] Rather than postulating a *general* mechanism of autocatalysis, it would be more plausible to interpret long term and increasing species loss as stemming from *causes* that continue to operate and augment one another.

From this perspective the intended scope of Norton's argument is reduced to much more modest dimensions. If downward spiraling is no longer regarded as an *autocatalytic* process, then it is more difficult to demonstrate why *all* species that currently inhabit the earth should be saved. It may very well be that certain "ecological services" that are performed by systems and are important for human well-being could be adequately carried out by a fraction of the existing species. The loss of species we are experiencing today would then not necessarily be a step in the direction of a global catastrophe but might mean instead that if we succeed in stopping it in time, we could experience a transition from a state of *maximal* diversity to one of *reduced* but still *sufficient* diversity. The question that then arises is, of course, how much diversity is *sufficient* for the future of humankind? Or expressed more cautiously, "How many species are required for a functional ecosystem?" (Woodward 1993, 271).

Unfortunately, even with respect to the second, more modest question there is no clear consensus among ecologists. Instead, several contradictory hypotheses have been formulated (Lawton 1994, 368; Vitousek and Hooper 1993, 6), of which the following two are particularly noteworthy, the *rivet hypothesis* and the *redundant-species hypothesis*. According to the rivet hypothesis, which goes back to the parable proposed by Ehrlich and Ehrlich (1981, xi), *all* species contribute to the integrity of the whole like rivets or screws in an airplane. Just as every loss of a rivet reduces the flying capacity of a plane by a small but significant amount, continual species loss gradually impairs the function of an ecosystem step by step. According to the redundant-species hypothesis, on the other hand, species diversity is pretty much irrelevant for maintaining life supporting systems on earth. The only really important thing is to maintain the biomass of the primary producers, the consumers, the decomposers, and so forth. (Lawton and

Brown 1993, 255). This can be achieved with a relatively small stock of species. Most species are redundant as far as "the functioning of spaceship earth" is concerned; like "passengers" they are completely superfluous (Walker 1992, 20).[79]

It is understandable that people interested in protecting species usually prefer to justify their goals with the rivet hypothesis. This seems to better suit our intuitively felt obligation to save as many species as possible than the redundant-species hypothesis. However, the lesson we have learned from the history of the stability-diversity hypothesis is that ecological evidence is not always on the side of nature conservation and that it can therefore be counterproductive to prematurely and exclusively rely on an ecological theory that is supposedly certain but in reality turns out to be both controversial and ambivalent. For the sake of credibility the rivet hypothesis should be employed with caution in discussions about species protection. After all, there is enough evidence that seriously curtails its validity and supports the idea instead that at least in some ecosystems and in the biosphere as a whole some species are indeed redundant.

Let me describe three examples of such evidence. First, various results from paleontology indicate that most of the terrestrial and marine ecosystems that existed in the past 600 million years were significantly less diverse than contemporary ones (Sepkoski et al. 1981; Benton 1985). And nothing suggests that the life-supporting systems of the planet functioned less efficiently in these cases. Granted, the climatic and biogeochemical cycles of the biosphere were quite different from those that exist today, but according to estimates by Lawton and Brown (1993, 257), this can hardly be attributed to the *lower number of species* that prevailed. Second, the large differences in species diversity that can be observed in forests of the temperate zone indicate that most ecosystems have more diversity than is necessary for maximal productivity and stability. For example, 876 species of trees and bushes can be found in East Asia, 158 in North America, and in Europe only 106 (as a result of the most recent ice age) (Schuh 1995, 34). Friedmann and colleagues (1988) report on a lichen ecosystem in the snow-free zone of the Antarctic in which six species apparently suffice to perform the essential functions of the system. Moreover, since this system existed as such as far back as the Tertiary Period, it must be quite stable (see Woodward 1993, 272). Third, the results of several model studies and experiments contradict the thesis that species diversity is *always* significant for the proper functioning of ecosystems. Analyses such as those of Vitousek and Hooper (1993, 6) as well as Swift and Anderson (1993, 36, 37) suggest instead that although an increase in plant species diversity from zero to a certain *threshold level* does improve ecosystem functioning, an increase in diversity beyond that level usually has only very little effect. As anecdotal as they may be, altogether these results do not seem to confirm the *extreme* version of the rivet hypothesis, according to which *all* of the currently existing species are truly relevant for "proper functioning" of the biosphere and its ecosystems.

What are the consequences of this insight for how we judge current species extinction? Even in a rigorous anthropocentric context it would be too simple to construct a cheap argument for playing down or even justifying human-induced species extinction from the mere existence of redundancy. Even though we can rightly assume that there are threshold levels above which a reduction in diversity does not significantly impair ecosystem functioning, no ecologist can say exactly where the threshold level might lie and how large or rather *how small* the margin is within which species extinction is currently taking place. It is still "not possible to quantify the minimum number of species which makes a functional ecosystem" (Schulze and Mooney 1993, 504). Woodward (1993, 288) even considers it probable that we will never find an answer to this question in the future. As the results of several investigations suggest, redundancy is not only species specific (which is to be expected according to the keystone species concept). It is also often time, location, and process dependent (Lawton and Brown 1993, 267). In light of this evidence it would probably be futile to wait for a generalized ecological theory that would someday permit us to calculate on paper, black on white, exactly how great the risks associated with species extinction are. As in the case of the keystone species concept, the limited analytic and predictive possibilities of ecology prevent it from giving its undivided support to nature conservation.

Thus in order to ecologically justify *comprehensive* species protection at a holistic level, it seems that all that is left is a variety of the "insufficient knowledge argument," namely the *unknown threshold level argument*. Since we don't really know how great the security margin of species redundancy in the biosphere is, and since it looks like we might never be able to determine it in all relevant respects, we should refrain from *any* further reduction in species diversity to prevent the possibility of overstepping such a threshold and incurring catastrophic consequences for the life-supporting systems of the earth. According to Ehrlich (1991, 175; 1993) anything other than such a strategy of caution would be "a single vast irreversible experiment." In the end we would know whether communities with significantly fewer species are capable of performing the ecosystem services we need, but we would have to accept having possibly destroyed the basis for human well-being or even survival in the meantime. In deference to future generations we can by no means assume responsibility for a risk of this kind whose course runs toward infinity.

As far as its factual content is concerned, I consider the "unknown threshold level argument" to be conclusive for justifying *comprehensive* species protection. However, for two reasons I doubt whether it is capable of convincing very many people *within the context of anthropocentric utility argumentation*. First of all, the cautionary deferential attitude it requires is quite clearly contradictory to the basic position of anthropocentrism and its fundamental claim that nature is basically at the disposal of human beings. In Chapter 23.a I shall demonstrate more precisely that this inconsistency is not only of theoretical significance but also a serious obstruction to implementing species protection in practice. In addition, when a cost-benefit analysis is conducted, and this is the final real test of the plausibility of utility arguments, the "unknown threshold level argument" will probably not come out on top. While Norton's "global downward spiraling argument" at least claimed that there is a fundamental relationship between species elimination at a local level and a catastrophe at an ecosystem or global level, the "unknown threshold level argument" lacks such plausibility. Due to lack of *specific* evidence (with respect to individual cases) it remains vague and abstract. Let me demonstrate this once again with Schreiber's (1978) book on protecting birds titled "Save the Birds—We Need Them." According to the threshold argument its title would have to be "Save the Birds—It Is Highly Probable That We Will Not Need *All* of Them in the Future, But Since We Don't Know Just How Many We Might Need, Let's Behave as if We Really Need All of Them, Just To Be on the Safe Side." It doesn't take much to understand that this kind of argumentation wouldn't be very effective in a political contest. According to Walker (1992, 20), it is "regrettable" but nonetheless "most likely that global biodiversity concerns will ultimately reduce to a cost-benefit analysis." But what counts in a cost-benefit analysis is not what we don't know but rather *positive* knowledge. "Without knowledge of redundancy, or more broadly, the relationships between levels of biodiversity and ecosystem function, we cannot estimate either costs or benefits." In view of this situation Pitelka (1993,

483) advises all ecologists to continue to try to *quantify* the relationship between loss of diversity and loss of ecosystem functions, in spite of obvious theoretical and practical difficulties. In his opinion, quantification would provide the most effective arguments for protecting species diversity. "Properly presented, such information could be convincing to both policy makers and the public and could make biodiversity a much higher priority issue than it currently is." Westman (1977) has a similar position.

Let us assume that ecology might someday really succeed in making the progress required in this area (which I still consider to be dubious, as discussed above). For reasons rooted in theory of science it is still questionable whether this approach would be adequate for dealing with the problem of species extinction. All the statements about the relationship between diversity and ecosystem functioning discussed so far are based more or less openly on a *holistic version* of ecosystem theory (discussed in Chapter 12), which places greater emphasis on the role of metabolic cycles and energy flow than on interactions between populations. A primarily process oriented and functional approach of this kind readily evokes the impression that the biosphere and its ecosystems are something like cybernetic machines designed to fulfill certain purposes, namely ecological system functions. That this perspective sometimes comes pretty close to the perspective of engineering is indicated by the images used to illustrate the relationship between the "whole" and its "parts." An ecosystem is an "airplane" (Ehrlich and Ehrlich 1981, xi), a "car" (Schulze and Mooney 1993, 499), or an "artificial respiration machine" (Norton 1988, 204), and its species are "screws" or "rivets," "drivers" or "passengers."

On the one hand it is certainly legitimate to look at nature from the perspective of a technician once in a while in order to better understand certain relationships. As I already explained in connection with the "verum-factum principle" (Chapter 12), this methodological approach seems to be almost fundamental to experimental science. On the other hand, the discussion of the terms "ecological health" and "nature's economy" revealed that there are limits to a holistic system perspective (in the sense of theory of science) as well as to the application of the term "function" to ecosystems. In addition, this can lead to serious misunderstandings and false estimates (see Chapters 12 and 13). The critical problem in particular is that it is incorrect to apply to ecosystems a term that was developed in connection with organisms, namely the concept of *teleonomy* (indirect goal orientation governed by a program). I will explain this further in Chapter 29.c. In order to avoid such mistakes when presenting ecological justification for species protection, it is important to keep in mind that the ecosystem and diversity concepts involved (like any scientific model) are not exact replicas of "reality" but *methodological projections* that more or less simplify matters (see Chapter 5). Closer inspection might perhaps lead to the insight that the "engineering model" differs quite distinctly from the "original" in nature so that there is no reason to overestimate its significance for species protection. On the contrary, in my opinion there are three inherent risks to this model.

First, by its very nature it fails to account for the *uniqueness* of species (see Chapter 4.e). Since this model is exclusively concerned with the function of a species (and usually only with its metabolic and energetic contribution to a hypothetical "whole"), it implicitly transports the idea that one species can be substituted by another (see Ehrlich and Mooney 1983). A species is simply a species, just as rivets are simply rivets. Granted, the keystone species concept has served to modify this abstract view of system ecology and redirect attention to the individual species by differentiating between large and small rivets, so to speak. But the fundamental idea of substitution remains unaltered by a distinction of this kind based on *considerations of function*. Large rivets can be replaced by other large rivets, and small ones by other small ones. In raising the case of the feather winged moth (*Pterophorus monodactylus*), a moth with unusual hind wings similar to those of a bird, Dahl (1989a, 68, 69) presents an example of an animal species that could easily be replaced in any ecosystem, in spite of its distinct individuality. Thus he writes, "For quantitative ecology the feather winged moth is altogether useless. However, that is not a shortcoming of the feather winged moth" but rather of a kind of ecology whose cybernetic terminology is unable to adequately represent certain decisive aspects (see Trepl 1983, 10).

Second, the functional and cybernetic approach is not capable of adequately accounting for the *dynamics* and *instability* of nature (O'Neil et al. 1986, 48f.). As described by Hengeveld (1994, 6) from a standpoint more closely associated with population biology, "species do not evolve by a steady process of co-adaptation within stable communities with a fixed composition but erratically and freely, being continually jolted about in environments that vary capriciously in time and kaleidoscopically in space. Populations are always on the move, flowing from one locality to the next, splitting and merging along the way." In view of this "turmoil" (Hengeveld 1994, 6) far removed from "ecological equilibrium," comparing an ecosystem to a car seems to be quite inappropriate, even if some parallels might exist. For this reason Schulze and Mooney (1993, 499) cautiously point out that contrary to a car the functional role of a component of an ecosystem can vary depending upon the activities of other components in the system. Thus something that might be a rearview mirror one day may change into the cover of the gas tank the next and to a steering wheel two days later. In view of this plasticity it is up to the reader to decide whether comparing an ecosystem to a machine causes more misunderstandings than reliable knowledge.

Third, thinking in technological terms as it is commonly found in modern systems ecology tends to cause researchers in this area to overemphasize the concept of the "ecological niche" while not directing enough attention to *historically contingent* diversity in ecosystems. As Schulze (1993, 275) emphasizes, species do not develop within *clearly defined* niches but rather in *open* ones. In other words, in the course of evolution they can succeed in using resources that were previously not available. "The evolution of species is by no means always associated with a particular function for the ecosystem." In order to underline the importance of the historical dimension for understanding ecosystems, Zwölfer and Völkl (1993, 315) compare an ecosystem with a cathedral instead of a car, "a cathedral, . . . the structure and decor (or rather the inherent cultural information) of which have grown in the course of centuries." This analogy clearly indicates the implicit dangers of systems ecology for the project of species protection. "From the standpoint of a technician, if the point is to maintain a cathedral, it would suffice to concentrate on the basic functional elements of the structure and forget about the rest that has been added in the course of history. But then sooner or later a cathedral would look no different than a factory" (Zwölfer and Völkl 1993, 315).

This analogy, which graphically summarizes the limits of *ecological* utility arguments, brings us to the third major branch of argumentation in anthropocentric ethics, to *aesthetic* justification for species protection. Having briefly outlined the most important aspects of this approach in Chapter 21, I shall now proceed to examine whether reference to aesthetic utility can be applied to justify and promote *general* species protection, or whether it is only applicable in certain cases.

22.c. Aesthetic Arguments

While the usefulness of economic and ecological arguments for saving species is seldom questioned, the status of aesthetic argumentation is highly controversial even among professed advocates of anthropocentrism.

On the one hand you may find *positive estimates*, according to which human interest in the beauty, comforts, and uniqueness of nature is "perhaps the most significant anthropocentric argument of all in favor of nature conservation" (Birnbacher 1989, 411). Support for this view is provided by three sources in particular, two empirical ones and one derived from ethical theory. One of the empirical sources has to do with the assumption that nature conservation activities "are primarily motivated by such interests." This view is expressed, for example, by Hampicke and colleagues (1991, 40) in their analysis of willingness to pay in matters of species and biotope protection. Another is the certainly quite plausible prediction that people's basic need for natural beauty, quietness, and relaxation "will probably increase rather than decrease in the future" (Birnbacher 1989, 411). In Singer's (1993, 271) opinion "the appreciation of wilderness has never been higher than it is today, especially among those nations that have overcome the problems of poverty and hunger and have relatively little wilderness left." According to the third source of aesthetic argumentation derived from *ethical theory*, the aesthetic approach is the most appropriate way to rationally master the ecological crisis, because *in effect* it grants autonomy to nature (the only anthropocentric approach that does so) without resorting to controversial metaphysical premises such as assuming that nature has intrinsic value (Früchtl 1991, 34f.).

In very striking opposition to these three pluses are the *negative experiences* of many conservationists and philosophers with the minimal significance of aesthetic arguments in political debates. According to Singer (1993, 271), "Arguments for preservation based on the beauty of wilderness are sometimes treated as if they were of little weight because they are 'merely aesthetic'." Norton (1987, 114) and Trepl (1991, 429) express similar views. But what exactly is the reason that such little store is set in aesthetic argumentation? One possible answer to this question is captured by the widely spread view that nonmaterial utility is fundamentally less important than *material* utility. Myers (1979, 46), for example, considers aesthetic argumentation to be "a prerogative of affluent people with leisure to think about such questions." In my opinion, a perspective that views economic and ecological matters as being of prime importance while aesthetic needs are regarded as a superimposed luxury is untenable. Just as economic and ecological utility arguments are not *all* directed at fundamental requirements for survival, so also is there no reason to exclude the possibility that aesthetic experiences in nature might be something essential for some people. As Leopold ([1949] 1968) correctly observed, "There are some who can live without wild things, and some who cannot."[80] Beyond certain indisputable minimal standards of material requirements it seems to be extremely difficult to distinguish between necessities of life and sources of happiness above and beyond them.

As this difficulty clearly demonstrates, the major problem of aesthetic argumentation is not that immaterial needs are *fundamentally* second to material ones but rather that enormous individual and cultural differences exist with respect to its orientation and significance. It is much more difficult to make generalizations about aesthetic things than it is in economic and ecological matters. This is not only true for aesthetic interests but also for the aesthetic judgments upon which they are based. Ever since Kant published his *Kritik der Urteilskraft* (Critique of Judgement) in 1790 there have been repeated attempts in philosophy to identify universally valid rules regarding our perception of the beautiful and the sublime that are capable of being rationally reconstructed. But none of these attempts at "aesthetic objectivity" has been very convincing (Jung 1987, 26f.). Since the possibilities for justifying an aesthetic judgment are very limited, in spite of a certain amount of structural universality, aesthetic views ultimately remain *subjective*.

That an aesthetic viewpoint is subjective doesn't, of course, mean that it is therefore completely irrelevant. As if objective and completely rational reasoning were the only kind of reasoning to be employed in nature conservation! (See Bierhals 1984 for a discussion of this topic.) But in the context of *ethical argumentation*, whose aim is to define ways of thinking and acting that are universally binding and that therefore requires more substance than personal feelings and interests, such subjectivity is without a doubt a critical deficit. It is particularly difficult to argue for support of the aesthetic interests of *future* generations when the limited universality of aesthetic judgments makes it almost impossible to estimate what these interests might be like. Claims of this kind based on *personal* aesthetic experience can only be shared by contemporaries with similar subjective feelings.

This basic weakness of aesthetic arguments can be seen very clearly when species protection comes into play. Obviously some endangered species are aesthetically appealing to almost everyone, as, for example, the Californian condor, the panda bear, or the European kingfisher. As far as these species are concerned, aesthetic judgments are usually quite similar, and the general interest that this generates is large enough to guarantee that costly protective measures are taken. But such exemplary cases of species protection notwithstanding, it should not be overlooked that the majority of species on earth are not fortunate enough to be among the favorite snapshot objects of human beings (see Zwölfer 1980). According to recent estimates 98 percent of the animal world consists of arthropods, which means that primarily invertebrate animals and in particular insects are the ones affected by worldwide extinction (Müller-Motzfeld 1991, 197). But the aesthetic notions of most people are not such that they react enthusiastically toward these creatures, except perhaps for the case of "useful" bees or particularly colorful butterflies. According to Müller-Motzfeld (1991, 205), insects are not only generally regarded as "not being worthy of being protected" but often even as "pests." If this is true, then it doesn't seem to be very realistic to expect more than marginal support for species protection from aesthetic argumentation.

Of course it might be objected that people's lack of interest in insects and other invertebrates is not the result of carefully developed aesthetic considerations but more often due to pure ignorance. Anyone who takes the time to study this group of animals more closely can hardly resist being profoundly moved by their aesthetic appeal. If beauty is thought of in more than a superficial sense[81] and the "beauty of interest" that scientific studies of nature evoke is also taken into consideration (Ehrlich and Ehrlich 1981, 38), then there is no such thing as a species that is not aesthetic. In this broader sense it would in principle be possible to attribute instrumental value to *all* species on the basis of aesthetic argumentation and thus also to secure their protection.

In response to this notion I first wish to point out that not all experts in aesthetics share the view that pristine nature is in principle beautiful and that those who are scientifically and aesthetically educated are therefore incapable of developing negative aesthetic attitudes toward it.[82] However, it is beyond the scope of this book to discuss these views in detail. In fact it is really even not necessary since I am not interested in a philosophical examination of aesthetic judgment but rather in the *practical* question of the extent to which views of this kind *really do* affect the way people deal with nature and their interests in species. From this perspective it cannot be denied that the aesthetic dimension with its inherent possibilities for mutually enhancing sensual experience and scientific knowledge bears considerable motivational potential for species protection. Relevant factual information, training in sensual perception, and contact with nature mediated by the senses can extend and reinforce interest in species and thus make a significant contribution to the aims of species protection. It is certainly important to recognize this and pursue the consequences it entails, but it is still a moot point whether *general* species protection, which conservationist intuition demands, can be secured by these means. To me it seems to be unrealistic to assume that "extended aesthetic interest" can ever be mobilized strongly enough to come even close to attaining this goal.

My apprehensions about the possibilities of extending the aesthetic approach to this extent stem from two observations. First of all, systematic scientific knowledge is certainly useful for developing aesthetic perception, but by no means sufficient. As Leopold ([1949] 1968, 174) remarked in a similar case, having a Ph.D. in ecology doesn't guarantee that a person will "see" nature the right way. "On the contrary, the Ph.D. may become as callous as an undertaker to the mysteries at which he officiates." In my opinion, however, another matter is more significant than this one, a morphological and systematic problem easily overlooked by those involved in widespread activities focused on protecting aesthetically conspicuous species of birds and mammals. It is the fact that the differences between many plant and animal species are so small that careful examination and profound knowledge of systematics is necessary in order to even identify them as separate species. If it were only a matter of differentiating between a monarch butterfly and a brimstone butterfly or between a blue spruce and a white pine, the problem might be able to be solved by additional efforts in biology classes. But in view of more than three hundred species of Thysanoptera, to mention only one of more than thirty endemic orders of insects in Middle Europe, the matter appears much more difficult. In cases like this, no manner of "nature conservation education" oriented toward the interests and possibilities of average people can meet the challenge. Even biologists with a university degree and armed with a microscope and books on systematics would probably find it difficult to accurately classify one of these tiny insects that are usually only a few millimeters in size. Since invertebrate animals can often only be distinguished from one another on the basis of minimal differences in external properties as, for example, a slightly different copulation organ, for some groups there are only a few specialists who are at all capable of classifying species in these groups.

The consequences of this situation for extending aesthetic argumentation are obvious. If species protection is to be achieved on the basis of intellectual and aesthetic interest on the part of human beings, all those species will be out of luck whose particular properties can only be recognized by a few specially trained scientists. Aesthetic and intellectual interest in a *particular* species requires that we know something about its specific properties and are able to recognize them, because without this knowledge, one species with similar properties would be as good as another. If, however, only very few people can be expected to be able to identify many invertebrates, then the amount of aesthetic interest attached to them will also be quite small. If, for example, the true bug *Corimelaena scarabaeoides* (an inconspicuous soil insect that has no common name in German or English) would become extinct, only a few heteropterologists (specialists for true bugs) would notice and lament its loss.[83]

From scientists in particular we sometimes hear the objection that extinguishing species is not just a matter of the aesthetic and intellectual interests of a few biologists but that in addition an act of this kind deprives science in general of potential knowledge. Even inconspicuous species that can barely be identified by nonspecialists are "raw material" for research in evolutionary biology and contribute to a better "understanding and historic view of ourselves and our world" (Barrowclough 1992, 124). The extinction of species is therefore no less reproachable than tearing pages out of a book that has not yet been read.

With respect to global (irreversible) species death, this argument is plausible, but its possibilities *in the context of utility argumentation* should not be overestimated. As so often in the case of utility argumentation, it may look quite convincing at first, but its supporters tend to forget that utility values are always in competition with one another (see Chapter 22.a). In the context of utility considerations, the value of species for systematic scientific knowledge cannot be judged as something separate from other utility values (i.e., as a kind of intrinsic value). In a cost-benefit analysis it must prove its worth when weighed against the potential for knowledge in other disciplines and against possibilities for utility outside of science. As Gunn (1980, 26) emphasizes, "scientists do not have overriding interests in furthering their knowledge at the expense of everyone else, especially where there is no reason to believe that the knowledge will ever be of any use except to the career of the scientist." Thus it would be necessary to determine just how great the social status of research in evolutionary biology and its aesthetic and intellectual appeal really are before reaching a political decision. Regardless of what such a discourse process might look like, it would probably not be easy for defenders of species diversity to prove that the value of scientific and intellectual interest in species is great enough to warrant maintaining *all* the species currently living on earth. We will more likely come to the conclusion that the relationship between a certain number of species and the aesthetic and intellectual benefits that can be gained from them is one of *marginal utility*. If the 10 million species that currently exist were reduced to half that number, this would probably not have much effect on the total aesthetic and intellectual appeal involved. Ninety-nine percent of the population would probably derive the same amount of aesthetic utility from 100,000 species as from the total number that currently exists (with the exception of specialists in biological systematics).

This points to a basic weakness in aesthetic argumentation, one that was already discussed in the context of economic and ecological utility argumentation, namely the *problem of substitutability*. If the aesthetic dimension is primarily regarded as a kind of utility when justifying species protection, then strictly speaking it is really not the species themselves that are at stake but rather their *effects* on human beings. However, as "aesthetic resources" of this kind many species could be substituted by others or even by artificial products. Birnbacher (1988, 75) refers to the first kind of substitution when he writes that the death of a species is not "a real loss" if all of its economic, ecological, and aesthetic functions can be assumed by other species. "Even if butterfly A cannot be replaced by butterfly B as far as its unique characteristics are concerned, the same amount of aesthetic satisfaction which A evoked before becoming extinct can be evoked by butterfly B in its place." Instead of regarding this line of thought as evidence for a weakness in the utilitarian approach he favors, Birnbacher concludes instead that the goal of *comprehensive* species protection cannot be ethically justified. In my opinion, the consequences of such an instrumental view of the aesthetics of nature are so far-reaching and contrary to intuition that I cannot simply pass over them lightly and return to anthropocentrism's home base. Let me point out three such consequences.

The first consequence is that from this perspective there would be no good reason to give priority to protecting *rare* species of animals and plants that are in danger of becoming extinct. Assuming that economic and ecological aspects were not decisive criteria, then the red lists of endangered species would have to be ranked primarily on the basis of aesthetic significance rather than rareness. As Gunn (1980, 34) has shown in great detail, "rarity" alone is not enough to merit granting a species particular value, "because rarity is not a quality." At the most it can function as an "intensifier of value" (Gunn 1984, 313). If the value of a species did indeed depend primarily on its rareness, similar to the value of certain stamps for collectors, then this would result in the paradoxical consequence that it would be in the interests of proponents of species protection to make sure that the species *remained* rare. However, this is obviously not the case.

A second consequence of the aesthetic-instrumental approach would be that conservation agencies would be required to optimize habitats with respect to their aesthetic utility and recreational value (e.g., for hunting, fishing, animal watching, etc.). Once again assuming that there are no conflicting ecological reasons, the communities that inhabit such areas would have to be manipulated by population control, introductions, and cultivation procedures in order to increase the number of "aesthetic" and useful animals and plants while lowering that of annoying ones (e.g., hornets, thistles, etc.). And it would be none the worse if such measures would cause extremely shy animal species that are relatively uninteresting for the recreational value of an area to disappear or even become extinct.

It would come as no surprise when as a third consequence certain aesthetic functions of live animals and plants were eventually substituted not only by other animals and plants but increasingly also by imitations, electronic media, or even computer simulations (cyberspace). Artificial plant decorations and plaster of paris deer in people's front yards indicate all the lucrative possibilities open to development here. Indeed, Tribe (1976, 62) reports that the trees on the grounds of a hotel in San Francisco are equipped with loudspeakers emanating recordings of bird songs, and that the city authorities in Los Angeles had nine hundred plastic trees and bushes in pots installed along the centerline area of a street where no natural trees and bushes can grow.

These examples, which have elicited numerous discussions in the conservation movement as well as among social scientists and philosophers,[84] have not been cited in order to demonstrate that a primarily aesthetic-instrumental relationship toward nature automatically results in technical reproductions of nature. Nevertheless, it is not so easy to refute such developments as dangerous flaws in our relationship to nature solely on the basis of utility considerations. At any rate, the strong fixation on human desires and needs associated with aesthetic argumentation suggests that it can provide no fundamental reasons for rejecting surrogates mediated by technology. Nevertheless, it cannot be denied that the possibility of substitution is a serious risk connected with defending nature and species protection from an anthropocentric position. If I can see black grouse close up on television, why should I be interested in placing an area under protection in which black grouse occur, especially if I have very little chance of actually catching a glimpse of them there or, worse yet, if I might not even be allowed to enter the area at all?

In response to the *superficial* consumer standpoint evident in this view, advocates of anthropocentrism often point out that technically reproduced or "fake" experiences with nature are never capable of providing the same kind of aesthetic pleasure as real ones. Elliot (1982) supports this thesis by referring to the analogy between restored nature and forged artwork. In both cases how we judge the object depends upon what we know about how it came into existence. These thoughts are certainly true for the schooled observer and critical friend of nature, and for this small group of people it might indeed be a convincing argument for nature conservation. However, the *majority* of people in industrial societies—and they are the ones who are politically relevant in the long run—can hardly be expected to succeed in distinguishing between different qualities of nature to the extent required for protecting it. According to a survey conducted by Schmidt (1993, 22) German citizens are capable of recognizing and naming "on the average just about five species of naturally occurring animals and seven wild plants." In view of this level of knowledge, which, by the way, is inferior to that of many so-called "primitive" cultures, it seems to be quite naïve to regard aesthetic interest in individual species as a significant driving force for species protection.

Of course one could reply that the strength of aesthetic argumentation does not necessarily depend upon ecological knowledge or a highly developed ability to discern species. The thing that many people enjoy about being in nature is usually not the possibility of seeing *certain* species of plants and animals but rather the awe-inspiring impression of species diversity *as a whole*. Therefore it would be possible to protect individual, even unknown species *indirectly* by protecting nature as a whole, of which species are a part. In the following I shall refer to this argument as the *holistic-aesthetic argument*.

In order to adequately weigh this argument it is first necessary to determine exactly what kind of interest in nature is involved when people use it. Is it the number of species or the (relative) "unspoiledness" and wildness of nature that appeals? If *species diversity* is its focus, it certainly is not strong enough to guarantee the subsistence of *all* the species in a particular habitat. Since there is most likely an upper limit to the amount of aesthetic pleasure that can be derived from species diversity beyond which greater diversity fails to produce greater pleasure, nature could probably be enjoyed just as well with a reduced level of diversity. Moreover, it should be recalled that species diversity is only one of several different competing goals in nature conservation and has no unquestioned value as such. This conclusion, which was drawn in Chapter 11.c and illustrated by referring to the high degree of species diversity in the city of Berlin, is, of course, also valid for aesthetic considerations. If holistic-aesthetic pleasure were a function of the number of species in a particular area, then Berlin ought to be a Mecca for nature lovers. After all, in greater Berlin you can find more species than in many excellent nature reserves (Reichholf 1993, 184).

If the object of holistic-aesthetic interest is *pristine nature and wildness* rather than species diversity,[85] the consequences for species protection are not much better. Like the "argument of extended aesthetic interest" the "holistic-aesthetic argument" is *theoretically* capable of covering a large number of species, but on the battlefield of competing interests, the strength of both is limited. Just as there are not enough people interested in Thysanoptera to justify the protection of all the species of this group of insects *for aesthetic reasons*, I suspect that there are also too few "wilderness lovers" to ensure protection of enough natural landscape for *general* species protection simply *on the basis of aesthetic argumentation*. Experts agree that for effective species protection it is by no means sufficient to preserve a few nature reserves and national parks as "islands of wilderness" within an otherwise completely rationalized industrial and agricultural landscape. What we need instead, in addition to changes in many other areas, is an extension of the wilderness idea beyond the narrow limits of isolated nature reserves (see Remmert 1990, 143, 161f.; Mader 1985).

The claim that an extension of this kind would not be very popular among most people is at first glance astonishing. Doesn't the extremely large number of visitors in national parks show that there is a holistic-aesthetic need for relatively pristine nature? Unfortunately this impression is superficial. Visits to a national park are not always an indication of professed interest in wilderness but often hardly more than carefully dosed out consumption of an exotic and romantic backdrop for heightened experience. As the recent controversy about expanding one of Germany's national parks, the Bavarian Forest, demonstrates, this backdrop is rarely appreciated when it extends up to your own front door (Keller 1995, 33). "Wilderness is nice as a romantic image in one's mind or in the virtual world of the media, but for most people the real thing is probably scary, bothersome or even repulsive" (Gerdes 1993b, 136).

The person who made this statement, a conservation official for the city of Bamberg, emphasizes his view by pointing out an area in which individuals have to "own up" regarding their relationship to nature, and that is *their own yard*. According to Gerdes (1993b, 136, 137) a brief look at the yards in newly developed residential areas exposes the hypocrisy of all the sentimental claims people make about loving nature that have become part of socially correct behavior nowadays. "Where fruit trees or other biotopes once stood you now find *lawns* which no longer issue the scent of wilderness but extend the monotony of carpeted living rooms outside. Wherever it is financially possible nature is domesticated, modified, and 'refined' by breeding and technology. Even a bed of perennials that's not carefully sectioned off can make a sensitive yard owner nervous." The following remark of Meyer-Abich (1984, 133, 134), a nature philosopher, indicates that urban aversion toward the lush growth of uncontrolled nature is not just a reflection of the diffuse fears of narrow-minded yard owners but extends far into all levels of society, even that of philosophers. "Nature suffers, and it suffers particularly from the fact that not everything in the world we perceive through our senses is natural and that a lot of things that occur in it are not good. We should be especially wary of calling it natural when something 'goes to weed,' especially yards."

At this point I cannot explore the psychological aspects that may be the source of such attitudes and whether or not, as Singelmann (1993, 112) surmises, they reflect deeply rooted imprints of "collective unconscious experience" that might once have been of survival value for people who farmed the land 10,000 years ago. However, there seems to be a considerable amount of agreement that humans' aesthetic feelings for nature are to a large extent governed by *cultural* factors or at least modified by them. The influence of culture is revealed by alterations in the so-called *basic landscape models* of nature conservation and landscape management. As Beierkuhnlein (1994, 17) mentions, at the beginning of the twentieth century the main model of conservation was the "original landscape" (Urlandschaft), even though it existed only hypothetically, while the model preferred by contemporary landscape planers is that of a "cultivated Middle European landscape." Bibelriether, the former director of the Bavarian National Park, regards this orientation as the decisive problem for promoting the idea of wilderness. "We are all imprinted by cultivated landscapes which we have looked after and cared for. We've lost our connections to natural forests" (cited in Keller 1995, 51). In view of this nonexistent or disturbed relationship to wild nature, it is probably an illusion to rely mainly on holistic-aesthetic interests for achieving species protection.

In order to uphold holistic-aesthetic argumentation in spite of these objections, it could be argued that it is always possible to alter the cultural context so that it is more conducive to this way of thinking. In other words, through appropriate schooling, education, and propaganda people could be brought to see the attractiveness of wild, uncultivated nature. I cannot deny this possibility. But as encouraging for species protection as this idea might at first appear to be, its ambiguity becomes evident when examined more carefully. If aesthetic feelings can be manipulated by culture, it is not only possible to *enhance* aesthetic sensibilities toward certain qualities of nature. The opposite option is also possible, namely to *throttle* or *simplify* aesthetic interests people have in experiencing nature. Leopold ([1949] 1968, 46) drew attention to this option in the forties of the last century when he commented on the reduction of indigenous plant species diversity with the following resigned and ironic words: "It might be wise to prohibit at once all teaching of real botany and real history, lest some future citizen suffer qualms about the floristic price of his good life." This statement opens our eyes to the tormenting idea that from a strict utilitarian standpoint it might be appropriate under certain circumstances not only to accept a reduction in aesthetic interest in experiencing nature but even to welcome it. One might eventually even have to *demand* such a reduction in the event of a conflict between aesthetic use and some other kind of use of nature if people are not prepared to do without the competing kind of use.

As far as the specific case of wild nature is concerned, one such conflict comes immediately to mind. Not only would protecting wild nature require giving up almost all the ways we have of using it. It would also require limiting tourist activities and thus also aesthetic experience itself. In view of this high price, no wonder the question is often raised whether we can really afford the luxury of undisturbed nature (Töpfer 1991). More precisely the question is the following: Do we really *want* to pay the price you'd have to pay for undisturbed nature? Wouldn't the opposite be better, that is, to reduce the aesthetic interests of humans with respect to nature to such an extent that we would be satisfied with "monotonous nature," with a form of nature capable of withstanding intensive and multiple use as, for example, in recreation areas, commercial forests, monocultures, and mowed lawns? These questions are not easy to answer if the individual aesthetic needs of humans are regarded as a starting point capable of being manipulated, and if nature is considered "a kind of dependent variable that can be altered in keeping with human needs" (Töpfer 1991, 494). According to Krieger (1973) it is even difficult to seriously object to plastic trees if people are satisfied with them and choose them rather than natural trees in a cost-benefit analysis.

It would be too simplistic at this point to merely deny that such a reduction in the aesthetic needs of humans with respect to nature is possible. The ability of humans to adapt to bare, inhospitable, and uncomfortable living conditions is amazingly large. As Bischoff (1993, 58f.) remarks in this connection, there have always been cultures that live on the edge of mere survival in vast regions of sand, stone, or ice and like it that way. Similarly the fact that people inhabit large cities seems to indicate that it is indeed possible to spend your life in a noisy, polluted, cement-and-asphalt wasteland and subjectively still feel good. Although enthusiastic city dwellers are always grateful for a few spots of green, it is usually enough to have a park around the corner, a weekend trip to the countryside, or vacation on the Bermuda Islands. Under these circumstances a profound need for as much untouched, wild nature as possible in the immediate vicinity such as that required by the holistic-aesthetic argument can hardly be expected to any significant extent.

However, even if this shows that it is indeed *possible* to reduce people's aesthetic needs for nature and that many would not even find that this detracts from their individual feelings of well-being, the question is whether a development of this kind is really *desirable*. As Willard (1980, 303) maintains, it is not enough, especially in ethics, to consider the value that the majority of people attribute to nature *in fact*. We must also think about the value that *should* be attributed to nature. With respect to the latter problem, almost all the (moderate) anthropocentric ethicists in environmental ethics would probably agree that it would be very bad morally if people's aesthetic sensibilities for nature were reduced in the manner described above as a possible (and sometimes even necessary) result of anthropocentric utility logics. In their argumentation in support of maintaining aesthetic interest in nature most philosophers attach a great deal of significance to wild nature devoid of human control. According to Seel (1991, 907), for example, dealing with *"wild* nature" (which he considers to be a necessary condition for *"aesthetically beautiful* nature") is not just one of many other things life has to offer but rather a "unique and indispensable sector of the human world." This sector of wild, beautiful nature is unique in that it can be experienced as a kind of island of autonomy, self-sufficiency, and non-functionality in an otherwise completely need-oriented and strife-torn world. In its "function as an island of nonfunctionality" the beauty of nature is a unique and indispensable corrective element for the goal-oriented individual and collective ideals of what constitutes a good life. If wild nature is further reduced or even destroyed, according to Seel (1991, 907) this means the destruction of an "authentic dimension of successful human activity."

As I mentioned at the beginning of this chapter, many advocates of an anthropocentric position regard this kind of aesthetic argumentation as the most appropriate because it acknowledges the autonomy of nature in the course of aesthetic experience but does not require that intrinsic value be attributed to nature in the context of ethical theory (e.g., Birnbacher 1980, 130f.; Früchtl 1991, 346f.). I can follow this thinking only to a certain extent. I agree with Seel (1991, 907) that it would be irresponsible toward future generations to "eliminate non-instrumental relationships to nature in everyday life" as would occur if the aesthetic dimension were reduced. Nevertheless, I still consider this argument to be consistent only if its proponents also assume that when nature is *perceived* as being autonomous in the course of aesthetic experience it is indeed truly, *objectively* autonomous. As a *purely utilitarian argument* "within the limits of pure anthropocentrism" it seems to me to be not only contradictory but also counterproductive.

Let us first look at its *contradictoriness.* Many anthropocentrists regard the relationship to nature generated by aesthetic experience to be "perhaps the strongest argument for the autonomous character of nature including even those parts which have no soul" (Birnbacher 1980, 130). According to Früchtl (1991, 347), "Anyone who experiences nature as beautiful cannot avoid granting it autonomy." Birnbacher (1980, 130) shares this view when he writes that nature only appears beautiful to us "when it confronts us with its existence as such, when it is not of direct functional significance." While Birnbacher (1980, 131) thus admits that it is essential to assume the existence of intrinsic value of nature in order to *experience nature aesthetically*, at the same time he contests such a concept of intrinsic value *in the area of ethics.* If from an aesthetic perspective nature appears to be a subject, something that exists as such, so Birnbacher argues, it must be recalled that "objectively speaking this perceived autonomy is only an illusion." In reality the value of nature is solely by virtue of its utility for humans, which among other things includes aesthetic utility. In this connection Birnbacher (1980, 132,133) coined the term *aesthetic resource*. "To the extent that a human being experiences a need to grant the things in his surroundings autonomous value of an aesthetic or metaphysical kind which is phenomenologically independent of any human needs, then nature becomes a resource for him in the sense that it permits metaphysical penetration, religious contemplation, an aesthetic-erotic relationship. This means that aesthetic resources are also resources." I doubt whether such a concept of the term "resource" is really wise and recommendable in the context of environmental ethics, because after all, then *everything that has value* is a "resource." In the following chapter I will return to this matter. Regarding aesthetic argumentation suffice it to say that by converting it to a resource argument two conceptual paradoxes arise. First, "the autonomy of nature is confirmed and negated at the same time" (Birnbacher 1980, 131). Second, the express nonfunctionality of nature becomes functionalized, if only at a second-order level.

What about *counterproductiveness*? The counterproductiveness of such contradictions for aesthetic perception of nature and thus also for species protection based on aesthetic argumentation is not difficult to comprehend. If *in the course of aesthetic experience* I have no choice but to perceive a red admiral butterfly sunning itself on a stone as beautiful as a subject, in other words, beautiful by virtue of being an end in and of itself, and if then philosophical reflection tells me that this impression of autonomy is *"objectively speaking* an illusion" (Birnbacher 1980, 131), this observation will destroy the pleasure of aesthetic perception unless I succeed in cleverly repressing it. Just as something can only function as a placebo if the patient really thinks it is medication, a reflecting observer who aesthetically perceives the autonomy of an object must assume that such autonomy *really exists* if she is not to regard herself the victim of a clever delusion. Therefore, when aesthetic argumentation is interpreted from a rigorous anthropocentric standpoint it undermines its very own foundation, namely aesthetic experience. Obviously this also holds true for the resource perspective proposed by Birnbacher. Even in the field of aesthetics it may be legitimate to draw upon the metaphorical context of economics in this manner to deal with *certain problems* (e.g., determining willingness to pay for species and biotope protection), but it is nevertheless important to be aware of the limits and risks associated with it. If the resource perspective is regarded as the *decisive* or even as *the only "realistic"* approach above and beyond the narrow scope of economics, this eventually will almost automatically lead to a distortion and impairment of aesthetic perception and thus also to annulment of the aesthetic argument. Just as love and friendship would eventually be destroyed if these areas of experience were viewed simply as "psychological resources" from a purely instrumental standpoint, so also would conscious reduction of a red admiral butterfly bathing in the sun to its instrumental role as an "aesthetic resource" eliminate the specific quality of this act of observing nature. If the term resource is not to eventually *decrease* the value of something whose value it purports to *increase*, at least its claim to absolute validity, its claim to be the only valid value category for nonhuman nature, must be abandoned.

If anthropocentric-aesthetic argumentation thus proves to be contradictory and counterproductive *within its own argumentative boundaries*, this still addresses only part of a much greater problem, one that was already revealed in connection with economic and ecological argumentation. The problem is that utility argumentation is inconsistent and more or less counterproductive both with respect to *altering attitudes* toward nature as well as with respect to *motivation* and *intuition*. This is the thesis that I wish to develop in the following chapter. In my opinion it is the most significant *practical* argument against staying on the straight and narrow path of anthropocentrism in environmental ethics.

23. Psychological and Sociopsychological Aspects

23.a. How Attitudes toward Nature Are Formed

In order to bring the psychological aspects of utility argumentation into closer perspective it is useful to analyze the *formal structure* of the previously presented criticism of economic, ecological, and aesthetic utility argumentation. This reveals three different categories of objections. The first of these includes objections that deny such argumentation any legitimacy whatsoever. This was particularly the case of arguments based on invalid generalizations as, for example, the stability-diversity hypothesis. The second category includes arguments that are not factually wrong but only of limited scope when cost-benefit analyses are applied or because of marginal effects on utility. Arguments of this kind such as the "extended aesthetic argument" with its broader understanding of beauty are capable of securing the protection of *certain* species but they are not sufficient for justifying *general* species protection. Finally the third category encompasses "enlightened" ecological utility arguments, that is, arguments (such as that of the "unknown threshold level") based on the fundamental limits of our ecological knowledge and the risk of global ecological disaster associated with them. Arguments of this kind are both factually correct and theoretically capable of justifying *general* species protection. Nevertheless, when it comes to balancing needs and interests they don't seem to be very convincing. Why exactly is this? As indicated previously, I regard the reason for this to be a psychological one associated with the *anthropocentric context* in which such arguments are usually presented.

Before examining this idea more thoroughly it is important to recall that a "convincing" philosophical argument is usually not the same as conclusive "proof" (as, for example, in the sense of mathematical deduction) but involves greater or lesser plausibility instead. However, the plausibility of a thought is not an isolated and absolute entity but something that is strongly influenced by the worldview context in which it is discussed. Whether or not an argument appears to be conclusive depends (among other things) upon the (often only implicit) premises, estimates, and attitudes in which it is embedded. One and the same argument can appear more or less convincing depending upon the *general attitude* of the recipients as well as their *views of the world and humankind*.

It is not difficult to discern that an anthropocentric view of the world

181

and humankind is a rather unfavorable context for asserting the plausibility of "enlightened" utility arguments. While all ecological versions of the "insufficient knowledge argument" suggest *in and of themselves* that we deal with species and ecosystems with great caution and restrict our activities as much as possible, the *anthropocentric context* of this argument, with its claim that nature is in principle at our disposal, points in just the opposite direction. All interventions in nature, even the extinction of species, seem to be justified in this context, at least in essence, as long as they do not impose on the interests of other humans. Since it is, so to speak, "normal" to have nature at our disposal in the context of an anthropocentric worldview, it is necessary to justify any *restrictions* to these basically unlimited options. If someone demands caution and self-denial from humans for the benefit of other species, he or she must demonstrate, if possible, by means of cost-benefit analyses that the utility of species protection is greater than the costs involved (or the utility of competing programs). However, enlightened ecological utility arguments are not compatible with this basic rule of utility argumentation. It is impossible to include insufficient knowledge as a factor in a cost-benefit analysis without having some idea of the potential utility or damage that may be involved. Thus it is not surprising that even though the scope of a utility argument of this kind is greater than all the others, it plays only an insignificant part in public debates about species protection. Proponents of species protection who rely on anthropocentric argumentation seem to sense that they must present *positive knowledge* if they wish to successfully compete with all the other interests opposed to species protection. If you feel you have to argue anthropocentrically, then you have to present so-called *hard facts*. However, this has three serious psychological consequences for humans' concept of themselves and their attitude toward nature.

The first negative side-effect of anthropocentrism is that it almost automatically provokes an attitude that was discussed in the first part of this book as one of the causes of the ecological crisis, namely the *attitude of scientism*. If as a proponent of species protection I am forced to provide strict economical or ecological evidence for every ethical matter I address, I can't really maintain much interest in an enlightened understanding of science and its limitations at the same time. Any doubts I might have about the methodological scope and accuracy of my scientific evidence would weaken my possibilities for accomplishing what I intuitively consider to be ethically right and important. Therefore I would have to either repress these doubts or at least keep them in the closet. Let the public believe that scientists really know how ecosystems "function" and what values should be attached to biotopes and their species. Then maybe the recommendations and warnings of those in conservation will finally be taken seriously!

Even though a pragmatic attitude of this kind on the part of many conservationists is understandable considering the desperate state of conservation, it is obvious that by assuming this attitude their credibility is at stake. Think about the uncertainty that arose when all the very self-confident predictions publicized in the 1980s about the death of seals on the North Sea coast or the phenomenon of "Waldsterben" (death of the German forest) gradually had to be revoked. Apart from this there are two other reasons why it is not wise to encourage expectations too great for science to fulfill. First, it should be recalled that blind belief in expert knowledge does not always operate to the benefit of species protection. It can also benefit the opponents. As discussed in previous chapters, it is often not difficult for advocates of "progress" to provide data that prove the insignificance of a particular species or area and thus to "scientifically justify" its loss. Second, it is also important to remember that scientism inevitably fosters *technical optimism*. Anyone who believes that ecology is basically in a position to analyze processes in nature in such great detail that it can unequivocally determine the usefulness of species and biotopes will also trust scientists to be able to precisely *manage and control* them sooner or later. In the first part of this book I already showed that belief in our ability to completely control nature is not only untenable but also truly disastrous for overcoming our ecological crisis. In light of these thoughts and in view of the relationship between utility argumentation and scientism outlined in the present chapter, it seems quite reasonable to conclude that from a *psychological standpoint* anthropocentric argumentation is a burden rather than a boon to species protection. Instead of promoting a badly needed change in our attitude toward nature encompassing more care, error-proneness, and intellectual modesty, compulsive recourse to supposedly hard facts in anthropocentric argumentation helps to consolidate the attitude of hubris and power hunger that has proved to be one of the more profound reasons for species extinction.

But there is a second negative side-effect of anthropocentric ethics that goes along with the risk of human hubris. Since species are regarded exclusively as resources in this way of thinking and thus forced to prove their worth in cost-benefit analysis, anthropocentrism contributes to a distortion of human value judgment and thus *unwittingly to a devaluation of nature* as well. This psychological effect was pointed out by Tribe (1976, 71f.) and Kelman (1981). Their criticism applies to *economical* approaches in particular, that is, to attempts to attach a monetary value to biological diversity and to make such estimates the basis for political and ethical decisions. Although the monetary value approach may seem attractive to many economists, planners, and conservationists because it allows completely different and opposing values to be represented and weighed on *one and the same* scale, it is this very act of standardization that constitutes a psychological problem. If the "total composite value" of a bluethroat is set at $678.56 (Vester 1984), as a result of this calculation its *special value as a living creature* fades out of sight. By virtue of an economic calculation it is practically relegated to the same category as an upper-middle-class CD player. Just what the morally relevant difference between a warbler and a CD player is and how the *particular* value of a warbler can be *theoretically* justified, a question that might trouble a radical reductionist in this connection, does not have to concern us right now. For the *practical* purposes discussed in this chapter it suffices to note that this kind of value truly *exists* in the minds of many people and that the rationalizing argumentation of economic approaches causes it to disappear, to the disadvantage of species protection.

As a result anthropocentrism fails to recognize the sensibilities and feelings of moral obligation of those people who tend to intuitively grant nature value of its own, which like the value of health and human life cannot be *directly* measured with economists' monetary value.[86] In this connection Kelman (1981, 39) has shown that an intuitive tendency of this kind to differentiate between different categories of value is of considerable practical significance. "The very statement that something is not for sale affirms, enhances, and protects a thing's value in a number of ways." And the other way around, if people are prepared to consider the existence of an animal or plant species to be a matter of cost-benefit analysis, it automatically reduces the value of these species. This is nicely illustrated by the following analogy from Kelman's work (1981, 38): Cost-benefit analysis is something like trying to use a thermometer to measure the temperature of a small amount of liquid. As soon as you put the thermometer into the liquid, the temperature of the liquid becomes distorted by that of the thermometer.

Although such *immediate* effects on the evaluation of nature are bad enough, you also have to take long-term negative effects on people's *percep-*

tive abilities and *value sensitivity* into account. Is the idea completely absurd that establishing cost-benefit analyses in environmental ethics might ultimately cause people to lose the capacity to differentiate between the value of a bluethroat and that of a CD player? After all, in the context of anthropocentric and utilitarian thinking they are manifestations of *one and the same thing*, human interests. In this sense Sagoff (1981) cautions us to beware of any kind of ethics for which value judgments reflect only *subjective* preferences, desires, and needs. Ethics of this kind not only makes it impossible to subject the *contents* of value judgments to rational discourse since everyone knows best what he or she prefers. In the long run it also subverts moral judgment. If every value that is perceived is only a "more or less abstract indication of personal interest" (Tribe 1976, 72), there is no place for a specific feeling of moral obligation toward nature. All that is left is human self-interest and a resource to be divided up as wisely as possible.

This leads to the third negative side-effect of anthropocentrism. Since it permits species protection to be justified exclusively on the basis of utility, it legitimizes and reinforces the *predominance of utility thinking*, which is thought to be one of the deeper causes of the endangerment of species. A survey of the history of species extinction by humans shows that utilitarian thinking poses an *immediate* risk to many species. Species such as the giant auk, the Madagascar ostrich, or the passenger pigeon did not become extinct by accident or because they were useless. They became extinct *for the very reason* that their usefulness was recognized and readily available (Werner 1978, 149). Nowadays it is the vast alteration of habitats that threatens species diversity, but this too is taking place under the banner of utility thinking. Species are disappearing because humans lay claim to all habitats as their supply depots, their agricultural space, their construction sites, or their "playgrounds," that is, as resources that *only* exist to be used by humans. The main goal inherent in such an exclusively instrumental attitude toward nature is the satisfaction of human needs. And if pursuing this goal is the very thing that has caused so many species to become extinct, doesn't it seem almost absurd to recommend just such a goal as a remedy for species extinction? Is it really possible that utilitarian thinking can provide solutions to the very problems that this kind of thinking has evoked?

Advocates of utility argumentation will reply that the reason for species extinction is not human utility thinking as such but rather the fact that this kind of thinking has not yet been presented in a comprehensive and sufficiently well examined manner. In the future, superficial and shortsighted utility goals must be replaced by enlightened and farsighted thinking. What we have to do is to use natural resources wisely and "sustainably." But as plausible as this response might sound, it is still not in a position to dispel the central problem of utility thinking, namely the "evolution of needs" (Marsch 1973, 20). Planet earth provides us with only a limited amount of space, and at the same time the claims to this space and the interests in using it continue to grow. Granted, these claims include numerous interests in maintaining biological diversity, but if the volume of *competing* desires, particularly aspirations for a higher standard of living, greater mobility, maximal comfort, and costly recreation continue to increase excessively, any more subtle and far-reaching interest in nature will automatically be backed up against the wall. As well justified as it might be, species protection will eventually become "a luxury and finally an impossibility" (Norton 1987, 130).

However, it is important to realize that this dilemma cannot be solved solely by *technical* means. Increased efficiency, recycling, and "ecological reconstruction of industrial society" are definitely very important, but "wiser" use of nature will not suffice to prevent "ecological collapse" (Vorholz 1995, 25), let alone save our planet's species. Instead it seems to be imperative that human beings set limits to their continually growing claims and give up the idea of realizing every imaginable use of nature that appears to be lucrative. In the words of Cobb (1972, 82) what we need is "a new asceticism, an *ecological asceticism*." Just what such an attitude might entail and how people can be motivated to assume it is described in greater detail in Chapter 32. At any rate, it seems clear to me that it requires a certain amount of *altruism*, that is, a certain degree of willingness to refrain from superficial use of nature on the basis of insight and out of concern for the well-being of nature and future generations.

It is highly doubtful that anthropocentric utility argumentation is capable of generating the psychological atmosphere in which such an attitude can develop and flourish. The results of my criticism of the basic contents of cost-benefit analyses in the previous sections indicate just the opposite. Since cost-benefit analyses require "hard facts," long-term arguments that emphasize the necessity for altruism with respect to future generations are automatically at a disadvantage. For a cost-benefit analysis aspects such as those inherent in the "unknown threshold level argument," which are difficult to calculate but morally portentous, can hardly be taken into consideration. The fatal consequences are that in public debates those arguments that come closest to being "ethical and altruistic" ones are either not brought up at all or "only presented when nothing else seems to work" (Amberg 1980, 74). Arguments that directly appeal to personal interests, however, are usually at the head of the list of reasons given for wanting to protect species.[87]

Now, of course, it is understandable that proponents of species protection tend to lean to those arguments that they feel are the most effective at the moment. A strategy of this kind corresponds to what Birnbacher (1982, 14f.) and Vollmer (1987, 93) recommend when they maintain that if several different reasons are available, you should choose the one with which you can convince the greatest number of people. Nevertheless, as plausible as this strategy might seem at first, it overlooks one important point. If you choose an argument that seems to be the most effective for short-term purposes, it may prove to be counterproductive *in the long run*. This has to do with the fact discussed above that the plausibility of an argument is not a static thing. Arguments can also have an autocatalytic effect and reinforce basic attitudes and worldviews that, in turn, determine the extent to which an argument is considered "convincing" in the future. Let me illustrate this with a problem of species protection. If proponents constantly appeal to personal interests, this not only generates the false impression in public opinion that "egocentric" arguments are basically the "best" ones. It also consolidates the predominant attitude that utilitarian thinking *per se* is the most reasonable way for an individual to interact with his or her surroundings. As if it were completely normal to put a price tag on all the things on earth. The more securely this ideology becomes rooted in people's minds, the more difficult it becomes to reach anyone with altruistic, ethical arguments. The price you pay for strategically following "the agenda" of the economic and instrumental way of thinking that currently predominates is thus a big one. By using this strategy you run the risk that in the long run the *only* arguments with which you can validly justify comprehensive species protection, ethical, and altruistic arguments, will eventually lose all their power.[88]

23.b. Motivational Aspects

"That's all very well," an anthropocentric skeptic might reply to such remarks, "but isn't it a complete illusion to place your stakes on altruism or perhaps even on the supposed intrinsic value of nature in a competitive society geared to utility, and then hope that the receptiveness for this kind of argumentation will grow? If you want to motivate people to protect species as quickly as possible, isn't it more expedient, even essential, to appeal to enlightened self-interest? After all, you can't deny that self-interest is an elementary, perhaps even the strongest motivating factor for human action!"

In order to test this objection let us attempt to determine the meaning of the term "self-interest" more precisely. If you look at it more carefully, you find that it is used in two different senses, that of individual or personal interest on the one hand and that of the collective self-interest of humanity as a whole on the other. As far as protecting species is concerned, it should now be obvious that it is usually not enough to appeal to *individual* self-interest. Only very few species can be guaranteed protection on the basis of egocentric argumentation, even if you include the representatives of the next generation or the one after it in the concept of individual self-interest. This means that ethicists in environmental ethics who favor an anthropocentric position must also go beyond the narrow limits of individual self-interest and resort to altruistic, ethical argumentation if they wish to further the very demanding aims of *general* species protection. Returning once again to the question of the most effective kind of motivation, what is at stake is not "self-interest" versus "altruism" but rather "altruism *with respect to future generations*" versus "altruism *with respect to nature*." If you weigh the different kinds of reasons presented for protecting species against *this* backdrop, the purported superiority of anthropocentric utility argumentation is no longer as obvious as it first appeared to be.

On the contrary, even advocates of (moderate) anthropocentrism admit that the question of motivation is a particular problem of their concept of ethics (e.g., Wolf 1987, 166; Birnbacher 1987, 72, 73). Moral obligations toward future generations can be sufficiently well *justified*,[89] but the good reasons presented as justification are usually only barely or not at all *motivating*. "[The] problem is that motivational drive decreases whenever not just care for immediate offspring but rather respect for undefined beings which may exist in the distant future is involved" (Wolf 1987, 163). Birnbacher (1988, 188f.) has identified three factors that he considers to be responsible for this effect: (1) the feeling of supposed powerlessness regarding any influence on future developments, (2) failure to recognize similarity between the present and the future, and (3) temporal distance. Experimental evidence from studies by Ekman and Lundberg (1971) that indicate that the degree of emotional involvement decreases exponentially with perceived temporal distance seems to show that the time factor does indeed play a decisive role. Since motivation is very strongly affected by emotional involvement, it is no wonder that apparently neither moral insight nor love of humanity are sufficiently effective motives for perceiving the need to assume responsibility for the future. "General love of humanity is a much too artificial and academic feeling to be of very much use as a motivating force in everyday life" (Birnbacher 1988, 200).

These psychological circumstances are a fundamental problem for *any* kind of ethics of things of the future (Birnbacher 1979a, 120), but in the case of species protection the situation is aggravated by the fact that highly complicated factual matters are involved in addition to temporal distance, which are difficult for the general public to understand. This is particularly the case of ecological argumentation. Recall, for example, the individualistic ecological version of the "insufficient knowledge argument," according to which the extinction of a rare, noninteractive species should be prohibited since a species of this kind may suddenly increase in number under altered environmental conditions and assume systemic functions of other species that used to be frequent but have since decreased in number. It is clear that such a complex and abstract line of thought steeped in theory is incapable of evoking very strong motivational powers. Even if it can be reconstructed theoretically, the morally relevant gist of the matter, the effects of the extinction of species X on the well-being of a distant generation Y, is so vague and uncertain that it can hardly be expected to generate significant emotional involvement.

Even if it is also difficult to evoke concern for invertebrate animals and plants, the conditions for generating concern for nature in general are definitely more favorable in the context of *nonanthropocentric ethics*. In this context *nature itself* has moral standing that justifies protecting it in and of itself. Therefore it is not necessary to take an abstract and "unemotional" cognitive detour via respect for future ecological consequences. It is no longer necessary to conduct complicated and often controversial disciplinary discussions that are known to tend to confuse the general public (Naess 1986, 22). The moral gist of the argumentation, which addresses the injustice of further jeopardizing or perhaps even irreversibly extinguishing an endangered species, does not have to be deduced by means of a highly theoretical line of thought. It is *immediately* evident. Species exist here and now; it can be clearly shown that they are endangered; and this can be perceived by sensory experience, at least in principle. The possibility of direct sensory experience is highly significant for consolidating moral values both psychologically and emotionally and thus also for motivation (Schurz 1986, 250). Of course, we are still left with the basic psychological problem of all environmental ethics, namely the spatial and temporal distance between the *actions* of individuals (causes) and their collective *consequences* (effects), but this is not a *specific* argument against nonanthropocentric ethics. The serious problem of emotional alienation is one that affects both anthropocentric and nonanthropocentric positions. However, the difficulties of anthropocentrism are further compounded by the fact that alienation, which is bound to occur anyway, is enhanced by the complicated cognitive exercises that are required in order to consider distant effects on future generations. Schurz (1986, 250) considers it an "empirical fact" that "such highly abstract ethical argumentation" is not enough for most people. "If their ecological conscience is to be aroused, their ecological values require direct psychological and emotional support."

The following text suggests that Birnbacher (1988, 201) shares these sentiments. "It is . . . not completely out of the way to conclude that the natural conditions for sustaining human life in the future can only be maintained effectively if the goal is to sustain nature in and of itself rather than to sustain future generations of human beings." But although such observations lead Schurz (1986, 250) to the conclusion that nonanthropocentric ethics should be preferred for *practical* reasons, Birnbacher (1987, 72, 73) continues to hold on to a (moderate) anthropocentric position for *theoretical* reasons. In order to fill the motivational gaps in this position he merely suggests that anthropocentric ethical principles be supplemented by nonanthropocentric "guiding principles." As an example of such a guiding principle he suggests Albert Schweitzer's ([1923] 1974) concept of "reverence for life." According to Birnbacher, guiding principles such as that of Schweitzer can provide motivation in instances in which ethical principles alone do not suffice. They have the *purely practical function* of making sure that what anthropocentric ethics has recognized as being right is put into action. But for Birnbacher guiding principles of this kind are not capable of meeting the standards of *theoretical competence* required for deciding what is right and what is wrong.

I seriously doubt whether such a distinction between "valid" ethical authority and "useful" motivational means is capable of successfully solving the problem of the motivational gap in anthropocentric reasoning. A guiding principle that cannot be ethically justified can hardly be of any good in the long run. If anthropocentric ethics tells me that "Albert Schweitzer's doctrine of 'reverence for life' is seriously deficient both as a general principle of ethics and as a principle of environmental ethics" (Birnbacher 1987, 70), then even a large portion of pragmatism will not be able to help me to derive motivational power from these teachings. As I have already explained in connection with aesthetic argumentation, the placebo effect of a medicine ceases to work as soon as the real essence of the substance is exposed. In view of these consequences, shouldn't every advocate of nature conservation who finds himself motivated by Schweitzer's teachings be strongly advised not to deal with questions of ethical justification if it is clear from the very beginning that ethical reflection will show him how theoretically deficient his guiding principle is? To me it seems quite obvious that from a psychological standpoint Birnbacher's double strategy must lead to inconsistencies that undermine the goals for which people in nature conservation are aiming. This explains quite clearly why Ricken (1987, 3) recommends that we not only pay attention to the criterion of "ontological economy" when choosing a procedure of moral justification but also observe the principle of *coherence*. "It is not enough for ethics to justify norms. It must also promote an ethos, an emotional attitude, which motivates us to really do what is morally right" (Ricken 1987, 20). As the discussion has shown so far, anthropocentric ethics is not sufficiently capable of achieving this.

23.c. Intuitions of People in Nature Conservation

Just how much coherence is lacking in anthropocentrism will become more clear when we take another look at the relationship of this position to another source of moral judgment that was the starting point of my discussion of a pragmatic ethical approach, namely that of *moral intuition*. To put it more precisely, it is advisable to take a look at the intuitions of people involved in nature conservation, that is, the intuitions of those who feel a kind of moral obligation toward nature and species protection and feel it strongly. After all, moral intuitions might be ignored or they may even be lacking all together.[90] I have already outlined the *substantive content* of intuitions in nature conservation, namely to achieve comprehensive and regional species protection, but the question of the *basic motivation* involved has not yet been addressed. What motives and convictions are associated with them?

If you study the official publications of people and organizations involved in nature conservation (e.g., Thielcke 1978; Schreiber 1978), you get the impression that their motivation for nature and species protection is derived almost exclusively from utility thinking. At the top of the list are economic arguments, ecological ones, and once in a while also aesthetic arguments. However, although such lists suggest that nature protection is pursued mostly for the sake of human beings, this impression is deceiving. In personal discussions and after studying publications dealing with the topic of motives and reasons for advocating nature protection (e.g., Bierhals 1984, 119; Meyer-Abich 1984, 50; Trepl 1991, 429) I have found support for the idea that the arguments presented officially are not the ones that really motivate people's activities in conservation. It seems instead that the majority of people involved in nature protection do it *for nature's own sake*, thereby intuitively attributing intrinsic value to nature.

In my opinion three things indicate that the basic motivation to protect nature is *nonanthropocentric*. The first such indicator stems from Routley's (1973, 205f.) so-called *last-people argument*, a reasoning experiment that deals with the consequences of anthropocentrism. It describes various scenarios in which human beings have lost the ability to propagate for one reason or another. The question then is whether or not the very last people on earth, who no longer have any responsibility for future generations, are entitled to exploit all life on the planet to their own advantage before dying themselves, or perhaps even to (painlessly) destroy it. According to Routley and Routley (1980, 120f.), the way a person responds to this problem reveals the extent to which he or she strictly advocates an anthropocentric standpoint or a nonanthropocentric one (see, however, Lee 1993 for a critical review of this argument). Birnbacher (1980, 132) offers an anthropocentric answer (under slightly modified conditions). He writes, "If we knew with certainty that planet earth would be uninhabitable for human beings forever from the year 2000 onward, there would be no ethical or aesthetic reason why we shouldn't leave it behind as a dump. "I am quite certain that almost everybody involved in species protection would intuitively take the opposite stance. The second indication has to do with the purely empirical observation that many people in nature conservation admit that they *don't feel quite right* about expressing their reasons for assuming responsibility for nature in terms of anthropocentric utility argumentation.[91] According to Stone (1988, 43) you can sense that official arguments based on utility "lack even their proponents' convictions."[92] A third indication finally has to do with the many grassroots activities in the field of species protection for which it is very hard to imagine that they are motivated by anthropocentric ethics, try as one may. Think about the controversy involving the hardly spectacular snail darter mentioned earlier (Ehrlich and Ehrlich 1981, 182), the efforts to protect the Antarctic from development (Tügel and Fetscher 1988, 30f.), or the controversy between sheepherders and supporters of species protection in southern Norway about a few wolves that hardly anyone ever sees but that attack sheep once in a while (Naess 1984, 266). In all these cases it would be completely farfetched to claim that the conservationists are acting "for the sake of their own species," let alone for the sake of personal interests. Since there are no prospects for material profits and often not even prospects for the immaterial pleasure of being able to see what's being protected, people in nature conservation like Stern (1976, 94) "can only shrug [their] shoulders" when others accuse them of secretly pursuing personal interests. Of course Stern (1976, 88) finds it no less erroneous when activists in nature conservation *themselves* constantly maintain that they are "protecting nature *from* humans *for* humans." "Let's get this straight once and for all. The promises we're making about how people will eventually be able to enjoy themselves in nature reserves are just not true, because we are not protecting nature primarily for people looking for recreation" (Stern 1976, 93).

If most people in nature conservation share this attitude, why do they continue to give priority to utility arguments and keep their real reasons a secret? A very obvious explanation is certainly the strategic idea that this is the way to achieve the most significant political effect (see Trepl 1991, 429). Some authors suspect, however, that more subtle psychological mechanisms also play a role. Stern (1976, 88), for example, believes that it is "fear of the stain of misanthropy" that causes activists in nature conservation to conceal their real feelings. And Ehrenfeld (1988, 213) suspects that people are afraid they might be laughed at if they failed to express their fears and concerns in the generally accepted terms of cost-benefit analysis. The idea implicit in such conjectures is that the socioeconomic atmosphere has a strong influence on people involved in nature conservation and the way they reason, an idea supported by the following thoughts of Meyer-Abich (1984, 50): "At least in western industrial societies unselfishness does not conform with the system and easily nurtures vague suppositions that proponents have not yet figured out what the conditions for success are in a highly competitive society. The simplest way to nip such suspicions in the bud is to consistently claim that everything you do is selfish, particularly your unselfish actions." If you accept this interpretation, you can hardly avoid the conclusion that when people in nature conservation employ utility argumentation, regardless of whether it is right or wrong, this kind of argumentation is really only a *pretext*. Psychologically speaking this phenomenon is equivalent to *after-the-fact rationalization* of nonanthropocentric moral motivation wrapped in the cloak of anthropocentrism.

You could, of course, think that a psychological insight of this kind is interesting from a theoretical standpoint but really not very significant for the practical purposes of species protection. What really counts is action, not whether or not what someone says is consistent with what he feels. However, this overlooks the fact that a discrepancy between reason and intuition can have serious effects on actions. The problem is that if a person constantly reasons in a manner that is inconsistent with her intuitions, this can eventually weaken her intuition and thus also the basic motivation for her actions. "If the right feelings are constantly suppressed by the wrong words and not brought out into the open, their compelling power will wane and in time they may even disappear" (Meyer-Abich 1984, 50). As Tribe (1976, 73) warns, by advancing utility arguments as a pretext, advocates of nature conservation maneuver themselves into the paradoxical situation of undermining the very feeling of responsibility that originally motivated them to become active in species protection.

If you look at analogies to other areas of ethics, it is easy to see that not only is there a *discrepancy* between utility argumentation and moral intuition. In a manner of speaking this kind of argumentation runs *countercurrent* to moral intuition. For example, everyone would consider it completely reprehensible to subject a criminal action to a cost-benefit analysis. "One does not consider the price if someone threatens to rape one's daughter" (Nash 1977, 12). According to Nash, if we really take the term "environmental ethics" seriously, we should think about and feel the increased danger to nature in the same manner. If this were the case, it would seem inappropriate if not absolutely cynical to attempt to solve problems of the life and death of species by means of economic analysis or other such estimates of utility.

By now it should be clear why the metaphors of species extinction mentioned earlier such as that of a library that has been set on fire or an airplane from which rivets have been removed have an unpleasant aftereffect on sensitive advocates of nature conservation, in spite of the strong impressions they convey. From the standpoint of their basic protective intuitions these metaphors fail to grasp the *gist* of the matter. The aspect of utility implicit in both metaphors reduces the growing problem of species extinction to a matter of sheer obtuseness. (After all, you don't saw off the limb you're sitting on.) In doing so, it seriously restricts the ethical dimensions of the problem. To use a term coined by Rolston (1985, 720), the utility standpoint is thus "*submoral.*" This term conveys the impression that anthropocentrism is *moral* to the extent that it considers the well-being of future generations of humans, but that it is also *amoral* since it denies any moral responsibility for the those who carry the brunt of the burden of species extinction, the species themselves. Because of this *interspecific amorality* anthropocentric ethics must be regarded as an incomplete form of environmental ethics. It is not capable of capturing the intuitions of conservationists in a complete and satisfactory manner.

One indication that anthropocentrism is indeed incomplete as maintained above is the observation that nonphilosophers who attempt to classify arguments for protecting species very often place anthropocentric and utilitarian arguments in a *different* class from ethical ones.[93] It seems to me that it is hardly possible to interpret this kind of classification any other way than to assume that anthropocentric and utilitarian argumentation is often not considered to be *genuine* ethical argumentation. However, a kind of ethics that the general public doesn't even perceive as a form of ethics will hardly be able to provide the badly needed impetus required for ethically reevaluating the way we deal with nature and species.

24. Expanding the Scope of Moral Responsibility

24.a. Moving Away from Anthropocentrism

"When the logic of history hungers for bread and we hand out a stone, we are at pains to explain how much the stone resembles bread." Leopold ([1949] 1968, 210) used these words in the 1940s to criticize the claim mentioned earlier in this book that traditional anthropocentric ethics and an ethics that takes the intrinsic value of nature into consideration are *essentially the same with respect to their effects*. More than fifty years have passed since then in which the ecological crisis has grown dramatically but a vast amount of knowledge and experience has been gained as well, both in theoretical ecology and in conservation practice. In the past chapters I have drawn on this information in order to demonstrate that the convergence hypothesis cited above is not tenable in light of the "touchstone issue" of species protection, even though it continues to be used with abandon by philosophers, economists, and decision makers in politics and administration. *Anthropocentric* ethics is neither capable of justifying *general* species protection, nor does it provide the social and psychological context that would be required for converting intuitively rooted moral concern for general species protection into deeds. In the case of *nonanthropocentric* ethics, on the other hand, this is possible at least *in principle*, to the extent that ethics is at all capable of achieving such ends. In nonanthropocentric ethics nature has moral standing independent of its usefulness.

However, the qualifying term "in principle" is important because nonanthropocentric ethics, of course, also has to deal with the problem of *weighing duties and interests*. When conflicts of moral concern arise, there is no guarantee that the requirements of species protection will *always* prove to be more significant than others. This situation is often used as an argument by advocates of anthropocentrism to demonstrate that anthropocentrism is inevitable and thus also functionally equivalent to nonanthropocentric ethics. Since it is always humans who ultimately weigh duties and interests in the context of nonanthropocentric ethics, any ethics of this kind must automatically revert to anthropocentrism. But this argument is not tenable. However, a more detailed theoretical explication of the still somewhat fuzzy concept of nonanthropocentric ethics would be necessary in order to refute it. Thus I will discuss this argument in depth later on (Chapters 25.b and 31).

Let me discuss a different objection at this point, one that allows me to summarize the criticisms presented in previous chapters regarding economical, ecological, and aesthetic utility arguments. According to this objection, my attempts to show that anthropocentrism is incapable of justifying *general* species protection are unrealistic since the phenomenon of species extinction we are currently experiencing rarely involves *selective* extirpation of individual species. Instead, the major sources of risk usually operate at a *global* level, in other words, they affect either several plant and animal species at a time (as, for example, when herbicides are used extensively against weeds in agriculture) or even entire habitats (e.g., when tropical rain forests are burned off). Thus it might be possible to achieve almost complete species protection without having to justify *individually* why each and every species should be protected. If we succeed in protecting those species and ecosystems that have been shown to be *useful*, a lot of other *useless* species will be protected as well under the protective "umbrella" of the useful ones.

I wish to present two reasons why this argument is dubious, an empirical one and a theoretical one. First, there are indeed many controversial cases in which the regional or global survival of individual "useless" species is the focus of debate rather than the continued existence of entire ecosystems. As the conflicts about the snail darter or spotted owl have shown, in conservation it is indeed sometimes an *individual* endangered species that might cause a construction or development project to be abandoned. By protecting an individual species a whole habitat achieves protected status, not the other way around.[94] Second, with respect to anthropocentrism the pragmatism of the idea of collective rather than individual species protection is more revealing than supportive. It reveals a relationship that is highly unsatisfactory *for any kind of ethical argumentation*, one that nevertheless is typical of utility argumentation, namely *contingency* (Katz 1979). Justifying species protection with anthropocentric arguments is more or less a matter of luck, because the arguments are based on the possibility that what it takes to protect one species might just *happen* to coincide with some currently sufficiently strong human interest in protecting a particular section of nature. Granted, these two things really do coincide once in a while, but *not necessarily*. The relationship between the needs of human beings and the ecological requirements of other species is not always one of "prestabilized harmony" (see criticism of the term "economy of nature" in previous sections). Moreover, in instances in which such harmony exists *at the moment* there is no guarantee that it will continue to stay that way. Human beings' preferences often vary quite strongly from culture to culture and may also change in the course of time. It is thus fundamentally precarious to base justification for species protection primarily on the *contingent* relationship between the vital needs of other species and the preferences of human beings. When an argumentative structure is pieced together out of contingent factors, there is always the possibility that it may suddenly collapse. This was clearly demonstrated a number of times in the course of discussing the "substitution problem" in connection with economic and aesthetic argumentation.

Of course, I don't want to go overboard with such a skeptical estimate. Even if contingency is a basic weakness of *utility argumentation*, this doesn't mean that *every utility argument* is untenable from the very start. Without a doubt there are instances in which common ground exists between human interests and the requirements of species protection that is quite stable in spite of contingencies and based on relationships supported well enough by science. It would be foolish not to take the significance of such relationships in economic, ecological, and aesthetic areas seriously. And it would also be a grave misunderstanding to draw the opposite conclusion from my criticism of apparently shortsighted anthropocentric arguments for species protection, namely the conclusion that anthropocentric arguments *against* species protection are better or based on more reliable grounds. Instead, opponents of measures for protecting species should be aware that the reason they are usually the winners in public debate is that in the context of anthropocentrism the burden of proof usually rests with those interested in protecting species. Opponents of species protection are not the ones who must explain why their economic interests or personal preferences justify endangering a species. It is the *advocates of species protection* who must show that there are significant human desires and interests in species that outweigh such preferences. As I demonstrated in previous chapters, with a constellation of this kind species protection usually stands to lose.

Spaemann (1980, 197) succinctly summarizes the consequences of such a constellation for nature in the following passage: "As long as humans interpret nature exclusively in terms of its functional possibilities for satisfying their needs and judge nature protection solely on this basis, they will continue to proceed with its destruction. They will continue to regard the problem of protecting nature as a problem of balancing needs and interests and always leave only that part of nature unscathed which manages to escape by the skin of its teeth. In the course of such detailed balancing, nature's proportion will become less and less." In order to circumvent this mechanism Spaemann maintains that ethically speaking only one avenue is left. We must abandon the anthropocentric perspective. Although Spaemann proposes this change for the sake of humankind, that is, in order to secure conditions for the survival of *human beings*, imagine how much more significant it is for achieving the even more demanding goal of *general species protection*.

What conclusions can be drawn from this intermediate estimate of argumentation in favor of protecting species? Even if for strategic reasons it may often be wise to select the argumentation you use to match the specific situation involved, the following basic principle still seems to hold true: If a change in attitude regarding the way we deal with nature is to become established on a long-term basis, advocates of species protection must find a way to reverse the order that usually prevails with respect to justification. Instead of justifying their cause with opportunistic utility considerations as they usually do, proponents must focus their efforts on the argument that has proven to be the most stable and ultimately decisive one in the ethics of interpersonal relationships as well, that of recognizing the intrinsic value of the "moral counterpart." Of course, it cannot be denied that argumentation of this kind is philosophically much more difficult than it sounds. Since only humans have been able to assume the role of a "moral counterpart" in traditional Western ethics so far, changing this order requires further development and reorganization of ethical theory. Several approaches of this kind have been proposed already, but they have not been able to gain consensus among ethicists. Even if they were to agree that the anthropocentric perspective must be left behind and that the scope of immediate human responsibility should be extended beyond the "closed society" of human beings, the question that still remains is *just how far*. Which objects of nature can and should be regarded as having intrinsic value? Although the "can" aspect of this question will be discussed in a theoretical context later on, in the concluding passages of my pragmatic approach I first wish to address the matter of "should."

24.b. Considering Nonhuman Interests

If you recall the classification of concepts of environmental ethics that have been proposed so far and that were described in Chapter 18, three different conceptual types of nonanthropocentrism can be identified that exhibit an increasing scope of moral consideration. Increasing scope means that the number of different kinds of objects of nature for which humans are thought to be immediately responsible also increases. For *pathocentrism* the limit is set at all consciously sentient creatures, for *biocentrism* at all living things, and for *holism* finally at everything that exists. Since a generally accepted, fundamental criterion for judging an ethical position is ontological economy, that is, avoiding metaphysical premises that are not absolutely necessary as much as possible ("Ockham's razor"), when considering extending the scope of moral consideration it seems appropriate to first examine the *pathocentric* approach. This approach has the *theoretical* advantage of being able to get along with the fewest number of assumptions that go beyond traditional anthropocentrism. But what about its *practical* capacity for guaranteeing species protection? Can this goal really be achieved by including *sentient creatures capable of conscious suffering* within the scope of immediate human responsibility?[95]

It can be readily seen that extending the scope of moral consideration in this manner would have only minor effects on the argumentative position of species protection. Since just barely 3 percent of all the species that exist belong to the class of sentient vertebrates, only a fraction of all species could profit from being recognized has having intrinsic value when such value depends upon the capacity of individuals for consciously suffering. For most of the species that exist, for example, for arthropods, plants, and fungi, pathocentrism provides no further *immediate* arguments for protection. Compared to anthropocentrism this position can only raise the *instrumental* value of these species somewhat by considering not only the interests of human beings but also those of higher animals in using lower animals and plants as resources. For the instrumental value of a tree, for example, not only the utilitarian interests of humans would be decisive but also the interests of the black woodpecker in using the tree as a source of food and places to breed. However, it seems doubtful that extending utilitarian considerations in this manner would be very advantageous for the plant and animal species involved. In pathocentrism as in anthropocentrism only *indirect* reasons for protecting species can be provided, namely the interests that individuals might have in them. If you recall the manifold flaws of such *indirect* and thus also contingent and complicated argumentation that have been discussed above, not much comfort can be found in conclusions such as those drawn by Weikard (1992, 120), who, after expounding on the pathocentric position that he favors, maintains that protecting species is "ethically no more significant than protecting the interests of any individuals." If this were the case, the goal of *general* species protection would certainly have to be abandoned. As a matter of fact, utilitarian and pathocentric ethicists such as Singer (1979a, 203, 204) and Elliot (1980, 29) seem to have come to terms with these consequences. Thus Elliot quite straightforwardly writes that "if [a] species is destroyed and [unwanted] consequences [for individuals] do not follow, I just cannot bring myself to think that this is wrong."

Since pathocentrism thus proves to be just as unsatisfactory as anthropocentrism for attaining the goal of general species protection, from a pragmatic standpoint it seems necessary to go a step further and to extend the circle of moral consideration to that proposed by *biocentrism*. After all, according to the biocentric position of environmental ethics at least *all living things* are recognized as having intrinsic value. With respect to protecting species, this means that not only higher animals but basically all animals, plants, and lower organisms can be defended by this type of ethics, regardless of how great their instrumental value is found to be. An irrefutable advantage of this even more comprehensive perspective is that it requires a basic change in human beings' attitudes toward nature. In any of its various organismic manifestations life is no longer *exclusively* a means for attaining human ends but instead deserves consideration *for its own sake* in the form of "respect" (Taylor 1986, 90f.) or even "reverence" (Schweitzer 1991). Compared to anthropocentrism this is undoubtedly a decisive qualitative shift. But as fruitful as biocentrism may be for initiating a sorely needed *change in our attitude* toward nature, when it comes to *providing arguments for protecting species,* it too appears to be unsatisfactory upon closer examination. Why?

The reason is that like pathocentric ethics, biocentrism is also a form of *individualistic* ethics. Only *interests* are morally relevant, and the only controversy that arises concerns the question of whether or not the unconscious efforts of plants and lower organisms to stay alive can be considered real "interests." Regardless of which side one takes in this controversy, it seems clear that *species as collectives* are incapable of having interests. Even though there have been some attempts to apply the concept of interest to ecosystems, communities of organisms, and species by referring to the self-regulation or "self-identity" of such collectives,[96] stretching the concept of interest in this manner is rejected by most philosophers.[97] In their opinion, if the concept of interest were applied to entire species, as in the case of ecosystems (see Chapter 12) the term could only function as a metaphor, the teleological content of which would probably prove to be more confusing than enlightening. Indeed, as Sober (1986, 185) notes, "What do species want? Do they want to remain stable in numbers, neither growing nor shrinking?" Contrary to the widespread myth of a "drive to preserve species" in nature or the purported "goal of maintaining species inherent in the existence of all creatures" (Kadlec 1976, 135), Sober (1986, 185, 186) remarks that according to the currently accepted Darwinian perspective it is completely untenable to think that species, communities, and ecosystems have adaptations that exist *for their own benefit*. "These higher-level entities are not conceptualized as goal-directed systems; what properties of organization they possess are viewed as artifacts of processes operating at lower levels of organization." In light of these comments, if one refers to "the interests of a species," this only makes sense if the sum of the interests of *all the members* of the species is meant, not the interests of the species as a whole.

However, the sum of the interests of individuals is very clearly not what those involved in species protection are concerned about. Of course it is impossible to protect a species without protecting its individual members, but the aims of species protection are more far-reaching. In order to better grasp them, the following intellectual exercise may be useful. Imagine two biotopes, one of which will have to be completely destroyed in the course of a development project that is apparently of paramount importance. Biotope A is a monoculture consisting of two hundred densely grown fir trees while biotope B is almost barren but contains the very last twenty specimens of an indigenous species of horsetail. Which of the two biotopes should be sacrificed for the sake of the construction project? Let us assume that (1) there is no way to get around the dilemma of having to destroy one of the biotopes, (2) that the instrumental value of the plants is not important, and (3) that the intrinsic value of a "primitive" horsetail is no greater than that of a tree (and indeed, there is no plausible reason to assume any such superiority). From a biocentric standpoint the decision would be quite obvious. Biotope B would have to be sacrificed, because if biotope A were destroyed, ten times more living things would be lost than in the case of biotope B. From the perspective of individualistic ethics it is completely irrelevant that destroying biotope B would not only lead to the loss of twenty individuals but to the extinction of an entire species as well. If only the "will to live" or rather the inherent orientation toward survival of individuals is morally relevant, there is no reason to rank the inherent drive of twenty horsetails higher than that of twenty fir trees. The *intrinsic* value of an individual (as opposed to its *instrumental* value) is basically independent of *how many* other individuals there may be on earth with the same properties. A citizen of the tiny country of Monaco has no greater intrinsic value and deserves no greater respect for human dignity than a citizen of China.

Even if this illustration from the perspective of biocentrism appears to be conclusive, it is also clear that not only proponents of nature and species protection but "general intuition" as well would still tend to find it wrong to destroy biotope B. According to their convictions it is definitely much worse to lose the last twenty members of a species than two hundred individuals of a widespread species.[98] Whether or not it is possible to rationally justify intuitions of this kind that apparently grant a species "a greater right to survive than particular individuals" (Lenk 1983a, 834) will be discussed later on. In addition, the delicate question of whether and how the value of a species can be *weighed against* that of individuals must also remain unanswered at this point. This intellectual exercise simply shows that questions of this kind only crop up if one assumes that some *additional, independent value* is attached to a "species" that is more than the sum of that of the individual members of a species. However, a kind of "intrinsic value of a species," which seems to be indispensable for any kind of nonanthropocentric justification of general species protection, can clearly not be defended from the individualistic standpoint of biocentric ethics.

24.c. Ethics beyond Interests

In light of the considerations presented above it looks like we have to go one last step in order to justify species protection from a nonanthropocentric perspective and extend the scope of moral consideration all the way to that of a *physiocentric* or *holistic* position. According to the holistic position in environmental ethics not only all living things but also inanimate matter and entire systems have intrinsic value. Since not only individual interests are what count in this type of ethics, and species and ecosystems are also considered to be *direct* objects of human responsibility, this type of ethics is the first and only one of all those discussed so far that is in principle capable of justifying general species protection in keeping with the intuitions of its advocates. Only in the context of holistic ethics is it possible to defend *all* species for their own sake, regardless of how useless, incapable of suffering, or rare their members might be. In contrast, with pathocentrism and bio-centrism you run into argumentative difficulties when dealing with *rare* species, because when competing with others they have only small sums of cumulative individual interests to offer.

This weakness of purely individualistic ethics is nicely illustrated by an example that Norton (1987, 161) describes: "When the blue whale became exceedingly rare, species preservationists supported laws for protecting them, even though they were well aware that these laws would increase pressure on more abundant species such as the sperm whale. They could not justify their support on the grounds that blue whales, taken individually, have weightier interests, as members of the two species have essentially equivalent levels of consciousness, and consequently, their interests should be accorded essentially equivalent treatment." The laws indicate, however, that in everyday life individual blue whales were indeed privileged. But how this can be justified? Let us assume that you want to resort to nonanthropocentric argumentation in this case or feel that you have no other choice. (Remember Clark's 1973 computer simulation that showed that for economic reasons it may be better to opt for the extinction of blue whales.) Then holistic ethics is the only recourse you have.

Advocates of other ethical schools of thought may agree but nevertheless point out that specific cases in nature conservation do not justify extending the scope of moral consideration all the way to holism. Granted, they will say, you cannot justify general species protection with an approach based solely on the concept of interests, but species protection is just one of many different problems dealt with in environmental ethics. There are many other areas (e.g., the greenhouse effect, nuclear technology) that not only can be mastered completely satisfactorily with arguments based on interests but for which such arguments may even be superior to arguments based on intrinsic value (Birnbacher 1982, 16). To depart from the concepts of rights, interests, and the value of the individual simply because of a few special cases in ecology is both unnecessary and politically dangerous considering the fundamental role these concepts play in traditional ethics (Johnson 1984, 359).

Allow me to respond to this objection. First, extending the scope of moral consideration all the way to holism doesn't mean that arguments based on interests are thus rendered irrelevant. Since the concept of holistic ethics that I have in mind encompasses the other types of ethics in the same manner as the outer layers of a bulb enclose the inner ones (see Figure 2, Chapter 18), it conserves the convincing moral arguments of anthropocentrism, pathocentrism, and biocentrism without altering them. Utility arguments are also legitimate in the context of holistic ethics. However, from the perspective of holistic ethics the rank and weight of different types of arguments varies. According to a system of classification proposed by Norton (1987, 177), the kind of ethics I am proposing would be termed "*pluralistic* holism." In this context both *individuals* and *systems* composed of individuals have intrinsic value.[99]

Second, I object to statements in which species protection is described as only a marginal problem of environmental ethics. Both practical and theoretical reasons can be presented for considering it a major challenge of such ethics. The *practical* significance of this topic becomes evident when you consider the global dimension of species extinction. What is currently taking place is a gigantic process of destruction comparable only to the great climatic and cosmic catastrophes that mark the history of life on earth. Due to the irreversible quality of this process, many ecologists consider this "quiet process of death" (Ehrlich and Ehrlich 1981)[100] to be all together one of the most serious and disturbing symptoms of the ecological crisis. Although the practical relevance of species protection seems to be obvious in view of these estimates, its *theoretical* relevance for environmental ethics is often underestimated. In many publications on this topic it is either mentioned only in passing or not at all.[101] The significance that ought to be attached to it in environmental ethics, however, becomes evident when you think about the fundamental role that species play in the functional operations and dynamics of ecological systems. The species is the basic unit of evolution and a fundamental concept in both evolutionary biology and ecology (Mayr 1982, 296; Willmann 1985, 5). As I shall explain later on, the species is not just an arbitrary systematic unit (or class). In the opinion of most biologists and experts of theory of science it is a historic individual whose temporal and spatial boundaries can be determined by objective means. According to Mayr (1982, 296) species are "the real units of evolution." As far as environmental ethics is concerned, I am convinced that a concept of this kind can only lay claim to normative competence in matters of how to deal with nature if it is able to account for the fundamental role of species in evolution and the organization of ecosystems. An ethics that is not capable of grasping *this* particular biological and ecological dimension and its meaning for human beings' relationship to nature does not really deserve to be called "*environmental* ethics" or even "*ecological* ethics".[102]

In the past years biology and ecology have shown quite convincingly that species and ecosystems are not just a collection of individuals in time and space that share certain characteristic properties. Instead they are supraorganismic *wholes* with emergent systemic properties and therefore *more* than the sum of their parts (see Chapter 7). Many ecologists are convinced that a proper conception of such wholes requires observing them at both a microscopic and macroscopic level (Cody and Diamond 1975; Gilbert and Raven 1975; Odum 1977). In light of these views and the scientific insights of many people involved in theory of science that indicate that the atomistic worldview of ontological reductionism is no longer appropriate, a type of environmental ethics that continues to regard species and ecosystems as nothing more than a class or collection of individuals must be regarded as anachronistic. An ethical position such as pathocentrism or biocentrism, which recognizes only individual interests, fails to face the realities of ecology in its efforts to incorporate accurate descriptions of the world. However, since normative aspects of ethics depend upon appropriate consideration of descriptive aspects (Vossenkuhl 1993a), chances are that an ethics that is based solely on an atomistic worldview will also be fallacious in normative respects.

The following specific case involving the population dynamics of large herbivorous mammals shows that apprehensions of this kind are legitimate. Endangered species such as the African or Ceylonese elephant have to make do with very limited habitats, usually national parks, because of increased expansion of human beings. If an elephant population suddenly increases at one location within these reserves, the elephants cause irreversible damage to the vegetation of their habitat by overgrazing since they are unable to migrate elsewhere (Kurt 1982, 57). If the elephant population is relatively small to begin with, destroying the source of their subsistence can cause not only a decrease in population size but even total collapse (Laws 1970). Is it under these circumstances justifiable to reduce the population to a more stable level by shooting some individuals in order to prevent the population from destroying itself? Advocates of species protection would usually respond positively to this question as long as they saw no chance for the population to regulate itself for the time being (Laws 1970; Caughley 1976). Those who favor individualistic, nonanthropocentric ethics based on interests, on the other hand, would seem to have no other choice than to respond negatively for principle reasons. How can you justify shooting individual elephants *on the basis of the interests of individuals?* If you justify the infringement on animal interests generated by herd culling solely by arguing that it serves to prevent even greater suffering, namely collapse of the population, you would also have to advocate regulating the *nonlethal* dynamics of the population in this manner as well. After all, the suffering of an individual in a population that is not threatened by extinction is no less than what it might suffer in a population on the verge of death. Since the loss of the entire population is as such irrelevant from the perspective of the individual, purely individualistic ethics can provide no *direct* argument for why intervention should be allowed *in this particular case.* It seems to me that this example shows that in some cases interest-oriented ethics and holism lead to very different conclusions and furthermore that ethics that is exclusively interest-oriented is inadequate for dealing with ecological problems.

Along this same line of thought Callicott (1993, 359, 360) warns his readers not to uncritically extrapolate from the individualistic principles of *interpersonal* ethics to the hierarchical systems in which nature is organized. According to Callicott you cannot simply assume "that what is right and wrong in the human moral community is mutatis mutandis also right and wrong in the biotic moral community." In assuming this, one completely overlooks "the very different structure and organization of the biotic community." Of course it is possible, in a manner of speaking, to simply superimpose the individualistic, interest-oriented model of ethics upon ecological problems (e.g., von der Pfordten 1996). Drawing on an analogy of Goodpaster (1979, 29), there is no reason why you can't explain the orbits of the planets by using the old epicycle theory of a geocentric worldview. But the risks that arise when an inappropriate paradigm is applied are obvious. When the individualistic model "is the only model available, its implausibilities will keep us from dealing ethically with environmental obligations and ideals altogether."

But it would be mistaken to think that "the elephant problem" (Laws 1970) discussed above is only a special case of species protection and not at all representative. On the contrary, it must be assumed that the individualistic model of ethics will lead to similarly counterintuitive consequences and results that contradict the basic goals of species protection when applied to most of the other problematic areas of conservation (e.g., the problem of succession, introducing exotic species, reestablishing indigenous species, maintaining genetic diversity) (see Hutchins and Wemmer 1987). This is by no means surprising. If you recall Remmert's (1984, 303) reference to "the schizophrenia which evolution has pre-programmed in individuals," it is clear that implausibilities are *inevitable*. Contrary to what the individualistic approach maintains, what a species needs in order to survive does not always coincide with optimizing the interests of the individual, let alone with the interests of human beings. In my criticism of "ecologism" I already pointed out this fundamental conflict between the individual, the population, and whole systems that is inherent in nature. And just as I showed that it is impossible to rely on ecology to *resolve conflicts between different systemic levels* by referring to a holistic concept of nature's economy based on systems theory, so also is it impossible to rely on a theoretical concept of ethics based on individual interests to *resolve the moral conflicts* that arise due to tensions between different systemic levels. In his essay titled *What is conservation biology?* Soulé (1985, 731) very clearly concludes, "The ethical imperative to conserve species diversity is distinct from any societal norms about the value or the welfare of individual animals or plants." According to Soulé *animal welfare* is concerned with reducing suffering and sickness for individual animals, whereas the aims of *species protection* are to secure the integrity and continuity of natural processes. At the level of the population it is genetic and evolutionary processes that are decisive, because these are the processes that have the potential to secure the continued existence of biodiversity (see Frankel 1974). Since evolution as it occurs in nature cannot proceed without suffering and adverse effects for individuals such as famine, sickness, and being preyed upon, an irresolvable conflict remains between the intuitions associated with species protection and those upon which the urge to protect individuals is based. For our dealings with nature it is not useful to play down these conflicts in order to simplify matters in ethical theory. If environmental ethics wishes to be taken seriously, it must take ecological reality seriously, with all its variety and contradictoriness.

What conclusion does this discussion allow us to draw with respect to the philosophical debate about the "right" environmental ethics? In my opinion it seems sufficiently plausible to maintain that it is *necessary* to continue to develop ethics from anthropocentrism to pathocentrism to biocentrism and beyond to an ethics of pluralistic holism. If environmental ethics is to be a *competent* form of ethics in the future and not split up into special disciplines such as animal protection and species protection, it is absolutely essential that it include both an individualistic and a holistic dimension. In a comprehensive form of environmental ethics both human and nonhuman individuals as well as wholes must be given adequate moral consideration. I certainly do not wish to give you the impression that such a concept of multiple moral consideration with respect to different systemic levels is capable of *resolving* all the conflicts that exist between these levels. At most it can help us to *master* them. However, compared to all other versions of ethics holistic ethics has one very definite advantage. Since it openly addresses *all* the conflicts that exist, at least the chances are good that they will be resolved *as knowledgeably and conscientiously* as possible.

II. A Theoretical Approach: Can Holism Be Justified?

Even though the preceding chapters should have clearly shown that the anthropocentric position in environmental ethics does not provide adequate justification for the general protection of species and that there are therefore important *practical reasons* for extending environmental ethics toward holism, it is nonetheless evident that with these arguments alone only half the battle is won. Even if holism is *desirable*, this by no way means that establishing it is truly *possible*. On the contrary, it may very well be that the holistic position is the most effective one, but that it is based on theoretical premises that are untenable. Indeed, this is the most common criticism brought against holism by advocates of the anthropocentric position. Before I attempt to demonstrate that—contrary to this criticism—the holistic perspective is sufficiently justifiable, I wish to elaborate on three very basic objections that can be found in the literature and are aimed at refuting the possibility of any nonanthropocentric position in environmental ethics. These can be considered very *basic* objections since they question the fundamental sense and legitimacy of expanding the circle of ethical responsibility beyond anthropocentrism. Thus they are not only relevant for holism but for biocentrism and pathocentrism as well.

25. Fundamental Objections to Extension

25.a. *Opportunistic Theory Choice?*

According to the thesis of the first objection any attempt to demonstrate that an extension of ethics is a moral duty is inevitably grounded on a petitio principii, that is, a statement that itself has not yet been proven is presented as proof of another. The reason is that in order to rationally justify such a duty, which in the first instance is perceived only by intuition, one is forced to rely on the help of the ethics one hopes to extend.[103]

If one embraces this logic, one would have to conclude that my endeavor is unacceptable. Instead of first examining whether the goal of general species protection is in compliance with ethical theory—so goes the argument—I have simply assumed this intuitively rooted postulate to be an obligatory premise and made it a critical test of ethical theory. Frankena (1979, 20) maintains that this sequence is like picking out a certain cart and then looking for a horse to put before it: "One must, in a fundamental sense, have one's ethics first, before one can decide, on moral grounds, for or against conservation, etc." von der Pfordten (1996, 59) expresses a similar view when he warns us about the pitfalls of "opportunistic theory choice," that is, incorporating a metaethical question about the effectiveness of an ethical position in the normative-ethical justification of that position: "The justification must support the results [in this case the goal of general species protection] rather than generate a theory that is useful for such results." In Chapter 20 I already presented pragmatic and metaethical arguments in favor of the opposite procedure. At this point I shall attempt to defend the thesis of the primacy of reflected intuition inherent in those arguments with a historical and systematic argument.

The fundamental primacy of theory as opposed to pragmatic and intuitive considerations might be justifiable if ethics were a cognitive and normative system that is largely independent of empirical and historical contexts, a system whose scope of responsibility had remained constant over the centuries and had never been seriously questioned before the advent of the ecological crisis. However, this is obviously not the case. A brief look at the history of ethics reveals not only repeated radical changes in theoretical thought systems but also shows that the scope of validity of moral systems in earlier epochs was not always the same as it is today (Vossenkuhl 1993b, 6). Usually it was smaller. In ancient times, for example, many humans (e.g., slaves, children under a certain age) had no rights of their own at all (Weber 1990, 112), and no contemporary moral philosophers took exception to this condition.[104] Members of foreign tribes and lower castes were excluded from the core of morality, sometimes all the way into the modern period. Not until the period of Enlightenment did people become generally convinced that *all* humans are direct objects of moral responsibility, regardless of their place of birth, race, or nationality. Thus a scope of moral responsibility in anthropocentrism that encompasses all humans and is considered almost self-evident in current ethical theory must be regarded as a relatively recent philosophical achievement. According to an estimate by Tugendhat (1989, 928) it's more recent than one might generally think, having become established in general consciousness only since World War II.

In order to illustrate that in view of this background it is neither histori-cally nor objectively justified to regard the scope of the ethical system that happens to predominate in a particular historical epoch as an unquestion-able starting point for a metaethical discussion, Meyer-Abich (1984, 22f.; 1990, 60f.) distinguished between three different levels within anthropocen-trism, in addition to three outside of it. This leads him to define eight possi-ble ethical positions with varyingly comprehensive scopes of consideration:

1. Egocentrism: the autonomous individual in the sense of John Locke
2. Nepotism: one's own tribe (reference group)
3. Nationalism: one's own people or race
4. Anthropocentrism with respect to the present: all contemporary fel-low human beings
5. Extended anthropocentrism: future generations as well as the above
6. Pathocentrism: higher animals (capable of suffering) in addition to the above
7. Biocentrism: all living creatures
8. Physiocentrism: inanimate things and whole systems in addition to the above

If one tries to fit the predominant moral system of a society into this table, it becomes evident that it is necessary to differentiate between theo-retically accepted moral positions and morality exercised in practice. The members of a New York street gang, for example, will most likely not rec-ognize anything beyond level 2, even though theoretically the animal pro-tection legislature of New York suggests a level of morality corresponding to level 6. According to Meyer-Abich (1989, 141), "in its political behavior humanity has made it to about level 3, in spite of the increasing globaliza-tion of human activity, and in moral consciousness it has reached level 5 at the most." Judging from Meyer-Abich's manner of speech, it would seem that he not only regards the sequence presented above as a logical and sys-tematic system of classification but would also have us envision it as a his-torical course of development as well. Is this legitimate? Is there anything like an *evolution of ethics* from egocentrism to physiocentrism, as for exam-ple, the historian Nash (1977, 6) postulates?

Brief reflection reveals three aspects of this postulate that raise doubts. In the first place—as Meyer-Abich (1990, 61) himself has indicated—there has probably never been an "original state of nature" such as John Locke proposed ([1690] 1966, 4f.), in which moral responsibility was limited to the self. It seems that individual human beings can only thrive in social surroundings, in which the egoism of level 1 has already been "tamed." In the second place, the religions of certain native peoples show that contrary to the assumption of linear progression from level 1 to level 5, some societies have developed a very pronounced attitude of respect for animals, plants, mountains, and rivers (level 8) at an early stage in the history of humankind, *even though* the complete scope of anthropocentrism in these societies usually was neither achieved nor perfected (Teutsch 1985, 116). And in the third place, one has to take into account the lack of uniformity in the history of philosophical theory, at least with regard to the ethics of ancient Greece and Rome, when judging the apparently irregular development of these religions. Thus von der Pfordten's (1996, 89) results indicate that there were "various different intermediate positions and nuances" between the mostly anthropocentric Stoics at one end of the scale and the more biocentrically oriented Pythagoreans at the other.

Even though these points contradict the idea of a linear and uniform extension of moral responsibility in the course of the history of ethics, it can nonetheless not be denied that a *general tendency* exists. In the long run the particular scope of morality that was regarded by the philosophers and judicial experts of a certain epoch as an inviolable minimum has increased. Basic concepts of ethics such as the Golden Rule, which used to be applied only within a limited social or geographical domain, are now regarded as universally valid. According to Nash (1977, 7), the fact that tourists nowadays are able to travel around the world in relative safety testifies to this effect: The existence of ethics that encompasses all of humanity "permits an Istanbul businessman to pass through Detroit without being captured and sold into slavery, just as it permits a Detroit secretary to visit Istanbul without fear of being pressed into concubinage." Of course, this doesn't mean that racism, slavery, and other forms of discrimination corresponding to level 2 or level 3 morality have disappeared, but they cannot be justified by modern philosophical ethics with its minimal scope corresponding to level 5. It cannot be denied that this is an improvement compared to the ethics of earlier periods.[105]

How does this historical retrospective relate to the criticism of holism on the grounds of "opportunistic theory choice"? I feel that it at least refutes the claim that it is unacceptable for *methodological reasons*, so to speak, to extend moral consideration beyond level 5 anthropocentrism. If this claim were true, it would never have been possible for ethics to develop as far as it has. Because from the internal standpoint of class-oriented, nationalistic, or racist ethics such as those of levels 2 or 3, justifying an extension of moral responsibility to level 5 would also have to be considered a petitio principii, just as the attempt to justify an extension of ethics beyond anthropocentrism presented here must appear to an anthropocentric ethicist. However, the fact that an extension has indeed occurred several times in the past, regardless of theoretical and practical objections, demonstrates that in certain historical circumstances and under the pressure of carefully reflected intuitions it apparently can sometimes be justified and necessary to "hitch the cart before the horse" and adjust ethics to intuition. The analogy to the development of knowledge in science is obvious. Just as new empirical results can lead to a paradigm switch in science when these results simply cannot be interpreted within the context of the predominant paradigm, new descriptive knowledge or external pressure due to the consequences of continually ignoring this knowledge can make it necessary to further develop or restructure ethical theory. In both cases paradigm change obviously represents an exception to the usual procedure of interpreting and testing new data in the light of established theories.[106]

The fact that it was sometimes necessary to reverse this procedure in the course of the history of ethics, of course, still does not provide *sufficient* grounds for justifying such a reversal in the present situation. Justification for a further extension of moral consideration has to be shown to be plausible in itself, in view of the ecological crisis, new scientific results, and an altered perception of nature. This will be taken up in Chapter 27. But before that we have to take the hurdle of a second major objection, according to which it is *basically impossible* to escape an anthropocentric perspective.

25.b. Is Anthropocentrism Inescapable?

According the second major objection, any attempt to establish nature protection on the basis of intrinsic value of nature will be handicapped by an "incurable logical weakness" (Löw 1989, 158) because of the fact that an anthropocentric standpoint is supposedly inescapable for reasons rooted in epistemology and for systematic and methodological reasons.

The *epistemological* argument is clearly articulated by von Ketelhodt (1992, 13): "Anthropocentrism poses a difficult position. From the standpoint of

pure logic we cannot escape it. It is always humans who think and pass judgment." This observation is also the main focus of von Haaren (1991, 30f.)
when she argues that nature protection can never be anything but anthropocentric: "Humans could and can always see and define nature only on the
basis of their individual knowledge, which is variable in time. Even if we
were able to determine the exact needs of all living things, this would still remain a definition established within the dimensions of human thought.
Therefore the claim to protect 'nature as a value in and of itself' appears to be
hypocritical. Thus we have no other choice than to derive measures for nature protection from basic human needs." Is this conclusion valid?

First of all, probably no one would deny that only humans are capable of
thinking about nature and the right way to deal with it. And everything we
know so far indicates that only humans are capable of assuming responsibility for their actions. Nevertheless, this does not necessarily mean that an
anthropocentric standpoint is unavoidable. As Taylor (1986, 16f.) and von
der Pfordten (1996, 32) indicate, the question of who is capable of *ethical
responsibility* and the question of whether or not nature requires *ethical consideration* are "logically independent of one another and must be kept apart."
Of course there are various normative ethical theories (e.g., discourse ethics)
in which these two elements are indeed coupled because of the particular
structures of these theories, but this coupling has to be *specifically* justified.
It is by no means a logical necessity in the sense of being unavoidable, as
maintained above. The unique cognitive status of humans, that is, their status as the only known *subjects* of morality ("moral agents"), is not *in itself* an
argument in favor of a unique ethical status of humans, that is, in favor of
their status as the sole *objects* of moral consideration ("moral patients").

Since these two aspects of morality are often confused, resulting in misunderstanding, I wish to introduce a useful conceptual differentiation that
Teutsch (1988, 60) has made in distinguishing between anthropocentrism
on the one hand and anthroponomism on the other.[107] According to
Teutsch the term *anthropocentric* means "to regard humans as the pivot
point of existence, to relate everything to humans and to subordinate
everything to them as well. In contrast, the term *anthroponomic* refers to
the fact that humans can judge everything that exists only within the con-
straints of human cognitive faculties." By making this distinction Teutsch
points out that in spite of all criticism of anthropocentric concepts we
must always take into account "that human thinking is contingent upon
the human condition and that therefore the objects of human reflection

never can be seen and judged "as such" but rather only in the context of human cognitive capacities." Thus Teutsch agrees with the epistemological premise of the objection discussed above, namely that all thinking processes, even those involved in environmental ethics, are inevitably anthroponomic. However, he very decidedly refutes the conclusion that some draw that all ethical thought must therefore be exclusively oriented toward human interests. Anthroponomism is by no means inevitably coupled with anthropocentrism.

The fact that the relationship between anthroponomism and anthropocentrism is not one of necessity and that *anthroponomic nonanthropocentrism* thus is indeed possible is convincingly demonstrated by the example of animal protection. Here there is a general consensus that causing unnecessary pain to animals is forbidden *for the sake of the animal*. Nowadays moral duties with respect to animal protection are no longer derived indirectly by weighing possible human interests but rather directly by arguing that animals are in themselves worthy of moral consideration. At the same time it is obvious that what is good for an animal can only be recognized and considered within the context of human cognition and judgment. It is obviously impossible to determine directly, from the internal perspective of the animal, whether or not a caged chicken suffers and to what extent it suffers. Instead we have to rely on observations made by humans, scientific theories, and analogies. Because of the hypothetical nature of such judgments, specific moral duties may be questioned in certain cases, but such skepticism is usually expressed with respect to particular factual relationships, not with respect to the moral obligation to prevent animal suffering *for the sake of the animal*. In conclusion the example of animal protection shows that even though what is good and right for nonhuman nature cannot be determined directly but rather via anthroponomism, it is still in principle possible and common practice to consider such forms of nature in light of their intrinsic value. In this case the fact that such estimations can only be made by means of indirect constructions and analogies is not an insurmountable obstacle for ethicists.

Were we to conclude that the hypothetical nature of such judgments is an insurmountable obstacle, this would also lead to major repercussions for the ethics of relationships between humans. Here too a necessary requirement for making ethical judgments is the ability and motivation to place oneself in the position of another, that is, to weigh the moral claims and interests of a fellow human being by indirect, hypothetical means. The discrepancies that sometimes arise when members of different races and genders evaluate social norms testify to the fact that assuming the perspective of another is by no means a matter of course.[108] Obviously verbal representations help us to indirectly draw the conclusions required for making moral judgments and facilitate estimations that are for the most part objective. But the viewpoints that are relevant for evaluation cannot always be successfully verbalized. And, as Nelson ([1932] 1970, 165) has shown, language can also be deceptive. If we were forced to rely exclusively on verbal representations, all interests incapable of being expressed in words would have to be ignored, for example, those of infants, the verbally handicapped, the senile, and the mentally ill. The fact that the interests of these persons are indeed represented in ethics indicates that they are granted moral consideration *for their own sake* even when the flow of information from the moral object to the moral subject only occurs in rudimentary forms of communication. Verbal representation is thus neither a sufficient nor necessary prerequisite for considering the moral claims of another being.

In view of these considerations it is not very convincing when the possibility of direct responsibility for nature is rejected on the grounds that we have no certain knowledge of the moral claims of nonhuman nature since they cannot be communicated. Neither in our relationships with verbally impaired humans nor in those with higher vertebrates are we absolutely certain of these claims, but this still does not prevent us from feeling morally responsible. Granted, when it comes to reconstructing the interests of others by hypothetical means, these two groups are particularly favorable from the standpoint of epistemology. After all, infants and the mentally ill are still humans, and within the family tree of evolution higher vertebrates are our nearest relatives. The greater the evolutionary distance between an organism and humans, however, and the more its way of life diverges from our own, the more difficult it becomes to put oneself in its place, so to speak, and to determine its moral claims via analogy. While we may succeed more or less well when dealing with animals, things are certainly more difficult in the case of plants, some of which are not individuals in the usual sense (see Chapter 4.d). The problems become even more serious when the question arises about how to do justice to a river, an ecosystem, a species, or the entire biosphere. Since these wholes are not only unable to express interests but quite simply don't have them, any linear analogy constructed on the basis of the ethics of relationships between humans is bound to run into trouble. In these cases any attempt to draw analogies by means of pure intuition will probably be futile. Thus it must be conceded that many more specific questions about how to deal with whole systems in a morally acceptable fashion cannot yet be answered and that it is often not even certain whether a satisfactory answer can be provided *at all*. Considering the fact that the field of holistic ethics is still in the cradle, so to speak, and that ecology, which is supposed to nurture it by providing a descriptive foundation, still has to struggle with serious methodological problems of its own, this is not surprising.

In spite of these difficulties in the area of description, it seems to me that there is no reason to conclude that the project of holistic ethics is hopeless or to consider it an enterprise that can only be realized in the distant future. Regardless of all the problems and snags that arise when delving into detail, the holistic concept of intrinsic value still permits us to establish a plausible set of *basic norms* that could form the skeleton of a theory of holistic ethics. These norms would include a prima facie rule against the destruction of individuals and species and the requirement to employ as little and as cautious intervention as possible in dealing with systems that have developed autonomously. From the perspective of epistemology I consider these constructs for maintaining the good of nonhuman entities, that is, for guaranteeing survival of individuals and autonomous dynamic development in the case of wholes, to be self-evident, providing that there is such a thing as "the good" of a non-organismic whole. It would take a generous dose of skepticism to reject the (granted anthroponomic) idea that it cannot be in the interest of a tree to be chopped down, that it is bad for the autonomous dynamics of a river if it is channeled and that a bog does not do well when it is drained. I deliberately chose three cases in which it is not very difficult to determine what is morally wrong since we are dealing with the prospect of existence or nonexistence of each natural entity. By doing so I hope to demonstrate that even within the scope of responsibility engendered by biocentrism and holism, anthroponomism does not necessarily lead to *agnosticism*. Contrary to what von Haaren (1991, 30) has come close to reproaching as being "hypocritical," anthroponomism also does not inevitably result in *anthropomorphism*. Just as animal physiology and ethology contribute to animal protection by helping to guarantee that the inevitably anthroponomic ethical judgments we make are as objective as possible, so also can ecology and evolutionary biology make a valuable contribution to nature conservation. Obviously the claim of objectivity inherent in this process can only be realized to a certain extent. But if the only alternative to an anthroponomically determined "good" of nature is anthropocentrically determined utility, it seems to me that it is appropriate and well worth it for environmental ethics to attempt to define such a good. After all, an ethical standpoint is characterized as one that at least *attempts* to assume a perspective that is as objective as possible, or rather, *universal.*

Considering this very fundamental claim of all ethics, it is surprising that Löw (1990, 293) maintains that it is impossible to assume a biocentric position. "A human being who protects nature or restores it does so as a human, not as nature." I would reply, "Why does he do it as a human? Why not as a citizen of France, a Muslim, a workman, a farmer, or simply as an individual?" If you think back on the four levels of responsibility *within* the extended concept of anthropocentrism described in Meyer-Abich's table (1984, 22, 23), there is no simple answer. Meyer-Abich (1990, 84) has rightly pointed out that "to think of something *from one's own perspective*" can occur in a very different manner depending upon the extent to which one feels himself part of the world and depending upon the reference group with which one identifies. "'From his own perspective' a human being can think egoistically, nationalistically, anthropocentrically or physiocentrically, but by no means only anthropocentrically, as some claim, because egocentric thinking is something less than anthropocentric thinking, and physiocentric thinking goes beyond it." According to this view anthropocentric thinking is definitely not the norm, neither in ethics nor in epistemology. Although I will explicate this later on in the context of ethical theory, my thesis in the context of epistemology is that, if one assumes any position at all, an egocentric position is more obvious than an anthropocentric one.

If you look at things from the standpoint of epistemology and seriously consider the fact that other interests (both human and nonhuman) have to be construed hypothetically, you will realize that in reality it is not an abstract *Homo sapiens* who makes assumptions about the exterior world and passes judgment on it but rather *the ego*, the subjective core of an individual human being. Both philosophical epistemology and empirical neurophysiology have shown that the starting point and foundation of all knowledge is consciousness and experience (Wigner 1964; Eccles 1970, 152; 1973,191). Only subjective states of consciousness such as thinking, feelings, perceptions, and memories are *directly* accessible to us; all our estimates about the exterior world, on the other hand, even the conviction that other humans exist with whom we can share a common "standpoint as human beings," are secondary, derived constructs. Thus Wigner (1964, 252) deliberately and pointedly maintains that " . . . excepting immediate sensations and, more generally, the content of my consciousness, everything is a construct . . . but some constructs are closer, some farther, from the direct sensations." If this view is right, then from the standpoint of cognition theory there seems to be no real reason why the anthropocentric position should be given a special status. Even though an anthropocentric position and the constructs related to it may be more closely associated with our "direct sensations" than a biocentric or physiocentric standpoint, its constructs are nonetheless still just constructs. The only standpoint that is not comprised of constructs, or, more cautiously formulated, the one that relies the least on constructs is the egocentric standpoint. From the perspective of epistemology this standpoint could be considered the best one in the realm of knowledge and evaluation of the world. In this connection it is interesting that hardly anyone considers an argument of this kind to be a convincing argument in favor of ethical egoism. Even though epistemology demonstrates that every human being is undoubtedly the center of his or her world, everyone (except an egoist) knows that from an objective or ethical perspective this is *not true*. It seems to me that this insight into the epistemological contingency of our knowledge of the world provides a plausible platform for refuting anthropocentrism's claim to exclusiveness in an analogous manner. Just as the key roll of subjective consciousness in epistemology fails to provide sufficient justification for egocentrism, so also does the inevitability of anthroponomism fail to serve as a convincing argument for anthropocentrism. *Epistemology* does not provide adequate support for anthropocentrism.

Because of this the "inescapability argument" is often reinforced by a *methodological and systematic* one in which reference is made to the problem of how to reach a decision in the case of conflicting duties and interests. According to this argument, anthropocentrism is inescapable, because "the coupling of our ethics to our interests is unavoidable" (Irrgang 1989, 47). What exactly this means is illustrated by two quotations by Bayertz (1987, 178): "We can only afford to abandon anthropocentrism when the interests involved are relatively weak ones (e.g., fur coats); whenever more vital interests are at stake (e.g., smallpox viruses), we have no other choice than to place these above the competing 'interests' of other parts of nature.

However, this means that we have to regard human interests as the critical ones, not completely arbitrarily, of course, but nonetheless human interests." Bayertz (1986, 231) summarizes these thoughts elsewhere in the following manner: "There is always an ultimate level at which decisions are made about whether or not we are willing to take the idea of equal rights for everything that exists seriously, and that ultimate level is anthropocentric." Since I will discuss the problem of how to reach a decision in cases of conflicting interests in greater detail later on, at this point I shall only be concerned with the question of whether the ultimate level of judgment is truly an anthropocentric one and whether this means that departure from anthropocentrism is therefore impossible.

In order to respond to this question in an intellectually satisfying manner, it is necessary to differentiate between "moderate" and "radical" versions of nonanthropocentric ethics. The cornerstone of a *moderate biocentric* position, for example, is a hierarchy of values or interests among all living things ("scala naturae"). In this hierarchy humans occupy the level corresponding to the greatest value, while microorganisms are relegated to the lowest level. Obligations toward a blade of grass are not as great as those toward a giant sequoia (Ricken, 1987, 18). When an attempt is made to balance interests on the basis of such a value hierarchy, a particular human interest is not always and necessarily more weighty than that of a nonhuman being, but the hierarchy makes it possible in principle to ethically justify sacrificing "lower" level creatures for the sake of "higher" level ones. At this point it is not yet appropriate to discuss the various philosophical proposals that have been made to justify such a value hierarchy of all living things. For the time being I shall simply point out that the process of establishing such a hierarchy can hardly occur apart from human interest in using the environment. As Vossenkuhl (1993b, 10) has argued, all attempts to assign value (in the sense of numerical credit) to nature can hardly be interpreted in any other manner than as attempts to justify the loss of animals and plants. "Assigning value only makes sense in the context of trying to justify losses. Of course, it is not nature toward whom this justification is addressed, but rather humans, to whom still other humans feel obligated to account for their behavior." Since the value ranking of moderate, nonanthropocentric positions in ethics ultimately serves human interest in self-exculpation, it is hard to get around Vossenkuhl's reproach that we are dealing here with a case of "masked anthropocentrism."

However, this reproach does not apply to *radical*, or rather, "absolute" ethical positions, which expressly reject both value hierarchies as well as any attempt to place moral claims in a relative position to one another within the chosen ethical system (Schweitzer [1923] 1974; Taylor 1986). Schweitzer would not agree that the case described by Bayertz as an example of the priority of human interests or that the fight against smallpox viruses is *ethically reconcilable* with his biocentric teachings of "reverence for life." For Schweitzer *any* action that would sacrifice the life of some other being for the sake of one's own life or good can be considered a fundamental infringement upon the basic principles of morality and can never be reinterpreted as being a morally acceptable one. If a violation of basic moral principles, as in the case of the smallpox virus, can be considered "excusable" nonetheless, then only because it is clear that this violation is *dictated* by the reality of life, which, of course, also includes the "survival instinct" of humans (Günzler 1990a, 98f.). Naturally this raises the question of where "the forces of necessity" begin and how the "ultimate limits to persisting in the maintenance and support of life can be defined." According to Schweitzer ([1923] 1974b, 388) the answer to this question cannot be determined *objectively*, but rather must be ascertained from case to case. The value hierarchies that may be employed in such instances are in his view "highly subjective measures" and therefore must be regarded as being outside the realm of ethics. At this point I shall not yet debate whether or not Schweitzer's ([1923] 1974e, 155) claim is right and reasonable that decisions reached outside of the context of *absolute* ethics are necessarily "arbitrary." Right now it is only important to note the following. When a balance of existential interests is attempted in the context of absolute ethics, the biocentric position does not automatically regress to an anthropocentric one, as Bayertz maintained, but rather, under the pressure of necessity the ethical standpoint is *abandoned*. Instead of reverting to anthropocentrism, in reality biocentrism reverts to egocentrism.

The fact that ethics occasionally capitulates to self-interest does not have to be considered a *specific* weakness of nonanthropocentric versions of ethics. The difference between anthropocentric and nonanthropocentric ethics in this respect is not a fundamental one but rather a quantitative one. Even within the context of absolute anthropocentric ethics (as, for example, Kantian ethics), it is possible to conceive of situations in which the contingent reality of the moral actor makes it difficult or even impossible for him or her to behave in a morally consequent manner. An example of such a *moral dilemma* might be the shipwrecked father of a family who manages to save himself and four family members in a lifeboat. Since the boot can *safely* hold only five persons and he does not want to risk the lives of his family, he forcefully prevents an additional shipwrecked person from entering the boot.[109] No one would come to the conclusion that the behavior of the father proves that tribal morality corresponding to level 2 in the long run is indeed "inescapable." Behavior of this kind shows simply that in this particular case anthropocentric morality corresponding to level 4 could not be maintained in the face of reality. Just as the example of the shipwrecked family does not suffice to exclude the possibility of anthropocentric ethics all together, so also does the example of the smallpox virus not provide irrefutable evidence that biocentric or holistic ethics are impossible for methodological or systematic reasons.

25.c. Refined Anthropocentrism?

Let us return to a third common objection raised against holistic ethics. The main thesis of this objection is that a departure from anthropocentrism is superfluous if we employ a definition of "human self-interest" that is broad enough. With a sufficiently broad definition of human self-interest *any* goal established in the context of nonanthropocentrism can also be justified with anthropocentric arguments. Even if one advocates abandoning an exclusively instrumental relationship toward nature, it is possible to derive this position from "well-meaning" human self-interest.

Bayertz (1987, 178) illustrates the basic idea of this thesis with a quotation from Spaemann (1980, 197), who justifies his appeal for nonanthropocentric ethics as follows: "When humans destroy nature, they also destroy the very roots of their existence. Therefore whenever we deal with nature we are also dealing with humankind. Nevertheless, or more appropriately, for this very reason it is necessary to abandon the anthropocentric perspective." According to Bayertz this form of argumentation proves once more "that the more avidly we try to show anthropocentrism the door, the more persistent it will be in slipping back in through the window." Bayertz argues that in Spaemann's new and highly sophisticated kind of anthropocentric argumentation the standpoint of human self-interest "is by no means discarded but simply shifted to a second, higher level of argumentation." While Bayertz continues to regard this kind of argumentation as a case of anthropocentrism in action, Meyer-Abich (1984, 66) rejects such an interpretation. He considers Spaemann's position to be a "nonanthropocentric" one. Thus we are confronted with the paradoxical situation that one and the same argument *on the one hand* is presented in order to demonstrate the necessity for abandoning anthropocentrism while *on the other hand* it is regarded as proof that in reality it is not really necessary to depart from this position. What is the source of this controversy and what does it mean in connection with the objection formulated above?

In order to get a better perspective on the matter it is necessary to define the term anthropocentrism more precisely. If it is understood to mean that human beings are the *reference point for all ethical justifications*, then this definition has to be further qualified to include the fact that reference to humans can be more or less strong. As von der Pfordten (1996, 21) has pointed out, it is not sufficient to simply talk about the existence or nonexistence of a referential relationship. In addition we have to account for gradual differences. And indeed in the literature reference is often made to "strong anthropocentrism" when talking about immediate and concrete human interests, while the term "weak anthropocentrism" is used to refer to more indirect and abstract relationships (see Armstrong and Botzler 1993, 275; Norton, 1984; 1987, 12, 13). According to this classification system economic or immediately relevant ecological arguments can be considered strongly anthropocentric, while aesthetic or enlightened ecological arguments can be subsumed under the category weakly anthropocentric. In this connection it should be recalled, however, that weak anthropocentric arguments are not necessarily weak arguments. It is only the inherent reference to human interests that is weak—and accordingly also the persuasive power of such arguments within the context of utility argumentation (see Chapter 23.a). The scope of weak anthropocentric arguments, on the other hand, as the test case species protection has repeatedly shown, is often greater than that of strong anthropocentric arguments. Although only a small sector of the entire spectrum of species can be defended with economic arguments (see Chapter 22.a), the "unknown threshold level argument" (see Chapter 22.b) permits protection to be extended almost without limits.

In view of these observations, it is not very surprising to find that *in philosophical discussions* weak anthropocentric arguments are relatively widespread. Since these circles cannot very well ignore the fact that the scope of the most common utility arguments is not very great, anthropocentrists who would like to see a comprehensive form of nature protection established tend to enhance their position with "enlightened-anthropocentric" arguments. A particularly elegant kind of enlightened argumentation in this connection is that of *multiple-stage justification*. According to this line of thought aspects of the relationship between human beings and nature that are noninstrumental (or at the most weakly instrumental) at a *primary* level may be reexamined at a higher, *secondary* level. When considered in consort at this level they once again attain instrumental value. A relevant example is the "aesthetic resource argument" of Birnbacher (1980, 132.f.) discussed earlier, with which he recommends protecting nature by arguing that humans have a need to attribute to the things of the world a kind of autonomous value that is independent of any kind of need (see Chapter 22.c). In the first (aesthetic) instance this argument is nonanthropocentric, in the second (reflected) one, however, truly anthropocentric. One can readily see that by employing such a multiple-stage argument it is possible to extend the term "self-interest" indefinitely. It is simply impossible to find anything that would not reveal itself to be "useful" when considered from a second level of reflection. Even actions and attitudes that are normally not associated with a self-interested perspective (e.g., worldviews, ethical attitudes, religious views) can thus be interpreted as being "grounded in anthropocentrism."

Thus a line of argumentation can also be found in the literature on environmental ethics that attempts to justify respect for nature primarily or exclusively on the basis of a human need for self-respect and dignity. A prerequisite for this viewpoint is that the term "dignity" be understood to "imply a force that exists beyond the human perspective, the acknowledgment of something deeper or greater" (von Ketelhodt 1992, 14). If this premise is accepted, then protection of nonhuman nature solely on the basis of superficial utilitarian arguments "would not be in accordance with the dignity of human beings" (von Ketelhodt 1992, 14). Although this line of argumentation does not exclude a priori the supplementary attribution of intrinsic value to nonhuman nature, many advocates of anthropocentrism point out decidedly that a "supplementary metaphysical assumption" of this kind is not necessary. In the long run it all boils down to human interests. *For their own sake* humans should not assign value to other things exclusively for their own sake. This statement pretty well summarizes the lines of thought found in the literature on environmental ethics that fits the term "refined anthropocentrism" (Meyer-Abich 1984, 65).

If a concept of human self-interest as comprehensive as that described above is selected, then it is obvious that this will affect how we usually differentiate between anthropocentrism and nonanthropocentrism. Two modifications must be taken into account. First, a dual-level perspective will cause the distinction between anthropocentrism and nonanthropocentrism to almost completely disappear. If it is not clear from the very beginning which of the two levels is the most important, contradictory views and misunderstandings may result. This is nicely demonstrated by the Spaemann argument cited earlier that is employed by both anthropocentrists and nonanthropocentrists to support their positions. Second, by stretching the concept of anthropocentrism to such an extent, the dividing lines between various positions in environmental ethics will be shifted. Thus in many specific instances of environmental ethics the difference between weak and strong anthropocentrism may appear greater than that between weak anthropocentrism and nonanthropocentrism. Norton (1984, 136) has demonstrated this using the so-called last-people example of Routley (1973) described earlier (see Chapter 23.c), which is frequently considered a kind of litmus test for distinguishing between anthropocentrism and nonanthropocentrism (see Lee 1993). Contrary to this view Norton maintains that this thought experiment is not useful for separating anthropocentrism from nonanthropocentrism but rather for distinguishing between strong anthropocentrism on the one hand and weak anthropocentrism or nonanthropocentrism on the other. According to his view of enlightened self-interest it would be more than plausible that the last anthropocentrists in Routley's scenario would also refrain from devastating the planet, because an act of vandalism of this kind would violate the "ideal of maximum harmony with nature" that they had internalized in the course of their life on earth (Norton 1984, 136). The decisive grounds for their considerate behavior would then not necessarily be some postulated intrinsic value of nature but solely their human value system, which they would not be willing to give up even shortly before their death. Doesn't this example show convincingly that "the right kind" of anthropocentrism can lead to the same consequences as nonanthropocentrism? Doesn't this prove that under these circumstances the position of nonanthropocentrism, which is ontologically more difficult to grasp, is indeed superfluous?

Now, on the one hand, it cannot be denied that *with respect to final consequences* "refined anthropocentrism" as described above and nonanthropocentrism overlap to a great extent. It would be a surprise if it weren't so, because one can almost always be sure that an act aimed at protecting nature that is performed for the sake of nature in an indirect manner will also benefit humans. One of the most important achievements of enlightened utility arguments has been to draw attention to such side effects. They clearly show that protecting nature is not only compatible with *a particular kind* of self-interest but, moreover, that "human self-interest" is not simply a matter of shortsighted and exploitative egoism of the species. An interpretation of the concept of self-interest that is so superficial that it completely ignores intellectual and spiritual needs of human beings not only reflects a very limited view of what humanity is. It would also fail to meet the explicit expectations of many enlightened anthropocentrists. On the other hand, of course, the question arises as to whether it is wise to employ a concept of "human self-interest" that in principle is infinitely expandable in order to define a concept of anthropocentrism that is so comprehensive that it practically "absorbs" the entire area of nonanthropocentrism by introducing the metaperspective of a second level of reflection. There are three reasons that make me dubious about the usefulness of a strategy of this kind.

First of all, it is important to remember that there would be a price to pay for a drastic expansion of the concept of anthropocentrism such as that which the introduction of an all-consuming, two-tiered concept of anthropocentrism would evoke. It would mean that the concept would become almost meaningless. We develop concepts in order to distinguish between things and delineate them from one another. However, if *each and every* act and attitude—regardless of whether it serves human purposes or not—can always be found to be in human interests when reflected at a higher level, then the term anthropocentric means everything and consequently also *nothing* anymore. This certainly can't be what people were thinking about when they coined terms like "self-interest" or systems of classification such as "anthropocentrism" and "nonanthropocentrism." Of course, it is also a fallacy to think that this conceptual problem is only a linguistic one and thus merely of *theoretical* nature. On the contrary, it is also a practical problem, because—and this is my second argument—expanding the concept of anthropocentrism would also have political consequences. The shift in the "argumentative watershed" (as illustrated by the last-people example) that would ensue would cause more confusion than clarification in political discourse. Since the most important line of demarcation in many specific and controversial cases would no longer be between anthropocentrism and nonanthropocentrism but rather between weak and strong anthropocentrism, misunderstandings are bound to occur. I fear the main danger is that refined anthropocentrism could be misused to justify a less refined form of anthropocentrism. In an intentionally ironic aside Meyer-Abich (1984, 68) predicted how this might take place. "Don't all philosophers maintain that environmental protection is inevitably always a question of human interests? . . . If this is the case, then it is only legitimate that we act according to our own interests in environmental matters. And we sure as heck won't let any of those philosophers dictate to us what our interests are supposed to be." I don't want to suggest that this line of argumentation is *logically* conclusive, but it seems to me to be *psychologically* plausible at any rate. In my opinion the strength of its psychological attraction results from what constitutes my third argument, the fact that any kind of refined anthropocentrism must seem inconsistent if it concedes the possibility of a noninstrumental relationship to nature at the primary level of consideration and then proceeds to refute it at the second and ethically decisive level. The reasoning that was presented in connection with my criticism of arguments that are both aesthetic and utilitarian applies in a similar manner to refined anthropocentrism as well. Arguments for the protection of nature based on more "profound" human interests such as interests in self-transcendence, self-respect, fulfilling duties, or maintaining dignity, harmony, and other ideals are only convincing if these ideals are accompanied by an objective concept of nature as being an end in itself. What is the sense of striving for "harmony with nature" if at the same time *direct* moral responsibility for nature is denied? As Regan (1981, 25) has noted, "An ideal which enjoins us not to act toward X in a certain way but which denies that X has any value is either unintelligible or pointless." According to his understanding of the term "ideal," this concept requires that the object to which actions are directed truly possess value of its own. Since it is difficult to refute this assumption, at least on psychological grounds, one might surmise that the abstract intellectual contortions of refined anthropocentrism really serve only one purpose, and that is to avoid assigning intrinsic value to nature. No thought seems to be too complicated when it comes to fortifying the anthropocentric position, if only it somehow or other fits into a concept of human interest. It should be noted that I do not contest the idea that indirect justification of this kind, which constantly refers back to human interests without embracing the concept of intrinsic value of nature at the same time, is *logically possible*. My thesis is that it is inconsistent and thus not convincing.

Support for this thesis can be found by examining an analogous form of ethical argumentation that is also logically correct but not very convincing, that is, the attempts to reinterpret altruism as a form of enlightened self-interest. Just as a kind of enlightened egoism *of the species* may be brought into play in order to refute the necessity of immediate responsibility for nature, it is also possible to contest the phenomenon of selfless behavior among humans by reinterpreting it as an enlightened form of *individual* self-interest. This is the pattern of thought found in psychological egoism and hedonism. If a person feels the need to help another, he or she does it for selfish reasons, so to speak, because, after all, this person will experience satisfaction in helping others. Since one can be quite certain that *every* action performed out of moral obligation is correspondingly motivated and accompanied by a "spin-off" product of personal gratification (see Schlick 1984, 97), it is easy to maintain that *these feelings* (or even expectations of a reward in heaven) represent the *real and sole* reason for that action, not respect for a moral other, as commonly believed. A claim of this kind is quite simply irrefutable. Nevertheless, or perhaps even for this very reason, it is doubtful whether such a claim can be considered a sufficient explanation of the action, and even less certain whether is appropriate as a foundation for ethics. This kind of claim might be of heuristic value as a model for stimulating methods of investigating specific problems of economics or sociobiology (keywords "homo oeconomicus" and "the selfish gene"), but in ethics, which purports to deal with much more than a small sector of reality, its reductionistic nature makes it appear unreliable and not very authentic.

At this point I cannot deal in any depth with the many arguments that have been brought forth against ethical egoism and hedonism.[110] In connection with the topic discussed in this section it suffices to refer to the objection against a complete reduction of interspecific altruism to enlightened anthropocentrism that has already been presented. If the term "self-interest" is defined so broadly that it is possible to interpret *every* action as being motivated by or performed for reasons of pure self-interest, then the term loses all of its semantic potential for differentiating between things. Even Maximilian Kolbe, who voluntarily chose to die in place of the father of a family in a concentration camp, would have to be considered an egoist—that is, an "enlightened" egoist. It is obvious that interpretations of this kind tend to obliterate differences that should not be effaced. As Wolgast (1981, 146) has shown, the special significance of a moral action is exactly that it *cannot be explained*. If a moral action is explained in terms of self-interest, the particular moral character of the action is abolished. Reducing phenomena to the same low level in this manner is certainly not very useful when looking for a source of orientation in everyday life. To understand what this means let me refer to a parallel line of thought that has already been discussed in the passages on epistemology in the previous chapter. Explaining morality in terms of self-interest seems to me to be similar to the thesis of the solipsist who maintains that the difference between a dream and awake experience is fictitious since both phenomena can be thought of as constructs of subjective consciousness. At any rate, a standpoint of this kind is just as irrefutable as the standpoint of *ethical* reductionism, according to which all moral phenomena can be reconstructed in terms of individual or collective self-interest. The common denominator of both is that from the perspective of logic they need not suppose the existence of an external (ontological or ethical) reality. But as the example of solipsism shows, not everything that is logically possible is also plausible.

If ethical reductionism so obviously lacks persuasive power, how can we explain the myriad attempts to interpret both interspecific and intraspecific altruism as a subtle form of self-interest? I agree with Stone (1988, 44, 45) that a fundamental problem of ethics is at the root of these attempts, namely the limited scope of all ethical reasoning and justification. The problem is not that it would be impossible to give good reasons why one particular action qualifies as being morally good and another as morally bad. *Ethical skepticism* of this kind that maintains that all moral judgments are inevitably relative and subjective is not difficult to refute.[111] In particular since Nietzsche ([1886] 1990), but not starting with him, the more profound and serious problem is rather a *general skepticism toward morality* (Nielsen 1984, 81). If morality is considered the propensity to reject egoism and assume an impartial standpoint, what on earth might motivate me as an individual to assume such an impartial standpoint? Why should I be moral? Moral reasons cannot be given for this problem since they presuppose the existence of the moral standpoint that is to be justified. What other choice is there than to revert to a "psycho-logical" appeal to self-interest (if no other possibility presents itself)? It can readily be seen that this represents a paradox of all ethics and earmarks fundamental limits at the same time. If any reasonable arguments at all can be proposed in favor of a moral standpoint, they must be nonmoral ones, and that means basically that they have to be oriented toward self-interest and pragmatic and psychological ones. A form of reasoning of this kind might be the following: "You are free to persist in maintaining a standpoint of pure self-interest, but you'd better think twice about whether you are willing to accept the possible consequences (whatever they may be). And think hard about whether or not your position complies with the view you have of yourself!"

Whether or not and to what extent reasoning of this kind is appropriate for convincing someone to assume a moral standpoint is a moot point among philosophers. I shall return to this problem in Chapter 28.b. But right now I would like to temporarily close and summarize this topic by pointing out a misunderstanding that can arise because of the two-tiered nature of this argumentation (moral standpoint at the first level, standpoint of self-interest at the second). It consists of the following: From the self-interested argumentation of level two which marks the borderline of all ethics, so to speak, the conclusion may be drawn that ethical argumentation as such is in principle based on nothing other than enlightened (individual or collective) self-interest. In my opinion Bayertz (1987, 178) falls victim to this fallacy in his interpretation of a critical argument proposed by Spaemann (1980, 187) and cited earlier on that it is necessary *for the sake of humankind* to abandon the anthropocentric perspective. Bayertz considers this an indication of the inevitability of anthropocentrism. In doing so he incorrectly converts a *borderline argument* concerned with the foundations of ethics to a *paradigm of metaethics*. Contrary to this interpretation Spaemann himself (1979, 286) does not regard his argument as a metaethical one supporting the inevitability of anthropocentrism but rather as a "functional argument in favor of nonfunctional thinking." In this connection he emphasizes that arguments "can only be functional." Of course, the question that arises is what argumentative thinking is capable of and what not. According to Spaemann (1980, 198) it is capable of a great deal: "Plato long ago demonstrated the maximum of all the possibilities of argumentative thought, which is that it can be stretched to its own limits, that is, to a threshold beyond which insights can no longer be deduced by argumentative or rather functional means."

26. Regarding the Nature, Claims, and Prerequisites of Justification

In the following chapters I shall attempt to theoretically justify extending the scope of moral responsibility to encompass a holistic standpoint, and, after all that has been said, there is no doubt in my mind that sooner or later this must lead us to the very threshold of insight that Spaemann has indicated to be the limit of all ethics. In attempting this I am not only concerned with questions about *which* behavior toward nature is morally right or wrong. Questions of this kind lie *within* the realm of ethics and can only be dealt with after others have been clarified. Before that a more fundamental question has to be considered, the question of *whether or not* it is at all possible to grant nature moral standing. In other words, *concerning our relationship to nature* it is necessary to refute a very fundamental version of skepticism, skepticism toward morality, or more precisely, to refute "class amoralism" (see Nielsen 1984, 88f.). According to this kind of skepticism the main question is, "Why should I assume a moral standpoint not only toward humans but toward animals, plants, landscapes, species, ecosystems, and the entire planet as well?" As discussed in the previous chapters a *moral* response to this question is impossible since it would require exactly what we have set out to test, that is, that nonhuman nature is already established as an object of moral consideration. If this is the case, what other kind of justification is possible?

Before dealing with the *content* of this question, a formal aspect must be clarified that repeatedly leads to misunderstandings and incorrect judgments in discussions of ethics, the *validity claims* of rational reasoning. When is it legitimate to speak of a *conclusive* argument, and just how conclusive must it be? Since it would exceed the scope of my venture to unfold a general discussion of this very fundamental problem of all ethics (for a detailed discussion see Frankena 1963, 94f; Forum für Philosophie Bad Homburg 1987), I shall restrict myself in the following to outlining the validity claims of my own argumentation in this treatise. To do this I believe light might best be shed on the matter by first addressing ideas discussed in the literature about the persuasive power of a "conclusive argument" with which I *disagree*. It appears that with regard to the validity of rational argumentation three obstacles must be circumvented in order to avoid asking too much or too little of ethics.

The first obstacle consists of demanding *absolutely conclusive proof* of one's own ethical position. In this respect the term "absolutely conclusive" would mean that anyone who could not accept my arguments would have to be considered irrational. He or she would have to be considered just as irrational as someone who denies that two times two is four. *Mathematical* conclusions derive their definitive certainty from the fact that they are analytic and grounded on axioms that are generally accepted as requiring no further justification. However, demanding definitive certainty in *ethical* argumentation would require determining *absolute justification*, that is, a series of arguments that lead to "statements [that] no reasonable creature can deny" (Köhler 1987, 304). Time and time again in the course of the history of philosophy many have sought a fundament of ethics of this kind. Kant ([1788] 1974) was convinced that it could be found in the concept of reason itself (or, more precisely, in the idea as such that something can be justified). This so-called transcendental justification approach was further developed by Fichte, as well as by Schelling in his earlier works, and is currently also being pursued in a modified form by advocates of discourse ethics or rather of transcendental pragmatics (e.g., Apel 1976, 1987; Kuhlmann 1985). However, at least with respect to claims of absolute justification it is noteworthy that none of these schools of thought has proven to be so convincing that it enjoys the general approval of all modern-day philosophers (Tugendhat 1989, 927; Gethmann 1987). Apparently neither reference to the "fact of reason" nor to any natural facts is able to provide so solid a foundation for ethics that it cannot be jolted by radical skepticism. Thus there is good reason for Tugendhat (1989, 927) to remark about the "profound helplessness of philosophers." He maintains that with respect to the question of how something can be justified "modern reflections on morality . . . have basically remained fruitless. . . . Nowadays we are all confronted with fundamental moral problems that we have to resolve one way or other, for which, however, we have no answers that we can justify sufficiently." Anyone who shares Tugendhat's skeptical assessment, will certainly understand that I by no means intend to provide an *absolutely conclusive* case for pluralistic holism in this treatise. Considering the fact that in the course of two thousand years of history anthropocentric ethics has not succeeded in establishing absolute justification for its moral claims with regard to relationships between humans, it would seem unreasonable to expect the much younger discipline of environmental ethics to present a set of supporting arguments of *such persuasive power*.

The second obstacle to be circumnavigated appears on the scene as soon as the first has been successfully mastered. If it is true that all attempts to provide absolute justification in ethics so far have failed, then it is only a short step to the claim that ethics *cannot be justified at all*. This belief is what noncognitivistic theories of ethics such as emotivism or decisionism have in common. While *emotivism* (e.g., Hume [1748] 1999; Ayer 1936; Stevenson 1937) maintains that moral statements are simply expressions of feelings, which themselves cannot be evaluated further, *decisionism* (e.g., Weber [1919] 1985; Hare 1952; Albert 1961) regards decisions as the final criteria of moral judgments. These decisions are subject to scrutiny by rational criticism at the most in only a limited sense (see Ricken 1989, 36). Even more radical skepticism toward ethics is manifested in Wittgenstein's philosophy (1963, 112, 115). In his opinion basic concepts of ethics "cannot be articulated at all." Things of this kind (like all things mystical and transcendental that elude articulation) will "reveal themselves," that is, they will crop up on their own. Although this view of the matter has some merit in that it points toward a basic problem of all ethics as discussed above, the fact that you can only go so far with arguments, it seems to me that this does suffice to contest the possibility of any kind of rational ethics. If one demands, like Wittgenstein (1963, 7) does, that we remain silent about all things about which we cannot "speak clearly," that is, about things that cannot be verified, then there are very few things (and trivial ones at that) about which we could speak philosophically. Even the many statements and conclusions of science would have to be considered obsolete, considering the highly metaphoric models of atomic physics, the paradoxes of quantum theory, the hypotheses of modern cosmology, or the problem of consciousness (see Heisenberg 1969, 279f.) And yet it is science that has shown that in regions outside the boundaries of ethics one can generally get along quite well with statements for which there is no guarantee of absolute certainty. Even scientific theories (so-called "natural laws") have not been proven in the sense of "absolute justification." At the most they can simply claim to be more or less plausible (see Chapter 2). Strictly speaking it is *only plausible* that the sun will rise tomorrow, but this plausibility is still sufficient for us to plan a picnic for the next day. If you keep this in mind, it is hard to understand why we can't be satisfied with plausibilities in ethics as well. Granted, a *plausible* justification is certainly less than *absolute* justification, but still it is also more than pure intuition *without* justification.[112] Of course, the question that immediately arises is the following: When is a justification "plausible"? Can criteria for plausibility be found?

The third obstacle that must be circumvented in this connection, is in my view the assumption that an argument or form of justification is *only* plausible if it meets with *general approval*. In this context Birnbacher (1981, 341; 1982, 14) refers to the requirement of "universalizability." In his opinion this is the case when the norms of ethics are grounded in values "for which one can be quite certain that anyone can comprehend and accept them so that the claim to general authority inherent in them can to some extent be realized." While he feels that the value premises of anthropocentrism (human interests) and pathocentrism (conscious feeling) meet this criterion, he thinks that the value premises of biocentrism and holism depend too greatly upon varying cognitive conditions to be able to meet with general approval. Wolf (1987, 167) maintains that moral consideration that extends beyond the pathocentric standpoint "can only be grasped in the context of religious beliefs." She feels that biocentric or holistic morality is not "based on opinions, feelings, etc. that we can assume to exist in all humans so that there is no chance of bringing it home to everyone." On the basis of these considerations Hartkopf and Bohne (1983, 68) have formulated the recommendation that if environmental ethics is interested in recognizing a broad range of value positions, it should not orient itself toward high ideals that only few accept. Instead it must embody an "ethical minimum" that is capable of both "capturing majorities here and now" and guaranteeing our future. In a pluralistic society, they say, the "ethics of environmental protection" must be firmly rooted in basic postulates and others derived from them that are "rationally comprehensible" and thus also "capable of being accepted by as many people as possible, independent of a particular worldview or a certain view of human nature."

As realistic and reasonable as this position might appear at first, it can hardly be denied that equating "plausible justification" with "being capable of capturing majorities here and now" would be disastrous in practice and theoretically unacceptable. In the first place, it is by no means certain that a so-called "ethical minimum" that the majority *nowadays* might agree upon is in reality sufficient to secure the *future* (which means the future of humans *and* nonhuman nature). If Hartkopf and Bohne think this simply goes without saying, then they silently presuppose a kind of "prestabilized harmony" between *future* ethical requirements and *current* ethical consciousness, an assumption for which there are no grounds. On the contrary, the ecological crisis seems to indicate that a gap between these two instances has developed that can only be bridged once again if moral sensitivity in the general public is developed even further, which means *raising the level of the ethical minimum*.

It is obvious that any attempt to increase ethical concern would surely fail if the plausibility of new value concepts could be disclaimed by simply pointing out that no majority supports them at the moment. If *being accepted by the general public* were the decisive criterion for plausibility, then no plausible argumentation in favor of extending ethics *could ever* become established. And it would never have happened in the history of ethics either. At the height of an ancient society practicing slavery, for example, an early advocate of rights for all humans would have had no chance to argue "convincingly" against slavery, because in the context of contemporary thought morality including more than one's own race would not have been able to appeal to "opinions, feelings, etc. that we can assume to exist in all humans," referring once again to Wolf's (1987, 167) wording as described above. It seems to me that this example from the history of ethics shows clearly enough that it would not be good for philosophical ethics to require that concepts of ethical value and their justifications be able to meet with general approval *in the same period in which they arise*. Instead of serving as a perceptive organ for new challenges and representing the avant-garde of ethical progress, philosophical ethics would function as a free rider of moral common sense. In this case its role would be little more than that of exoneration, justifying the moral status quo of society. In order to avert the danger of *ethical opportunism* that would ensue, Frankena (1963, 96) accordingly distinguishes between *real* and *ideal* agreement in ethics. "The fact that ethical and value judgments claim a consensus on the part of others does not mean that the individual thinker must bow to the judgment of the majority in his society. He is not claiming an *actual* consensus . . . , he is claiming an *ideal* consensus which transcends majorities and actual societies." In this connection Frankena emphasizes the fact that "one's society and its code and institutions may be wrong."

If in a given society it is possible for moral consensus to be "askew" with respect to certain value judgments, then, of course, it is not only necessary to rule out the idea of equating the *plausibility* of value judgments with the degree of general approval. We must also give up the idea of defining the term *rational justification* in such an exclusive manner. The reason is that from a superior, detached, or more advanced perspective the concept of rationality favored by the majority can also be *rationally* criticized. In retrospect what once seemed reasonable may be deemed unreasonable.[113] Although this historical perspective teaches us to beware of rash conclusions regarding our concept of rationality, in literature on environmental ethics it is not uncommon to find examples where rationality is equated unconditionally with the predominant opinion of the majority, especially when the majority opinion happens to be shared by the author. Thus, for example, Wolf (1987) and Patzig (1983, 340f.) have no qualms about describing their pathocentric positions as "being derived from rational moral consideration," as if rational thought could lead to no other concept of morality than a pathocentric one. They explicitly deny that farther-reaching concepts of morality can be justified with "rational arguments." Contrary to the theologian Auer (1988, 31), who favors anthropocentric arguments himself but still maintains that both anthropocentrism and physiocentrism can be upheld "in a rational manner since both undoubtedly are expressions of rationality," Wolf and Patzig practically propose a dual-class society. On the one hand we have anthropocentric or pathocentric ethics based on rationality and established exclusively on rationally conclusive arguments, on the other there is confessionally oriented biocentric or holistic ethics, which is forced to rely on religious or metaphysical premises or ones derived from other worldviews. The quotation from Hartkopf and Bohne (1983, 61) cited earlier reveals what the decisive difference between the two classes might generally consist of. Rational ethics, so the basic assumption, is characterized as operating "independently of a particular worldview or a certain view of human nature."

Of course this assumption is based on a fundamental error. There is *no* ethics that is not associated with some worldview or perspective on human nature. If ethics is defined as an attempt to reflect upon the right way for humans to deal with the people and things of this world from a universal standpoint, it is clear that reflection is impossible without a minimal hypothetical concept of what a human being "is," what the things of the world "are," and what the relationship between the two involves. At least *this insight* has emerged from all of the unsuccessful attempts to develop an absolute justification for ethics that is completely independent of empirical premises about the world and the "nature" of humanity. If Hartkopf and Bohne as well as quite a few other moral philosophers nevertheless continue to insist that the ethics they favor is free of premises resulting from fundamental worldviews, this only goes to show that the premises upon which their ethics is based have been internalized to such an extent that they simply are not aware of the contingency of these empirically determined premises. However, a closer look at the matter would lead them to recognize that the rational morality of anthropocentrism is also firmly rooted in premises determined by certain worldviews, that is, in those of the anthropocentric view of the world and of humanity (to be described in detail in the next chapter). Thus the options at stake in metaethics are not really "rational ethics" versus "worldview ethics" but rather "ethics A corresponding to worldview X" versus "ethics B corresponding to worldview Y" (etc.). An option *without* a particular worldview or view of humanity is not available.[114]

What is the significance of this observation with respect to the problem of justification? If each and every variety of environmental ethics is and *must* be based on some worldview, then it is obvious that the plausibility of the justification for a particular school of thought does not depend solely on formal criteria (e.g., conclusiveness, consistency, coherence, etc.). To a very great degree it will also depend upon whether the basic premises of the corresponding worldview are right. If these premises are wrong or obsolete, then, of course, even the most watertight set of arguments cannot prevent the conclusions derived from these premises from possibly also being wrong.[115] This logical conclusion illustrates how significant a worldview is for the justification of a particular form of environmental ethics. According to a number of authors it is no less than the very pivot point of any school of thought.[116] If one analyzes the controversies between the concepts of environmental ethics that mark the extreme endpoints of the entire range, anthropocentrism and physiocentrism, one indeed finds that the discussion "almost always [winds up] examining the view of human nature to which advocates of the position adhere or must adhere" (Strey 1989, 76, 77).

In this connection Meyer-Abich (1984, 22) is of the opinion that the particular view of the world and of humankind associated with a certain position in environmental ethics can only seem conclusive to someone who already embraces it. He illustrates this thesis by referring to the perspective an egoist has of the world. "Anyone who even vaguely asks why his own personal interests are not the only measure for his behavior demonstrates with this query that his view of humankind is one in which the social context of the individual does not exist. A person whose basic character is not rooted in the perspective of a social context, who is not aware of the fact that language and the ability to love are at his disposal and that he can only be truly human as a social being, can in no way be convinced that he should embrace certain considerations derived from this context." Exactly the reverse situation is true for someone whose concept of herself has always included the perspective that a human being can only be human among and in the context of other human beings. She would either simply not pose the question cited above in this form or reply that considering the interests of others is only natural. According to Meyer-Abich what is true for the worldview of an egoist or anthropocentrist is also true for the worldview of a physiocentrist. In the end his position is also based on an "existential decision at a fundamental level." A human being's behavior toward his or her environment always depends upon an "orientation that is in the broadest sense religious—or existential, that is fundamental to all arguments." Does this mean that the choice between anthropocentrism and physiocentrism is only a matter of making a *decision*, that in the long run the noncognitive theory of decisionism is the right one?

The answer to this question depends upon what one understands by the term "decision." It would certainly be incorrect to equate it with a "blind jump," that is, with a choice that can neither be reconstructed nor justified rationally. This surely cannot be what Meyer-Abich meant, because he did not just drop the discussion with the straightforward proposition of a "fundamental existential choice" but rather has attempted to support his case for the physiocentric option with various different arguments. Bayertz's (1987, 179, 180) criticism of these arguments to the effect that they can only suggest, promote, or enhance the motivation for the proposed choice but never "conclusively justify" it nor render it "rationally necessary" is beside the point. No worldview, not even an anthropocentric one, can be shown to be "rationally necessary." As Marietta (1995, 103) has demonstrated, philosophical worldviews are in principle "not wholly correct or wholly mistaken." They are probably rather "mixtures of sound learning and cultural nonsense." If this estimate is true, then it is quite clear that there can be no justification for worldviews in an *absolute* sense. Nevertheless, it seems legitimate to speak of *relative* justification when several different worldviews are compared with one another and the most plausible one is given priority after careful examination. Worldviews can obviously not be verified by this procedure, but by means of critical comparison it is at least possible to establish a kind of plausibility ranking. Of course, for this purpose it is necessary to agree on certain formal test criteria. If the discussants give rational criteria the highest ranking and if they also share a common concept of rationality,[117] they will most likely also agree with Taylor's (1986, 158, 159) suggestion and enlist the following criteria, "which have been found throughout the entire history of philosophy. They are (a) Comprehensiveness and completeness. (b) Systematic order, coherence, and internal consistency. (c) Freedom from obscurity, conceptual confusion, and semantic vacuity. (d) Consistency with all known empirical truths."

The last criterion in particular seems to be of great significance for the topic of this treatise, because through modern science our knowledge of nature and the position we occupy within it has developed more rapidly and to a degree unknown to previous generations. If consistency with all known empirical facts is to be maintained in spite of this fundamental transformation, it is obvious that plausible worldviews can no longer consist of unalterable, timeless systems. Instead they must be regularly tested for their "compatibility" with new discoveries about the world, and if necessary they must be "purged" of views that have become untenable (see Chapter 17). "Worldviews, like old houses, need to be refurbished now and then, and some of them need a complete remodeling" (Marietta 1995, 102).

Having clarified this matter, have we satisfactorily answered the question about the special character of "choice" between various different worldviews? Taylor's list of criteria somewhat mitigates the decisionistic aspects of such a choice by providing general measures for judging and correcting worldviews as well as for testing the plausibility of individual components of them. However, this does not mean that the necessity of a *personal* position is therefore completely superfluous. Knowing as much as possible about relevant *descriptive* data is necessary for forming a plausible worldview, but it is not sufficient for determining its content unequivocally. In the context of different worldviews the same descriptive data may be interpreted and weighed in very different manners, and this, of course, has serious *normative* consequences. Thus two people may possess the same knowledge about species, about evolution, and about the role of species in evolution, but this does not mean that both will reach the same conclusion regarding the question of what it means to extinguish a species. In light of the discussion of the "naturalistic fallacy" (Chapter 8), it is obvious that one has to reckon with the possibility of diverging views. From the knowledge of "pure" facts alone it is impossible to derive value judgments that are *logically compelling;* a "should" does not automatically follow from an "is." But how then can we reach a decision regarding the best worldview, which in the long run is a normative question, if not indeed through a "blind jump"?

I agree with Spaemann (1990) that the existential choice of a worldview and the basic values associated with it should be thought of as exactly the opposite of a *blind jump*. It should be considered an act of *enhanced perception*. In reaching a decision of this kind, if it is to be a truly rational decision, one should try to "see" and "hear" as attentively as possible, in order to attain a more or less clear insight into what "should" be (see Jonas 1988, 58). Of course, insight into what "should" be can only succeed if three prerequisites are met. First, faith in evidence generated by perception is required, which means believing that it is indeed possible to gain insights into what is and what should be.[118] Second, the best knowledge possible should be acquired. And third, it is necessary to be prepared to overcome the archaic and purely egocentric perspective of an "instinctive creature" and to regard the reality of external objects as the reality of autonomous subjects, not just that of objects of instinctive drives. At this point Spaemann (1990, 130) talks about "awakening to reality." According to this interpretation rationality is nothing but "totally aroused awareness—awareness that knows itself and sees itself as particulate reality on an absolute horizon" (Spaemann 1990, 116). "As long as life remains a captive of instinct, as long as it persists in a state of 'centrality' [i.e., in the egocentric perspective], the world will not appear as real. The rest of the world does not appear to be something of its own. It appears only as 'environment,' as an object of instinctive drive" (p. 119). With reference to Kant, Spaemann (1990, 133) emphasizes here that the requirement of real things that they be truly perceived as real cannot be deduced from any moral law. In the end it can only be perceived through an "act of autonomy." "I can decide anytime to regard some living thing which I encounter as only a machine, in other words, as a kind of reality that is not substantial. This will not disturb the comprehensiveness of my experience or my ability to identify objects in space and time. I can decide to behave toward the world as if it were a realm of pure objects." More concisely, I can refuse to *awake to reality*.

In my opinion Descartes' ([1637] 1956) philosophy of nature demonstrates that refusal to perceive reality as it is is not necessarily a sign of an individual psychological defect, but may instead be an option, a general worldview to which one consciously adheres. In the context of the sharp dualism between the mind (res cogitans) and the material world (res extensa) envisioned by Descartes there is only *one* creature that possesses subjectivity and therefore also reality, and that is man. Animals and plants are nothing but complicated machines. Since animals have no mind and consist simply of "extended matter," they cannot "really" feel pain. If a rabbit begins to scream in the course of vivisection, Descartes would see no basic difference between this phenomenon and the squeaking noise that arises when one cranks a bucket up from a well. Obviously this interpretation of animal suffering is "convenient," because it wipes out any moral scruples that people might have in dealing with nature. Both scientific progress and the triumphs of modern technology have profited from this interpretation. But still nowadays we cannot help renouncing Descartes' view of nature as being extremely out of touch with reality. Besides being so counterintuitive that one wonders how a genius like Descartes could really believe it, this view of nature has been so thoroughly discredited by what we now know about the evolutionary relationship between humans and other vertebrates and the neurophysiological structures they share that as a philosopher one can support it only blindfolded. In our day and age it must be considered as good as untenable.

The proposition that I intend to unfold in the next chapter is that in light of the knowledge we have gained in modern times about the world and human beings, not only Cartesian philosophy as a particular variety of anthropocentrism but the *anthropocentric worldview itself* have become unacceptable. When the list of criteria proposed by Taylor (1986, 158, 159) is used to evaluate the four worldviews that correspond to the four basic positions in environmental ethics,[119] the anthropocentric worldview is found to be the least plausible. A closer look shows that while the pathocentric and biocentric worldviews manage to smooth a few existing incompatibilities with empirical evidence, only a holistic worldview is capable of generating a reasonably close fit with both what the humanities have revealed about the unique status of human beings with respect to nature and what science has discovered about their status as part of nature. Of course, the plausibility arguments in favor of a *holistic worldview* presented in the next chapter by no means suffice to justify *holistic ethics*. They simply provide a platform upon which a justificatory structure will be erected later on. Clearly one could object that this represents a kind of naturalism, as Schäfer (1987, 23) does when he rejects any "attempt to construct a concept of ethical responsibility on the basis of a concept of nature." Nevertheless I don't believe this objection can be sustained in the case of the strategy pursued here, which involves *relative* justification at a very fundamental level, or rather, comparing the plausibility of basic premises. If I am correct in my estimation that no ethics exists that is completely detached from basic empirical assumptions about what nature and human beings are, then this objection either applies to *all* forms of ethics in the same manner or it is *irrelevant* for all forms of ethics.

27. From an Anthropocentric Worldview to a Holistic One

In environmental ethics the many different systems of moral thought that can be subsumed under the generic term "anthropocentrism" are far too diverse to assume that they all stem from *one and the same* view of the world and humankind. Nevertheless, according to Teutsch (1985, 8) they share at least one common property. They regard "the world as being oriented toward *humans*; everything is there to serve their purposes; everything else in the world is a means to be deployed by humans." Of course, advocates of anthropocentrism do not always *explicitly* articulate this view that human beings are the pivot point of the world and that the rest of nature is merely contingent "environment," but I maintain that it is *implicit* in their ethical premises and thought structures. In the following I shall refer to this as "ontological anthropocentrism." I suspect that for the most part this view stems from three different traditions: (1) a religious one, (2) one based on philosophy of nature, and (3) one derived from epistemology.

Regarding the first of these traditions, the *religious* one, in Western cultures for centuries the anthropocentric worldview was justified by referring to the Judeo-Christian tradition, which regards humans as having been conceived as "images of God" and commissioned to conquer the earth. Indeed, in Genesis 1:28 one reads, "Be fruitful, multiply, fill the earth and conquer it. Be masters of the fish of the sea, the birds of heaven and all living animals on the earth." This frequently cited quotation from the Bible can, of course, be contrasted with other passages in which the superior position of humans is further qualified or in which at least the claim of *unlimited* power over nature that is often deduced is convincingly refuted (see Liedke 1981, 109f.; May 1979, 159; Teutsch 1980, 26). Nevertheless this does not alter the historical fact that the mainstream of thought in Christian theology, similar to that of Cartesian philosophy, advocated a dualistic theory in which the world was divided into two parts: humans (with reasoning capacity, souls, and prospects of salvation) and nature (without reasoning capacity, a soul, or any kind of prospects for salvation) (see Liedke 1981, 73). The relationship between the two was usually interpreted as a God-given hierarchy of power. This view of the world led to the widespread attitude that the world was given to human beings "as a kind of inexhaustible cornucopia to be exploited at will and from which they could help themselves without any scruples as long as they didn't forget to say grace" (Teutsch 1980, 119). Indeed, just shortly before the churches began to become aware of the ecological crisis and to think differently about their relationship to nature the following statement was issued by the Second Vatican Council (1962–1965): "It is the almost unanimous opinion of all believers and non-believers that everything on earth should be oriented toward human beings as both the center and highest level of creation" (quoted in Auer 1988, 31, 32).

The claim to fundamental power over all of the nonhuman world expressed in this statement was, of course, not restricted to either Christianity or the other great monotheistic religions, Judaism and Islam. It can also be found in many systems of thought developed in the second major tradition, *philosophy of nature*, from antiquity to modern times. In particular it is typical of interpretations of the world that not only regard nature as an unwavering system of purpose and order but also regard humans as realizing the highest and final purpose of the internal organization of the world. Cicero (106–43 B.C.; 1997, 103–105), for example, did not doubt for a minute that "the universe itself was made for the benefit of gods and men" and that everything that exists "has been provided and devised for us to enjoy." He explained this view using the example of domestic animals. "It would be a long story to recount the services rendered by mules and asses, but they were undoubtedly created for the use of men. As for the pig, what role has it other than to become our food? Chrysippus in fact remarks that life was bestowed on it to serve as [pickling] salt, to prevent its going bad." In a more general statement Thomas Aquinas (1225–1274; 1953, 153) summarized the concept of *teleological* anthropocentrism so graphically outlined in the quotation cited above: "No one commits a sin if he uses something for the purpose for which it was intended. However, with respect to the order of creatures, the less perfect ones are there for the use of the more perfect ones." Since nature does not produce anything that is useless or superfluous, so Thomas Aquinas argues, it must be true beyond a doubt that it created all animals for the use of humans. It is only a short step from this viewpoint to a cosmological perspective as delineated by Francis Bacon ([1619] 1968, 63) in the context of his interpretation of the myth of Prometheus: "If we look for the reason for certain purposes, man [can] be regarded as the center of the world . . . to such a degree that if he were removed from the world, what remains would appear completely confused, with neither goal nor purpose and to no good. Because the entire universe serves mankind, and there is nothing from which he cannot derive use and the fruits of his efforts." In order to counter any thoughts that worldviews of this kind are merely something of the past, I shall simply refer to a publication by Schäfer (1987, 27, 28) as one example of several similar positions representing a kind of *modern* teleological anthropocentrism. Here too humankind is expressly understood as "the final purpose of nature."

However, since the idea of humankind as the physical center and only real purpose of the universe has been dealt a serious blow by modern science, a new trend in worldviews can be observed, an increasing shift from a teleological perspective to one based on a third tradition, namely *epistemology*. More and more often ontological anthropocentrism is based on reference to the preeminent position allotted to man as a knowledgeable subject in modern epistemology. Thus Kant ([1756] 1985, 78), the most prominent founder of epistemology, explicitly rejects *teleological* anthropocentrism in his early writings. He maintains that the idea that everything in nature is there for humans is nothing but an expression of human prejudice and vanity.[120] Later on, however, in his *Kritik der reinen Vernunft* (Critique of Pure Reason), Kant ([1787] 1976) presents a theory of knowledge that not only retains humankind in a central position in the world but also paves the way to an even more strictly anthropocentric view of reality. A fundamental element of his approach, which he himself compares to the Copernican turn in astronomy (albeit with reverse effects) is the proposition that our knowledge is not centered on objects that exist as such but rather the other way around, that objects (as perceived by the senses) must be somehow generated in accordance with the nature of our capacities for perception. Kant regarded this as a consequence of recognizing that we can never grasp the world "as such" but only as an "appearance," that is, through the filter of our cognitive structures, a concept that to date has not been questioned. If we believe we have discovered some laws in nature, this is because what we perceive and experience has been organized in the form of mental categories by the defining characteristics of our cognitive apparatus (e.g., space, time, causality, etc.). In this sense we do not really *discover* laws of nature but rather *construct* them. Thus if nature is considered "the essence of all possible objects of human experience," then humans are, so to speak, the lawmakers of nature. Although Kant still assumed that beyond what we perceive as nature by virtue of experience there is a nature *as such* that exists as something independent of us, this hypothesis has been abandoned by post-Kantian philosophy. Since it seemed to make no sense to believe in a world *as such* that exists apart from the cognitive categories of space and time and moreover is completely inaccessible by empirical means, there was no recourse for philosophies such as German idealism, phenomenology, and existentialism than to accept the world as one of appearances, and this world, as noted above, is a construct of human thinking. Everything other than humanity no longer enjoyed the status of being independent reality in a literal sense of the word.

In light of these developments in the humanities and the views of the world ensuing from them, it is not surprising that many modern-day philosophers and theologians still regard nature as being oriented toward humans and to a certain extent even as a construct dependent upon their reasoning capacities. Thus referring to the fact that only humans are capable of understanding, respecting, and destroying ecological interactions Auer (1985, 175) says that "nature realizes itself only in humans. Only in humans does it fulfill its purposes. . . . That is why the idea that 'only humans are capable of representing the universe' does not constitute an 'ignorant and arrogant power position' [as Amery (1974, 211) has claimed], but rather the very determining factor of human dignity, through which alone nature is capable of attaining value and dignity."

By presenting a rough outline of various religious, philosophical, and epistemological roots of anthropocentrism as I have, I do not mean to intimate that the anthropocentric worldview has developed exclusively from these sources.[121] I also do not suggest that anyone who favors anthropocentric ethics also entertains premises based on the worldviews summarized here. However, if he *does not*, he will have to address the question as to how consistent it is to explicitly reject many thoughts that have led to the establishment of an anthropocentric worldview and still implicitly or explicitly adhere to the core idea of this position. It seems to me that only two positions are consistent in this matter. Either one essentially rejects the religious, teleological, and epistemological premises concerning the world and role of humankind outlined above; then one can hardly continue to uphold an anthropocentric worldview. Or one feels compelled to adhere to this worldview because one still finds a large number of these premises accurate; then certainly it is necessary to ask whether they really comply with what we know today about the world and about human beings. In the following I wish to attempt to show that these premises cannot withstand a confrontation with such knowledge. I shall support my arguments by drawing on knowledge from three disciplines: (1) astronomy, (2) evolutionary biology, and (3) ecology.

If thinking about "worldviews" is not restricted to thinking about "earth views," it is impossible to avoid dealing first and foremost with the results of modern *astronomy*. The necessity for considering this perspective becomes obvious if you consider the fact that hardly any other results have so seriously jolted human beings' ideas about occupying a privileged position in the world than Copernicus' discovery that the earth revolves around the sun, not vice versa. "As long as people followed Ptolemeus in believing that the earth lies at the center of the universe, it seemed reasonable to suppose that this centrality somehow carried over to humans as well. However, as soon as the earth came to be known as one star among others, beliefs about the outstanding uniqueness of human beings could no longer be upheld" (Landmann 1976, 71). Furthermore, in the meantime post-Copernican astronomy has demonstrated that human beings' hopes that at least "their sun" and together with it humankind itself might occupy a central position in space have also been disappointed. Not even our sun is at the center of the universe but rather in the outer regions of a giant spinning galaxy about 30,000 light years away from its center. This galaxy, the Milky Way, harbors more than 100 billion suns, a truly astronomical number.[122] As Ferris (1981, 2) notes, our spiraling galaxy is "so large and abundantly populated by stars" that no one "would feel disappointed had it proved to be the whole of the cosmos. But it is only one among many galaxies." According to Barrow and Silk (1983, 204) the entire universe is currently estimated to contain more than 100 billion major galaxies. In addition there are an unknown number of minor galaxies, which all together probably contain just as many stars as the brilliant major galaxies. Thus all in all the part of the universe that is currently observable contains more than 20 sextillion (2×10^{22}) suns, not to mention all the planets that possibly circle about them as well.

It is hard to imagine that these dizzying results about the universe could have no effect on humans' view of the world and of themselves. Certainly great thinkers of centuries long past already described the earth as a tiny point in the universe (e.g., Marcus Aurelius, 121–180; 1990, 24) and expressed "great awe at the infinite expanse of space" (Pascal, 1623–1662; 1993, 189). However, the results of modern astronomy have confirmed and enhanced these abstract and vague premonitions about enormous distances and multitudes of suns in a manner that far exceeds all (relatively well founded) speculations of the past. If anything can be rightly found to correspond to Hegel's idea of a quantity being converted to a new quality, then it is this. Our present knowledge about the universe has absolutely nothing in common with the old, intuitive, and still widely favored image of a "star-studded canvas stretched above us."

In the context of this book it is obviously not possible to even come close to sounding out all the consequences that these observations have for the philosophy of nature. The goal of this chapter, as cited at the beginning, is merely to *compare worldviews*, in other words, to address the rather modest question as to which perspective is more plausible in light of new knowledge about the universe, an anthropocentric one or a holistic one. I believe that in this case the answer is not too difficult. In view of the immense dimensions, the rich dynamics and the billion-year-old history of the universe, the anthropocentric idea that humankind represents the center and final purpose of nature is clearly the hypothetical worldview that is least compatible with empirical results. It is absurd to postulate that the 100 billion galaxies with their many billions of stars, comets, and nebulae are "geared toward human beings" (Auer 1985, 177) or perhaps owe their reality to some act of knowledge on their part. A perspective of this kind certainly might be able to satisfy human longing for significance and reassurance, but nowadays it can no more be considered an expression of humanity's efforts to perceive and earnestly interpret reality as comprehensively as possible. In pre-Copernican times it might have been plausible, but it can by no means meet the test posed by new empirical results.

A holistic position, on the other hand, is certainly much more convincing. It maintains that not only humans represent reality in and of itself but *all of nature* as well, even all the inanimate components of nature that predominate in the universe. Their existence cannot be explained only as a function for some part of the whole or an act of cognition performed by such a part, regardless of whether this part consists of humans (anthropocentrism), conscious beings (pathocentrism), or living ones (biocentrism). Their "center" is within themselves. According to Brennan (1984, 44), a characteristic of the autonomy of all that has been brought forth by nature, as opposed, for example, to the tools and artifacts that human beings have produced, is the "lack of intrinsic function." Of course it is obvious that all the things and processes of the world interact with one another more or less intimately and may therefore exhibit many different functions. Nevertheless, these are always *extrinsic* functions, or rather, *roles*. For example, the sun and the moon play a role *for us* by illuminating day and night, but it would be naïve to conclude that they are primarily or exclusively there for that purpose. They are primarily there *for themselves*.

The second scientific discipline whose results support a holistic world-view is *evolutionary biology*. Its significance is indicated by the fact that it is often called the "backbone" or "bracket" of the entire discipline of biology. The basic idea of evolutionary biology is summarized in the proposition that all creatures that have ever existed in the history of the planet, including human beings, were not generated as separate entities through a single act of creation but rather arose in the course of a series of reproductive acts as the offspring of slightly different but common ancestors. Although even the pre-Socratic philosophers Anaximander (611–546 B.C.) and Empedocles (492–432 B.C.) proposed elements of such a theory (Störig 1981, 127, 137), Darwin (1859) was the first to present evidence for species diversification (inconstancy) and to propose an explanatory mechanism. Since then support for the Darwinian theory of evolution has been provided by innumerous pieces of evidence (particularly from genetics and molecular biology), and in the past century the theory has grown into a complex edifice of ideas (see Wuketits 1988). Naturally, differences of opinion regarding specific questions (e.g., concerning the significance of various evolutionary factors and their explanatory power with respect to certain phenomena) still exist among biologists, but nevertheless, experts nowadays no longer entertain any serious doubts that the evolution of species and human beings is a historical fact. According to Wuketits (1988, 10) in this instance the evidence is so convincing that it can be considered a "closed case."

Of course, neither the unqualified approval that evolutionary thought enjoys in science nor its great heuristic significance can alter the fact that the concept of humankind's affiliation with nature inherent in the theory has entered the minds of the general public only hesitantly and in a somewhat skewed manner. In addition to popular misunderstandings and incorrect biologistic interpretations (as, for example, social Darwinism or certain varieties of evolutionary ethics), the current worldview of most human beings is still pre-evolutionary. Thus it is fitting that Meyer-Abich (1984, 94) refers to "a blind spot in the perception of our common natural world in industrial societies." The reasons for this tendency to ignore the phylogenetic relationship between human beings and the rest of nature are most likely psychological ones. For many the theory of evolution is probably still offensive, even 150 years after its first appearance. The idea that we are part of an ancestral line together with animals, plants, and bacteria and that all existing species are our relatives, in a manner of speaking, continues to meet with considerable internal resistance. Nevertheless, this idea is irrefutable.

Representative of the many examples from molecular biology, physiology, and ethology that demonstrate the close relationship between humans and other vertebrates is the evidence that the sequences of active genes in humans and chimpanzees are 99.6 percent identical (Sagan 1994, 31).

This example itself shows that the idea of a radical discontinuity between humans and the rest of nature, which the anthropocentric worldview often maintains, is fictitious from a biological standpoint. In some respects there may be a considerable degree of divergence between humans and other organisms, but in general the differences are neither absolute nor regular. Thus the evolutionary distance between humans and chimpanzees is obviously smaller than that between humans and paramecia. And with respect to almost all characters the distance between a paramecium and a chimpanzee must be viewed as much greater than that between a chimpanzee and a human being. The finding of such gradual degrees of difference not only makes any claims to an absolute distinction between humans and the rest of nature appear invalid. It also makes other kinds of absolute distinction (as, for example, the one postulated by pathocentrism on the basis of the phenomenon of consciousness) seem questionable from the very beginning. If one mentally draws a line extending from the paramecium to the clam to the earthworm, the herring, the robin, the chimpanzee, and finally to human beings, it is impossible to select a single point along this line that one could justify against all other options as defining as an *absolute* cut-off point, regardless of all the classification possibilities that exist. At the most one could make a case for the *end* of this line representing such a cut-off point, since it designates the temporary end of a branch of evolution. However, this interpretation would require us to ignore the fact that any line drawn perpendicular to the historical course of evolution that ends with human beings is basically a subjective construct of the human mind. From the standpoint of evolutionary biology there are no lines that converge exclusively at human beings, but instead, what one finds is simply a tree of life with its trunk, its many different branches, and a countless number of branch endings.[123] Like any tree the phylogenetic tree of species has neither a single center nor a single endpoint. At the most one could identify a starting point and many temporary endpoints. If one takes these results of evolutionary biology seriously, it would seem to me that one can hardly get around a holistic (or at least biocentric) view of life, according to which all species and the entire phylogenetic tree pose a reality of their own that is not derived from anything else. Either everything belonging to this tree is a "center" with no exceptions, or nothing is. Compared to this view, ontological anthropocentrism, which takes the opposite stance by declaring *a single branch ending* to be the center toward which all other branch endings are oriented and must be oriented in the future, has to be considered an ontologically overloaded and implausible interpretive model. I would guess that it can be reconstructed nowadays only in the context of pre-evolutionary thought structures.

This view is further substantiated if not only the *form* of the tree of life is take into consideration but also the *time* in which it evolved. A brief review of natural history should make this clear. Although the universe is currently estimated to be 15 billion years old (Sagan 1994), according to recent calculations the earth is thought to have been in existence for approximately 4.55 billion years. Life is believed to have existed on earth for at least 3.5 billion years (Eldredge 1991, 54, 55). However, life consisted merely of single cells (bacteria, algae) for all of 3 billion years, until the first more complex multicellular organisms arose 670 million years ago. Aquatic vertebrates have been shown to have existed for 450 million years. After life gradually succeeded in conquering land about 400 million years ago and then proceeded to produce the first extended forest ecosystems some 350 million years ago, it had to survive several devastating phases of mass extinction before bringing forth dinosaurs 245 million years ago. These creatures became extinct 65 million years ago (probably as the result of a powerful meteorite that struck the earth), and this then cleared the way for the mammals that entered the picture as early as 200 million years ago but were only able to occupy a marginal position in conjunction with dinosaurs. Finally, something more than 2 million years ago *Homo habilis* arose as a relatively recent species of mammal and the oldest known ancestor of the family of humankind. *Homo sapiens*, corresponding to human beings as we know them today, on the other hand, is "only" one hundred thousand years old, whereby in general only the last ten thousand years are counted as the "history of humanity or human culture."

It seems to me that the results of paleontology briefly outlined above lead to three insights that all speak quite clearly against an anthropocentric worldview. The first is that one view that has remained characteristic of the anthropocentric position up until today must be considered untenable, namely the idea that nature constitutes a kind of *ahistorical* machinery made up of deterministically regulated objects that can be readily replaced and that the dynamics of nature, contrary to those of human history, represent a kind of "eternal recurrence of the same."[124] The opposite is true. If one looks closely and applies the right scaling system, nature, like culture, is a *historical process* that has always brought forth new and unique things in the past and will continue to do so in the future. In the course of evolution probably all together more than a billion different species have arisen (Mayr 1988a, 72)— each one a unique and irreproducible design of nature. In my opinion the thing that best demonstrates the enormous "creative" potential of nonhuman nature and should by no means be overlooked is the fact that in the end it is nonhuman nature from which human beings with all their unusual mental capacities and cultural possibilities emerged. A second important insight, which paleontological results disclose, is that when human beings "arrived on the scene," they did not find a barren planet just waiting to be "worked over." Instead a richly developed flora and fauna organized in complex, well-adjusted systems already existed on all the continents. Taylor (1981, 207) remarks on this as follows: "The earth was teeming with life long before we appeared. Putting the point metaphorically we are relative newcomers, entering a home that has been the residence of others for hundreds of millions of years." In view of this image, *reason* seems to almost force us to assume "that this home must now be shared by all of us together." The third insight is finally that the fraction of time that civilization has occupied in the entire history of nature is so very marginal that it would require a lot of help from mythology to still regard humans as the "center of nature." Just how effective paleontological results, in addition to those of astronomy, are in promoting the proverbial conversion of quantity to quality becomes evident when the history of nature is presented in the form of a so-called "planetary calendar" instead of in absolute numbers (see Mayr 1988a, 69; Zahrnt 1993, 115).[125] Here the 4.55 billion years of planetary history are projected onto the dimensions of a year. If we assume that the earth came into being on the first of January, then it wasn't until the 27th of February that the first organisms (procaryotes) appeared, and the first unicellular organisms with a true nucleus (eucaryotes) didn't develop until the 1st of September. The first vertebrates (fish) appear on the 21st of November, mammals on the 12th of December. On the last day of the year at 7:30 P.M. the oldest known ancestor of the family of human beings enters the picture. At 11:57 P.M. his successor, *Homo sapiens*, follows, who proceeds to colonize all the continents within a fraction of a minute. The beginning of the modern period, heralded by the rise of modern science, takes place two seconds before midnight, and the industrial revolution occurs half a second later. Modern technical civilization so familiar to us with its cars and colored TVs spans a period of less than a single second.

To me it seems that the inappropriateness of an anthropocentric worldview can hardly present itself any more vividly than in this time-lapse scenario. If one namely advocates an anthropocentric interpretation of the planetary calendar, then one must in all seriousness assume that the *real* history of the planet didn't begin until the last three minutes of the planetary year. The entire twelve months of the planetary year that preceded the 31st of December were merely a preliminary period in which the material and biological groundwork was completed for the history of humankind. Thus with reference to a quotation from Karl Rahner, Auer (1988, 36) maintains the following: "Only human beings with their capacity for reason and their autonomy are capable of making sense of the process of evolution and giving it direction." Does this mean that 4.5 billion years of planetary history, or rather 99.998 percent of the entire time in which the universe has existed were of no value and purpose and would have remained so if human beings had not come about? In view of the temporal and spatial dimensions described above it is almost impossible to accept such an incredible conclusion. In order to get around it, anthropocentrism has no other choice than the no less unbelievable option of assuming that in the context of *teleologically* interpreted evolution humankind always was and is the hidden and ultimate goal of the preceding millions of years in which life developed. If cosmic history cannot be accepted as an end in itself, it would at least have some indirect meaning and purpose in having brought forth humankind sometime or other. However, in light of modern scientific results demonstrating the fundamental nondirectionality of evolutionary processes (Mayr 1988b, 235), a finalistic interpretation of evolution of the kind proposed by Teilhard de Chardin (1961), for example, with his theory of cosmogenesis oriented toward humankind (the "omega principle") can no longer be maintained. It would require the additional assumption that with the creation of human beings evolution has finally reached its ultimate goal and is therefore more or less finished. No evidence can be found to support this speculation. On the contrary, one can reasonably assume that evolution is continuing and that human beings will one day meet with the same fate as any other species and become extinct, either because of biological circumstances or due to a cosmic or geological catastrophe.[126]

Would evolution without human beings then once again be meaningless and nature wilderness of no value or purpose? If you recall Routley's

(1973) "last-people example" (see Chapter 23.c), it seems clear that a holist would negate this question while an anthropocentrist would reply affirmatively. However, I doubt whether the anthropocentric position can be considered coherent. It seems to me that the assumption connected with it that (in a manner of speaking) meaning and purpose emerged out of nothing with the birth of humankind and will disappear into nothing when human beings die out is less plausible than the alternative assumption of *continuous* meaningfulness or meaninglessness. This idea is supported by thoughts that Spaemann (1990, 153) expressed when he wrote, "Anyone who regards reality in its entirety as the meaningless existence of *facta bruta* cannot suddenly regard it as meaningful when in the course of evolution some creature opens its eyes and in that very act this universal meaninglessness becomes aware of itself. Schopenhauer saw things correctly. In this case conscious life is only absurdity to the nth power. Thus the reverse is true. If there is meaning in the process of becoming aware of existence, that is, in consciousness, then this meaning must have been there before the awareness of it." In connection with the problem of the "meaning of life" (in more than a subjective sense) Nagel (1987, 98) comes to an analogous conclusion. It would only seem legitimate to refer to the meaningfulness of some *part* (i.e., individual life) if one also assumes that the *whole* (social relationships, humanity, the planet, the universe) is meaningful as well.[127]

A modern advocate of anthropocentrism will be tempted to reject these thoughts as "speculative metaphysics" and claim in return that her worldview does not require any help from the philosophy of nature because it is based solely on *epistemology*. That's all very well. Nevertheless, she should be aware that epistemological anthropocentrism leads to an interpretation of natural history that is no less disconcerting than the teleological one discussed above. If human experience is considered the *only* reality about which we can speak meaningfully, and there is no such thing as "a thing in and of itself," then this must mean that nature "came into being" with the first cognitive act of a human being. Cobb (1972, 105) pointed out the paradox of this idea: "It would mean that all the events that produced the planet and started the evolutionary development that culminated in man became real abruptly, either when man emerged or when man first learned about them. But the whole thrust of evolutionary theory is just the opposite. It teaches that there were billions of years of real occurrences, that man appeared very late, and that no sharp line separated him from his animal ancestors." I have to agree with Cobb (1972, 105, 106) that philosophy can no longer simply ignore this theory about the position of human beings in the universe that science has generated. If philosophy is really interested in resolving the paradoxical consequences of an anthropocentric conception of reality, it will have to incorporate the results of evolutionary biology in its system of thought. It will have to take into account that "what existed before man, and before the sensory perception of animals, was already something for itself" (Cobb 1972, 107).

The repercussions that this would have for worldviews is obvious. If nature must be granted the status of "being something for itself" *before* the advent of human beings in order to avoid the paradox mentioned above, then it is impossible to deny *presently* existing nature the same subject quality. The nonhuman world surrounding us must be accepted as reality as such and not as something exclusively oriented toward us (or some other thing). Nature is not only "environment" and a resource for human beings and sentient animals but rather common reality that exists *in and of itself*. This conclusion reveals how the results of evolutionary biology not only shed light on nature's *past* but also elicit far-reaching consequences for our understanding of nature *in the present*. These results reveal the untenability of the idea that still (at least implicitly) underlies current anthropocentrism that nature is not real *in itself* but only real to the extent that it functions as an object of human cognition or use. There may have been times when this idea was able to lay claim to a certain amount of plausibility, but since the rise of modern astronomy and evolutionary biology these days are over. The American astronomer Sagan (1994, 52) remarks on the old-fashionedness of ontological anthropocentrism as follows: "If you lived two or three millennia ago, there was no shame in holding that the Universe was made for us. It was an appealing thesis consistent with everything we knew; it was what the most learned among us taught without qualification. But we have found out much since then. Defending such a position today amounts to willful disregard of evidence, and a flight from self-knowledge."

This, of course, raises a question that already cropped up in connection with the theoretical problem of how to make a choice between different worldviews, namely, the option of *choosing to ignore*. One can either ignore empirical facts about the world or deliberately play them down in order to prevent them from destabilizing one's own familiar or comfortable worldview. In Spaemann's words, one can refuse to *awake to reality*. The somewhat resigned resumé of White (1967, 1206) indicates that this option is not only one that turns up once in a while among individuals but apparently represents a common collective process. He writes: "Despite Copernicus all the cosmos rotates around our little globe. Despite Darwin, we are *not*, in our hearts, part of the natural process." How can we explain this? In addition to the fact that these new results are difficult to grasp, another reason is probably that it is quite pleasant to live as if we had been placed in a nicely tailored garden in which it was left to us to do whatever we want with all the other living things around us. Why should we bother about the results of astronomy and evolutionary biology if these only serve to contest our privileged status in the world? What does everyday existence have to do with the dimensions of our own galaxy or the "planetary calendar"? It isn't easy to respond to these questions, because any response capable of being the least bit convincing requires a certain amount of interest in reality and self-knowledge on the part of the inquirer. If this interest isn't there, the discussion is useless. The only recourse is "final" argumentation similar to what was already proposed in favor of the moral standpoint. "You can go ahead and continue to regard the world with blinders, but you'd better think twice about whether you really want to deal with the consequences of consciously refusing to accept reality."

An argument of this kind would probably not have much of an effect if it only were based on the results of astronomy and evolutionary biology. However, the odds are better if a third discipline is brought to bear, *ecology*. Refusing to accept reality doesn't incur any serious penalties from reality if only facts from astronomy or evolutionary biology are at stake, but refusing to accept reality with respect to *ecological* facts does. In the past years the ecological crisis has repeatedly forced humans to recognize this. It has become clear that every major intervention in an ecological system that is performed without taking its autonomous regulatory rules into consideration will have direct or indirect repercussions for humans and their civilization. The idea that the world of autonomous humankind and the world of natural objects are *in principle* separate from one another, which for a long time was an unquestioned tenet of the anthropocentric worldview, has thus been drastically refuted. Contrary to this view of humanity as a "closed society," ecology has clearly shown that the entire biosphere must be regarded as a complex, hierarchically structured *community*, in which humans are integrated along with all other living things and the abiotic world via numerous different interactions. "We all live in systems and are parts of them" (Kreeb 1979, 140). Moreover, human beings' claim to a privileged role within these systems cannot be confirmed by ecology. On the contrary, as a species occupying a position in the upper section of the food pyramid humans are dependent upon other species, especially the producers and decomposers at the bottom of the pyramid, while the functioning of the system and the well-being of the other species only rarely depend upon the ecological contribution of humans. From an ecological perspective it can be said that human beings need an appropriate biosphere for their survival and well-being, but the biosphere doesn't need them.

A modern advocate of an anthropocentric worldview will not contest such results of ecology. He will no doubt recognize the many connections that exist between humans and other species and how we are integrated in higher-order systems, but at the same time he will point out that humans differ from all other species in their ability to obtain *insights* into these relationships. These insights enable and empower us to assume a *central* position in the biosphere, in spite of being integrated within it. It is high time for humans to assume the role of the captain of spaceship earth, he will say. What we need nowadays is "biosphere management" (Markl 1995, 207). In my criticism of technical optimism (see Part A.I) I have already gone to great lengths to demonstrate that such hopes for perfect control of nature and the self-confidence of humans that goes along with them are not only unrealistic but even dangerous. The evidence summarized in that section reveals that the epistemological capacities of the self-appointed captain would be greatly overtaxed by such a task, and furthermore that sooner or later spaceship earth would withdraw from such attempts to steer it. One of the reasons for this was shown to be the fact that the biosphere and its ecological subsystems are *decentrally* organized. Natural ecosystems and their organismic communities never develop by means of centralized regulation. Instead they regulate their functions solely via the interactions of their components, compartments, and factors. On the basis of this structural property of nature any attempt to govern higher-order systems by means of centralized regulation appears extremely dubious from the very beginning. It is fitting that Kornwachs and von Lucadou (1984, 153) warn us against regarding ecosystems as "entities that can be reliably influenced." For self-organizing, open systems they recommend instead always assuming "a certain amount of autonomy." Consequently they arrive at the same conclusion that was already outlined under the headline "ecological thinking" in Chapter 15. If humans wish to contribute toward establishing favorable conditions for further development on earth without running the risk of an ecological catastrophe, they will have to abstain as much as possible from direct intervention through design and planning without compelling reasons. Instead they will have to allow nature autonomous dynamics to as great an extent as possible. In the long run this strategy will produce the best results, at least according to most ecologists.

In the context of ecological results the anthropocentric view of humankind and the world has once again proven to be wrong. Both the traditional idea that nature *is somehow centered* around humans and the forcefully propagated claim of supporters of technical optimism that nature *should be centered* around human needs are not compatible with current knowledge about the decentralized, self-organizing structure of ecological systems and processes. What is absolutely wrong with this perspective is the old-fashioned way it contrasts humans with their environment. Himmelheber (1974b, 98) has emphasized the "short-sightedness of this standpoint": "As creatures of nature human beings are irreconcilably integrated in the biotic cycles of the planet; there is no utility or damage that is only of significance for them, no environment that is solely there for them." In light of this, the popular concept of so-called "human ecology," as honorable as its intentions might be, must be viewed with a great deal of skepticism. In this respect Himmelheber points out that the perspective of isolated human ecology is not only dubious from a *theoretical* standpoint but also involves a serious *practical* risk as well, namely, the risk of "miscalculating the problems that exist and deducing incorrect measures." In view of this practical risk Küppers (1982, 73) recommends that "we by all means disengage ourselves from an ecology informed by an anthropocentric standpoint that tends to separate the biosphere into humans and the rest of the biotic environment." The unity of all living things must be restored "through readjustment to ecological 'balance' by humans." According to Küppers history shows that "incidences in which we have turned away from anthropocentric worldviews belong to the greatest advances in cultural development."

It is now not difficult to recognize that extending an anthropocentric worldview to a *pathocentric* one can certainly *not* bring about the progress that Küppers hoped to see. The major fault inherent in anthropocentrism, the dichotomy between a postulated center and its supply of resources, is maintained in pathocentrism almost without any discrepancy. The only difference between the two positions is that in the first case a single *species* (humans) is declared to be the "epicenter" of nature while in the second a single *phylum* (essentially vertebrates, representing less than 3 percent of all species) is assigned this role. If one realizes that in doing this one accords the most important organisms for life on earth (namely plants and microorganisms) only a very marginal role in things, then one can hardly regard the pathocentric worldview as an adequate foundation for dealing reasonably with the complex systems of nature.

The *biocentric* worldview also proves to be inadequate in this connection. Even though it is more "ecologically enlightened" in that it closes the gap between "higher" animals and all other living things, it still retains two basic flaws of the pathocentric worldview. In the first place it remains a captive of the old dichotomy between a center and its periphery by granting a reality of its own only to *animate* matter but not to *inanimate* things. However, drawing a sharp dividing line between animate and inanimate matter is not only questionable from the standpoint of evolutionary theory. After all, life emerged from inanimate matter in the course of a long process. It also fails to take the fundamental role of geochemical factors in the self-organization of ecological systems into account. As Lovelock (1988, xvi) has shown, the biosphere of our planet is the product of the coevolution of living organisms and their material surroundings and both parts are tightly and inseparably coupled with one another. According to Lovelock (1988, 34) the evolution of rocks and air and the evolution of living things simply cannot be understood independently of one another. "There is no clear distinction anywhere on the Earth's surface between living and nonliving matter." While this remark is not meant to question the possibility and significance of attempts at classification altogether, an *absolute* discrepancy such as that implied by the biocentric worldview seems to be at least difficult to justify.

In addition, the biocentric worldview also seems to ignore a piece of information that is very fundamental to ecology, the fact that nature cannot be regarded solely as a collection of *individuals* but rather also as a community of hierarchically structured *wholes*. In spite of the observation that ecological wholes are characterized by so-called "downward causation" (see Chapter 7) and thus quite obviously represent more than the sum of their parts, such wholes have no reality of their own in the biocentric worldview, nothing except that defined by the sum of the individuals of which they are made up. Since I have already criticized this atomistic and individualistic view from a *pragmatic* standpoint in Chapter 24.c, I shall only discuss a *theoretical* objection at this point, the objection that wholes are simply constructs of the human mind. In a radical epistemological sense, of course, this objection is correct (see Chapter 25.b). However, in this case it not only applies to the existence of wholes but to the existence of their parts as well. Thus not only holistic and synthetic representations of nature are constructs but reductionist and analytical ones as well. Apart from this basic contingency of all knowledge, the victory of quantum theory over classical physics also provides good reasons to regard not the *parts* of a system as "original" but rather the *whole*. C. F. von Weizsäcker (1991, 25, 26) demonstrates this with two examples. "The hydrogen atom is a whole. It is not really made up of a proton and an electron. It is simply relatively easy to destroy it in such a manner that a proton and an electron remain. Strictly speaking an ice crystal is also not composed of water molecules. It can simply be broken down into them. And the world is not comprised of objects. It is only the limited mind of humans that separates the whole to which they themselves belong into objects in order to cope with it." In this sense Plato (427–347 B.C.; 1990, 80) seems to be right when he allows the "visitor" in his dialog *Sophistes* say, "What becomes always becomes as a whole. Therefore, one must not proclaim that there is either being or becoming if the whole is not put down as among the things that are." As this quotation readily reveals, the holistic worldview is not really new, but it has gained a great deal of plausibility through the results of modern science. It seems to me that the evidence from astronomy, evolutionary biology, and ecology presented here makes it sufficiently clear that of all four worldviews upon which the fundamental positions of environmental ethics are based the holistic one is the most convincing.

After having discussed worldviews and views of humanity exclusively from the perspective of science, at the end of this chapter it is important to point out that a *complete view of humanity* cannot be drawn solely from scientific evidence. Human beings are creatures of mind as well as of nature, and this combination makes them unique products of nature. No other being except for humans is at the same time both part of nature and capable of removing himself from it by virtue of his consciousness (Fäh 1987, 54). In order to do justice to this "double nature" of humans, results from both the natural sciences and the humanities must be consulted when attempting to explain it. The ecological crisis can really only be understood in the context of *both* dimensions, and it can only be mastered under the assumption that human beings are not simply governed by nature but are rather reasonable beings (within certain limits). It follows that it would be a grave mistake to ignore the mental dimension of humanity simply because it cannot be grasped accurately by empirical means.[128] In Chapter 7 it was not a coincidence that I criticized making scientific evidence the *final measure* of everything or reducing what we know about the world to what can be "positively proven" as examples of dogmatic scientism. The reason I have not dealt with the results that the humanities have produced concerning humankind's role in nature in this chapter is that I regard these results as compatible with all four worldviews discussed here. The idea of a unique position of human beings in the world developed by philosophical anthropology is not contested by a holistic worldview as I understand it. However, the aspect of our view of humankind that is indeed altered by modern scientific evidence is the *significance* attached to these results. Thus from the standpoint of holism the unique status of human beings can no longer be regarded as a convincing argument that humans are the *only* things in nature that are substantially real. The rule of thumb of holism is uniqueness, *yes*, center of the world, *no*.

It would be beyond the scope of this treatise to elaborate in detail the unique position of human beings. Instead I shall refer the reader to appropriate literature in philosophical anthropology.[129] Among the characteristics listed there that define the uniqueness of human beings (such as language, cosmopolitan perspective, eccentricity, biological insufficiency, etc.) two seem to be of particular significance for the purposes of this book: *autonomy and morality*. Human beings are the only creatures to which (conditional) free will and the capacity for self-government can be attributed and which therefore are able to and obligated to assume responsibility for other humans and for the nonhuman world. With this observation, which must be regarded as the indisputable premise of any ethical discussion, the line of thought of my inquiry will return to the problem of determining exactly *which* objects of nature can be considered objects of direct moral responsibility. Is the possibility of direct moral responsibility restricted to certain parts of nature, or is it possible to rationally justify a holistic standpoint that includes *everything* that exists in nature within the scope of moral consideration?

28. Justification for Holistic Ethics

28.a. The Universal Nature of the Moral Standpoint

In order to pursue the question of a justifiable scope of *environmental* ethics it would seem to be informative to first analyze how the current scope of *interpersonal* ethics is justified. Which arguments can be proffered in support of universal human rights as opposed to a morality of class distinction, the scope of which is restricted to a certain reference group or exclusively to the bearers of certain traits? Tugendhat (1994, 93) justifies this kind of "morality of universal respect" with a formal analysis of the concept of morality. He maintains that formal analysis shows that the concept of morality very automatically leads to Kant's categorical imperative. He explains these thoughts to a fictitious discussion partner by inviting him to imagine the two basic options that are possible in matters of morality: egoism and altruism. "Imagine that you have reached a fork in the road. One path leads to egoism. The convinced egoist acts exclusively according to the maxim, 'I will do only what pleases me.' . . . It is not that the egoist has no relationship whatsoever to fellow humans, but the relationship is a purely instrumental one. Other people serve as a means for satisfying his needs, and it follows that he sees himself exclusively as a person of power in his relations to others." According to Tugendhat, the other path, the alternative to egoism, is altruism. It means that we are considerate of others as well, but not just whenever it pleases us. What is decisive is that the altruistic alternative to egoism can in principle not be discriminating. "You see, to the extent that you might indeed determine which of your fellow companions you consider and which ones you don't, you would be acting according to your own discretion, that is, according to your egoistic perspective, from a position of absolute power, to determine the circle of those who are to be respected. That is why the alternative to egoism can only be to respect each and everyone whatsoever. But that is exactly the content of the categorical imperative." In the so-called second formula Kant ([1785] 1965, 52) expressed it as follows: "So act as to treat humanity, whether in your own person or in that of any other, in every case as an end, never as means only." Tugendhat (1994, 80) believes that this rule can be summarized in the imperative "Never regard anyone an instrument of your ambitions!"

As convincing as Tugendhat's argumentation in favor of a universal and egalitarian morality of interpersonal relationships is, it still cannot conceal the fact that it does not satisfy its claim to be simply a *formal* analysis of the concept of morality. As far as the question of scope is concerned, both Tugendhat's "morality of universal respect" and Kant's "categorical imperative" are clearly grounded in a preliminary decision *with respect to content*. Both intend the moral community to be limited *a priori* to beings capable of reason (Kant) or rather cooperation (Tugendhat). However, this preliminary decision, which means that only humans can be considered moral objects, contradicts the results of the analysis, namely, that a restriction of this kind is incompatible with a moral standpoint. In my opinion, the fact that Tugendhat (1994, 193, 376f.) does indeed perform such a discriminatory act by selecting the ability to cooperate as a defining criterion can only be understood in the context of a traditional anthropocentric worldview that regards humanity as radically separate from the rest of the world. In this two-class world, ethics is *from the very start* something that only applies to humans. After having demonstrated in the previous chapter that an anthropocentric worldview is clearly no longer acceptable, a concept of ethics as something that is inevitably anthropocentric automatically also ceases to be completely self-evident. In our day and age limiting the moral community to members of the species *Homo sapiens* is implausible because it fails to recognize three aspects of the relationship between humans and nature.

First of all, it ignores the fact that humans are part of an *ecological* community together with other living creatures, species, and systems. Contrary to what people until recently used to be able to talk themselves into, humankind does not exist as a closed society of human beings but rather in intimate association with animals, plants, ecosystems, water systems, and soil systems. In this community all members are reciprocally more or less dependent upon one another. If like Leopold ([1949] 1968, 202) we regard ethics in very general terms as "the tendency of interdependent individuals or groups to evolve forms of co-operation," it seems obvious to include everything that is potentially part of this community in these efforts. Leopold ([1949] 1968, 203) considered this concept of an ethics extended to include nature not only as an obvious possibility in the history of ethics but flatly as an "ecological necessity." The ecological crisis has now emphatically confirmed this estimate articulated more than fifty years ago. If ethics hopes to measure up to the new challenges posed nowadays, it has no other choice than to adjust the normative area of its conceptual edifice to the new knowledge in the area of description. Since we now know that we live not only in a social community but also in an ecological one, there is no plausible reason to exclude nonhuman members of this community from moral consideration *a priori*.

A second aspect that an anthropocentric concept of ethics ignores is that humans are part of a *phylogenetic* community together with all other creatures on earth. As evolutionary biology has shown, humans emerged from nature and are therefore more or less closely related to all other species of our planet. In a figurative sense the other species are our phylogenetic cousins, nephews and nieces, uncles and aunts. It was in this sense that Jimmie Durham (quoted in Pister 1979, 348), director of the International Treaty Organization and of native American origin, expressed the following views in a hearing about the extinction of the endangered snail darter in 1978: "To me, that fish is not just an abstract 'endangered species,' although it is that. It is a Cherokee fish and I am its brother." In light of what we now know from evolutionary biology, there is no reason to dismiss the feeling of unity with other forms of life expressed in this statement as mystical irrationality. Instead it seems to give pause to reconsider a concept of rationality that finds it necessary to explicitly exclude beings that are phylogenetically related to us from moral consideration.

The third thing that an anthropocentric concept of ethics ignores is that human beings are not only part of a community of other humans but that together with nonhuman nature they are part of a community of *all that exists*. Humankind has in common with all of nature the existential situation of being "thrown into the world," as Heidegger ([1927] 1977, 134f.) would say. What this metaphoric expression intends to say is that we are situated in this world within certain temporal and spatial coordinates that are far removed from our intentions, and that we do not know why we exist here and right now and in this particular form. Pascal (1623–1662; 1993, 189) expressed the mysteriousness of these existential facts in an inimitable fashion.[130] While it is true that humans are the only creatures that are capable of being conscious of their existential state of being "thrown into the world," this does not alter the fact that animals, plants, rocks, species, oceans, and the like also are in such a state. They too are characterized by distinct coordinates in a spatial and temporal continuum, by a historical state of existing as such that amounts to much more than simply being useful for something else. Once one realizes this, it is impossible to understand why human beings should not assume responsibility for nonhuman existence *from the very beginning.*

What conclusions can be drawn from this for argumentation in favor of universal morality? If Tugendhat's *formal* analysis of the concept of morality is correct in assuming that a moral standpoint automatically leads to universality, and if *with respect to content* his preconception of morality as something limited to human beings can no longer be considered justifiable, it is clear that his thoughts demand a more comprehensive kind of universality than one that only includes *humans.* If pursued rigorously they inevitably lead to pluralistic holism. As Tugendhat (1994, 93) rightly emphasized, a moral standpoint must not be discriminating. Once I have opted for the path of altruism, I may not on the basis of my egoistic perspective determine whom or what is to be included in my scope of moral consideration. This would correspond to a "path of power," not a "path of morality." Therefore, logically thinking the alternative to egoism must be to respect *all things without distinction.* A moral standpoint must mean that everything that exists in itself and "for itself" is included in the scope of moral consideration, that is, all such things must be respected for their own sake. It follows that Kant's categorical imperative would have to be reformulated: "Act so as to treat everything that exists in every case as an end, never as a means only." This principle of practical reason means the following: "No part of reality may be reduced to the status of being regarded exclusively as a means for attaining some individual goal unless its existence is defined by such a function from the very beginning" (Spaemann 1990, 227). Nothing that has not been made by humans is *only* a resource. All other things have intrinsic value and must be respected as ends of their own.

The terms "intrinsic value" and "end of its own" that I have repeatedly used in this treatise obviously require more detailed explication. To put it briefly they mean that something is not just a means for or of value for something else but rather valuable or purposeful *in itself.* Thus they do not refer to something to be aimed for or to be realized but rather describe "something that from the very beginning always forms the fundament of any kind of realization" (Spaemann 1990, 124). Granted, this sounds quite abstract, but that's not surprising. The reason is that concepts like "intrinsic value" or "end of its own," like the concept of "moral good," are monadic predicates and as such can neither be functionally deduced nor defined in terms of something else. If advocates of anthropocentric ethics find fault with holistic arguments in favor of species protection for being obscure about "what the intrinsic value of a species really means and what criteria can be used to define it" (Wolters 1995, 248), their criticism addresses this very difficulty. Since the term "intrinsic value" as opposed to the term "instrumental value" is nonreferential, it *cannot* be explicated in a satisfactory manner or defined unequivocally. What this criticism often fails to see, however, is the fact that this insufficiency not only applies to the controversial concept of intrinsic value of species but also to *any* kind of intrinsic value postulated by ethics, including the intrinsic value of humans, which is usually taken for granted. Intrinsic value of this kind cannot be defined satisfactorily either, that is, without the easily contestable deduction of values from facts. Thus Katz (1987, 240) seems to be right in his estimate that the meaning of the term "intrinsic value" is best specified negatively. The term must be regarded as a symbol of the ethically unavoidable conclusion that exclusive reference to instrumental value is inadequate. Ethics cannot avoid requiring that *another*, noninstrumental kind of value exists above and beyond instrumental value and belonging to the category of primary qualities, even if these qualities can only be described in abstract or indirect terms.

In addition to the limited possibilities that exist for describing the concept of intrinsic value, this term, as opposed to the term instrumental value, is also unusual in that it cannot be quantified or calculated in any manner *a priori*. When Kant ([1785] 1965, 58) writes that human beings "do not simply have relative value, that is, a price, but rather inner value, that is *dignity*," it is exactly this characteristic of the term intrinsic value that he addresses. With this remark he denotes all that is incalculable, sublime, and unconditionally worthy of respect, that to which in the end the idea of moral responsibility toward other humans refers. While Kant and traditional anthropocentric ethics clearly restricted the term "dignity" to humans and furthermore used it to distinguish humans from "brute creatures," in holistic ethics this kind of distinction doesn't seem to make much sense anymore. As a result, in more recent publications by ethicists who argue in nonanthropocentric terms one often finds expressions such as "the dignity of the creature" (Teutsch 1995), the "dignity of what has come into existence over a long period of time" (Ruh 1987, 133) or "the very own dignity of nature" (Jonas 1984, 246). One objection that is often voiced is that by stretching the concept of dignity in this manner we will eventually define more and more different things as being the same (Schlitt 1992, 181) and thus possibly undermine the dignity of human beings. Teutsch (1995, 34) counters this objection by maintaining that extending a concept doesn't automatically exclude differentiating within it at the same time. Thus reference is usually not made to some kind of abstract dignity as such but rather to "human dignity or the dignity of animals or living things or nature." Specifying the meaning of the term in this manner permits us "to appropriately address both what is common and what is different."

In spite of all that is obviously different, what is common about the dignity of humans, animals, plants, ecosystems, species, oceans, and cliffs? As the discussion of the concept of intrinsic value has clearly shown, because of the nonreferentiality of the term "dignity" this question can only be answered in an indirect, descriptive manner with the help of imagery. I tend to agree with Sitter-Liver (1994, 150), who maintains that what all uses of the term "dignity" quite clearly have in common is the "experience of the final inaccessibility of everything that exists." This is based on the existential insight discussed above that we humans as well as animals, plants, ecosystems, species, oceans, and cliffs always have and always will find ourselves in a state of having been "thrown into the world." They and we were already there or have come to be; we were neither brought here nor manufactured by other humans. As Sitter-Liver (1994, 150) has pointed out, "[human beings] lack the power to bring anything original into being. Whatever they do, they remain dependent on employing something they can manipulate, alter or rearrange." This insight also applies in cases in which humans restructure landscapes, breed organisms, or, more recently, produce transgenic organisms by means of gene technology. "Whatever exists never stems completely from humans but is always rooted in some other form of existence." If one recalls the proposition of the "verum-factum principle," according to which we can only *know* an object "to the extent that we are able to *make* it" (Habermas 1973, 32; Vossenkuhl 1974), then in the long run the existence of this other being that has not and cannot completely be made by us must remain obscure. What that old oak tree on the edge of the path *really* is must remain a secret, in spite of (or perhaps, more accurately, *because of*) all the basic biological information we have accumulated. We do not really understand the "nature" of such a tree. It is this incomprehensibility, this "condition of having to rely upon another" that gives not only the aesthetically appealing oak tree but *everything that exists* special dignity, "a final uniqueness that eludes human manipulation, in its self-referentiality a value of its own" (Sitter-Liver 1994, 50).

According to Burkhardt (1983, 417, 418) dignity of this kind carries with it an obligation to deal with nature as carefully as possible. "We tend to grant any person the right to destroy the things that he has produced himself and which he owns. That means that only a producer of something who is its sole owner at the same time has the right to destroy it or alter it." According to Burkhardt the fact that practically all religions regard the Creator as the owner of the earth shows that it is quite plausible to interpret the production of something as a criterion of ownership and rights of disposal. "All possible reservations aside, this is the least that one can learn from most religions. Humankind has no right to destroy what it has not created itself." The richness of nature that humans *came upon* must be regarded as inaccessible, at least in the first instance. Since the distinctiveness of this richness is not reproducible, it may not be diminished without good existential reason.

It is obvious that a concept of ethics that is so far-reaching that no area of nature is excluded from human responsibility presents problems. Because of this, holistic ethics, as young as it is, has kindled a great variety of criticisms. I shall deal with the most serious *practical* objections in more detail in Chapter 31. Right now I would like to discuss a prominent *theoretical* objection since it permits me to return to the argumentation in favor of universal morality presented at the beginning of this chapter and to substantiate my appeal for holism. According to this objection, articulated by Wolters (1995, 249), my approach lacks internal coherence. "When practically *everything* has intrinsic value, then in the long run *nothing at all* is of value. Just like everyday life morality requires making distinctions."

At this point there is no need to argue about the fact that acting in a moral fashion requires making distinctions. But it is another matter to maintain that a prerequisite for making these distinctions is *the absence of value* in certain areas of nature. This claim seems implausible, because it leads to the assumption that intrinsic value is, so to speak, inversely proportional to the frequency with which it occurs. Wouldn't it follow that the concept of personal dignity must have become more and more meaningless in the course of history as more and more human beings have laid claim to it and that nowadays *no* human beings possess dignity anymore, after all of them have attained it? One has to exaggerate Wolters' argument in this manner in order to draw attention to its inconsistence, which, in my opinion applies not only to his objection but also to any nonuniversal concept of ethics. From tribal morality (level 2) to biocentrism (level 7) (see Chapter 25.a) the basic assumption is that the world consists of a core area that is valuable as such (an end in itself) and a peripheral area that is valueless as such (a means to ends). Moral behavior toward what is valuable as such and unlimited exertion of power over what is valueless as such not only go hand and hand. One necessitates the other, something like the case of a national park whose exclusive protected status to a certain extent benefits from the ruthless exploitation of the land surrounding it. In a detailed analysis Birch (1993, 315) showed that this traditional view of ethics is founded on four premises. First, it means that with respect to any moral community there are and always must be insiders and outsiders, citizens versus noncitizens (slaves, savages, women, etc.), members of the "club of consideranda" versus the rest. Second, we can and must determine the criteria of membership. Third, we can establish these criteria in a rational and nonarbitrary fashion. And fourth, we must establish rules and customs that both uphold the criteria of membership and the intactness of the "club" and increase the good of its members.

One readily recognizes that all four premises are very dubious. If you take a brief look at the long list of criteria for moral considerability that are currently being negotiated on the philosophical market,[131] and if you also recall the history of ethics, in which more and more supposedly objective and definitive criteria were eventually exposed as chauvinistic and thus became subject to alteration at a later date, you will find little reason to trust in the success of being able to determine the boundaries of the "moral park" in an impartial and unprejudiced manner. Too great are the indications that *interests which guide the formation of knowledge (erkenntnisleitende Interessen)* are consciously or unconsciously involved. If the universal character of the moral standpoint such as that developed by Tugendhat (1994, 93) in his analysis of the concept of morality is taken seriously, it seems to me that with respect to the problem of criteria the theory of ethics offers only two different options. Either one accepts a pluralistic theory of value that allows drawing on various relevant criteria (from existence to life to the ability to suffer to personality) from case to case (see Stone 1993; Wenz 1993), or one deliberately leaves the question of criteria open since in itself it is apparently ethically problematic (see, for example, Birch 1993). In this connection Birch (1993, 317) rightly maintains the following: "Thus, the institution of *any* practice of *any* criterion of moral considerability is an act of power over, and ultimately an act of violence toward, those others who turn out to fail the test of the criterion and are therefore not permitted to enjoy the membership benefits of the club of *consideranda*. They become 'fit objects' of exploitation, oppression, enslavement, and finally extermination." It is clear that this certainly is not the path of altruism but rather a power play. If the moral standpoint is to be truly consistent and valid, it must grant ethical consideration to *everything* that exists.

In order to avoid any misunderstandings at this point, equality with respect to *moral considerability* certainly does not mean that everything on earth must be *treated* in the same manner. It would be absurd to draw the conclusion that trees must be treated like humans and humans like trees, because trees aren't humans and humans aren't trees. As a *methodological* guideline for just treatment of various different natural entities Teutsch (1987, 76f.) recommends the *principle of equality* originally developed for the ethics of human society. In essence it maintains that equals are to be judged and treated the same way in accordance with the degree to which they are equal, whereas things that are not equal should be treated differently depending upon just how different they are. What this might *in fact* mean in individual cases cannot be further discussed in the context of this treatise. A study of this kind would require a very comprehensive investigation of its own and can only be undertaken in a subsequent step.[132] What is important to me at this point is the insight that this *second* step can only be accomplished when the *first* has been made. First of all, *everything that has come to be naturally* must be included in the scope of immediate human responsibility.

28.b. Limits to Justification

Has the preceding argumentation succeeded in justifying pluralistic holism, and with it general species protection? If one applies moderate criteria for the strength of ethical justification such as those discussed in Chapter 26, then it seems legitimate to respond affirmatively to this question. However, it must be conceded that the justification developed here is relative and hypothetical in two respects. First of all, it requires assuming a moral standpoint. *If* one has already reached a decision regarding the "fundamental choice" between egoism and altruism and opted for the path of altruism, *then* according to the preceding argumentation this decision automatically leads to the standpoint of plural holism. In other words, as a *consequent* altruist one cannot respect other humans for their own sake and at the same time regard animals, plants, other species, and landscapes only as resources. Since one would otherwise have to *arbitrarily* determine what one treats respectfully and what not, in the end one would remain an egoist, even with respect to personal friends and acquaintances who would just happen to be fortunate that for the time being there is no reason to use them as instruments to their disadvantage. It follows that anyone who does not want to view himself as a person of power in his relationships with other humans must also avoid this role with respect to nature.

The argumentation developed in the preceding chapter is relative and hypothetical in a second respect. It will have to do without "final justification," as discussed above, and as I see it, just like any kind of ethical justification. When I propose that consequent altruism inevitably leads to pluralistic holism, the moral skeptic, as described in Chapter 25.c can still reply, "That's all very well, but what on earth could ever move me to become an altruist? For what reasons should *I of all people* be moral?" Contrary to what many moral philosophers think, this question is neither contradictory nor can the radical skepticism that it manifests be dismissed as "irrational" from the very beginning (see Nielsen 1984). If namely the skeptic's question is meant to indicate that he wants to be shown that his life will always and under all circumstances be better if he orients himself toward morality rather than his own interests, it will be impossible to give him a satisfying answer. Of course, there is a long philosophical tradition, supported by a whole series of results from anthropology and psychology, that has produced strong arguments for the proposition that in general a moral life is advantageous for the physical and psychological integrity of humans (see Fromm 1947). Nevertheless it would be a mistake to conclude from such a statistical correlation that a moral life is *always* and for *every* individual the best. It is by no means seldom that someone who has chosen the path of morality not only has to get along without certain comforts but may also have to suffer deprivation and persecution. This is even lamented in the psalms of the Bible. Thus Singer (1979b, 50) is quite right when he writes, "'Why act morally?' cannot be given an answer that will provide everyone with overwhelming reasons for acting morally."

But what if someone insists on an answer? In this case, according to Tugendhat (1994, 92) it is impossible to get around reformulating this question and directing it toward oneself. "What kind of person do I want to be? What is important for me in life? And what might be affected by whether or not I regard myself as belonging to a moral community?" More importantly the question is, Am I interested in awaking to reality? As Tugendhat (1994, 92) emphasizes, the answers to these questions are not rationally compelling for a personal decision to join the moral community. However, they can be supportive by making it clear what exactly might be involved with the decision. "There is only this relative compulsion such that if I want something and it is contingent on something else, that I then also have to want this other thing as well." If, however, one who is skeptical about morality is prepared to jetty membership in the moral community and everything that goes along with it, then there is no argument that could convince him or her otherwise. The discussion about ethics is then over. According to Tugendhat this demonstrates "to what extent autonomy is final." A decision for a moral standpoint cannot appeal to a compulsive "I must," in spite of what those with an authoritarian understanding of morality might like. In the end it must be content with a freely expressed "I want to."

The discovery that there apparently is no irrefutable final justification for ethics may be disappointing, but its significance for everyday life should not be overestimated. The fact that it is possible to doubt everything doesn't mean that it is reasonable to *persist* in such doubts. Regarding a possible "no" to a moral standpoint, I do not believe it is an option that many people would *definitely* choose if they take all the practical consequences into consideration and don't just deal with it as an intellectual game. Most people would hardly seriously contest the sense and necessity of ethics. They would neither be prepared to do without the existence of morality in everyday life, nor would they want to explicitly consider themselves egoists. These are the only people to whom the preceding discussion can be addressed. I hope that coupling *environmental* ethics with the widely accepted ethics of *interpersonal* relationships as I have done it has produced an argument that in their eyes is so strong that the ethical obligation to protect other species *for their own sake* can be regarded as reasonably justified.

29. Objections from Other Ethical Schools of Thought

The thesis that a *formal* analysis of the concept of morality automatically leads to pluralistic holism doesn't, of course, exclude the possibility of objections to the *contents* of such a universal concept. Advocates of other ethical schools of thought will claim that in spite of the implications of the previous discussion, it is by no means whim or arrogance associated with a position of power that causes them to ban certain parts of nature from moral consideration but simply circumstances that can be determined in a *rational* and *objective* manner. These circumstances make it seem inappropriate and misleading to even consider using terms such as "power" or "exploitation" when dealing with some nonhuman entities since their failure to satisfy certain decisive criteria makes it impossible to treat them in an immoral manner.

I do not wish to deny the *possibility* of such circumstances, but reference to a theoretical possibility is a weak argument. It must be *clearly demonstrated* that because of such circumstances exclusion of parts of nature is absolutely conclusive. In my opinion the prima facie universality of morality as discussed in the previous chapter makes it necessary to reverse the commonly accepted burden of proof. Then it is not the conservationist with his or her holistic arguments who must show which natural entities should be granted moral consideration for particular reasons or on the basis of certain characteristics. Instead, the representatives of a nonuniversal concept of ethics are obligated to convince others that a certain empirical property (as opposed to some other one) is absolutely essential for granting moral consideration. *They* must explain why it is necessary to depart from the egalitarian principle of universal consideration and exclude certain parts of nature from the scope of morality. In connection with questions of justice Tugendhat (1994, 374) points out the basic reason for reversing the burden of proof (unfortunately without drawing the conclusions with respect to his own concept of morality that to me seem imperative): "It is wrong when people try to force someone with an egalitarian position to justify this position. In and of itself an egalitarian position requires no justification. It is the non-egalitarian position . . . which must be justified. Equity and inequity do not balance out one another. This is evident from the fact that there is only *one* form of equity, while many different kinds of inequity exist. If a concept of inequity is proposed, it is only one among numerous others." Because of this, nonuniversal concepts of moral consideration require *specific* justification.

On principle I myself believe that it is highly questionable whether any justification of this kind can be proposed that is generally convincing. When the burden of proof is reversed, the advocates of a nonuniversal concept of ethics are the ones who must rely on the very process of argumentation, which they were so eager to criticize regarding other ethical concepts before the tables were turned. It is they who must then draw a normative conclusion from an empirical, nonnormative observation. It is a common strategy of proponents of restricted concepts of morality to denounce as an example of naturalistic argumentation any proposal to extend moral consideration that attempts to justify the intrinsic value of, for example, species or ecosystems on the basis of a particular empirical property (such as information, stability, homeostasis, etc.). Sometimes these attempts are quite clearly stamped as instances of a naturalistic fallacy (e.g., Wolters 1995, 248 with reference to Rolston 1994; Irrgang 1989, 54). In other cases in which the argumentative relationship between empirical and normative properties is explicitly described as weak and logically not completely conclusive, the "intersubjective validity or normativeness of such forms of justification" are questioned (e.g., von der Pfordten 1996, 129). This kind of knockdown argumentation that makes any theoretical attempt to justify extending the scope of moral consideration almost impossible presents its proponents with the same problem when the burden of proof is reversed. Then *they* are the ones who must bear the criticism of naturalistic argumentation. If they claim that there is a *logical* relationship between some empirical property and the necessity to exclude an entity from moral consideration, they can be reproached for having fallen victim to a naturalistic fallacy. If they merely claim that there is a certain amount of *plausibility* to their justification, they will become targets of the same objection they themselves once raised, namely that of insufficient intersubjective validity and therefore also arbitrariness.

It would be way beyond the scope of this book to try to examine all the arguments in the literature of environmental ethics that have been proposed for or against the exclusion of certain natural entities from the moral community. And I believe it is really not necessary since the voluminous debate between representatives of various different positions has already been summarized well elsewhere (e.g., Hargrove 1989; Pojman 1994; DesJardins 1999; Stenmark 2002). Therefore in the following sections I will

examine only *three* of the most important and frequent forms of argumentation presented by representatives of anthropocentrism, pathocentrism,
and biocentrism in favor of restricting the scope of moral consideration. I
am not so much interested in offering an exhaustive presentation of all the
aspects of these positions. Instead I would like to discuss in greater detail
and demonstrate the plausibility of the thesis developed in the previous
chapters that an arbitrary act is required to unequivocally establish any single criterion as decisive for moral consideration, and that this is therefore
not in keeping with a moral point of view.

29.a. Lack of Reciprocity?

The first objection to a universal concept of morality, one common to anthropocentric positions, can be summarized under the heading of "lack of
reciprocity." It maintains that *all nonhuman nature* must be excluded from
the moral community since its inability to reason and its lack of autonomy
preclude the ability to cooperate. "By virtue of its very form" morality is
thought to require "a community of cooperative beings", since for morality
it is essential "to demand goodness from one another" (Tugendhat 1994,
193). Since one cannot establish any kind of symmetrical relationship involving mutual respect and responsibility with animals, plants, species, and
ecosystems, it is also not possible to include these creatures and systems in
the moral community. Anyone or anything that is not able to assume duties
toward other members of the community is not entitled to moral rights.
Passmore (1974, 187) summarizes this view as follows: "The supposition
that anything but a human being can have 'rights' is . . . quite untenable."

I must admit that it is indeed rather problematic to extend the strongest
of all moral claims, the concept of *moral rights*, beyond the social sphere of
humans to nonhuman nature. Not only advocates of anthropocentrism but
many representatives of nonanthropocentric ethics as well doubt whether
an extension of this kind is wise or truly expedient for nature protection.[133] However, the question of possible "rights" of nature is not really
critical for weighing the objection of nonreciprocity. Even if one refuses to
grant nature moral rights, it does not follow that we also have no *moral duties* toward nonhuman nature. Contrary to what the nonreciprocity argument suggests, moral duties are by no means restricted to entities that are
themselves capable of assuming such duties (Rescher 1980, 85). Tugendhat's requirement of symmetry in this respect between the moral subject

and moral object is neither logically conclusive nor convincing at the level of facts. According to von der Pfordten (1996, 55) "there is no reason why the moral actor and the entity subjected to moral consideration must possess the same qualities. On the contrary, the basic situation of morality and ethics is in principle characterized by asymmetry with respect to activity and passivity, subject and object. Therefore it is legitimate to expect different things from the two parties."

The plausibility of this position becomes more apparent when one realizes that not only the relationship between humans and nature is asymmetrical but that asymmetry also occurs in interpersonal relationships as well. Granted, from a historical perspective it is conceivable that the roots of morality did indeed involve establishing reciprocal claims. Contract theory (contractualism) and the Golden Rule are evidence of this. Nevertheless, as von der Pfordten points out (1996, 296), "the entire development of subjective rights . . . can be envisioned as a gradual departure from such immediate reciprocity between the holders of rights and duties." According to contemporary thought, ethics and morality are particularly important in situations in which reciprocity *does not exist*, for example, with respect to people who are physically or mentally handicapped, those who are seriously ill, comatose patients, infants, and unborn babies. All of these people are unable to function as moral subjects. Nevertheless, or, more correctly, *for this very reason*, the need for direct moral responsibility is even greater in such cases. While such instances of asymmetry between the moral subject and moral object at one time may have been regarded as exceptions to the rule, the greatly enhanced power of humans generated by science and technology has caused lack of reciprocity to be the normal state of affairs in at least one area of ethics, the area of moral responsibility for things in the future (Birnbacher 1979a, 1988; Jonas 1984). All responsibilities toward future generations are unilateral responsibilities and therefore, as Tugendhat (1989, 935) himself admits, they cannot be dealt satisfactorily with either contractualism or with the morality of mutual respect.

If this significant and, for the survival of humankind, increasingly important area of ethics must and does indeed manage to get along without the symmetry of reciprocity, it is hard to understand why reciprocity is still considered to be absolutely essential for ethics with respect to nature. What both the ethics of responsibility for things in the future and environmental ethics share in common is the fact that they must deal with extremely asymmetrical relationships of power and control. Because of this,

in both cases it is justifiable and necessary to accept asymmetrical responsibility and overcome the widespread notion that ethics is exclusively and
always a "reciprocal matter." According to Lenk (1983b, 14f.) it is the very
act of *renouncing* reciprocity that makes a human being a truly moral, fair,
and humane person and accentuates his unique position in nature: "A
human being is characterized by the fact . . . that he can not only assume
responsibility for his own activities directed toward fellow human beings,
but . . . that by symbolic and projective means he can also bestow upon
other creatures moral quasi-rights, the right to exist, the right to continue
to exist, that he can and should assume duties for and toward them without receiving any services or duties in return. A human being is a creature
who is able to understand whole systems and venture beyond his anthropocentric limits to grant entire systems as well as ecological subsystems
and living partners in nature the right to exist."

Spaemann (1984, 76, 77) even goes so far as to maintain that only in the
context of this extended perspective is it at all justified to speak of "human
dignity." Because, as Spaemann argues, "as long as reference to human dignity is regarded as merely a manner of speaking by which the members of
the species *Homo sapiens* defend themselves against others, then this manner of speaking is of no real normative significance. In doing so the human
species behaves in principle just like any other species in nature, except for
the fact that its intelligence provides it with a much greater ability to assert
itself, which in turn permits it to gradually abandon any scruples." If, on
the other hand, human dignity means something that is "objectively" characteristic of human beings, then it must mean the ability of humans to voluntarily exercise restraint with respect to other species and self-organizing
natural systems; in other words, to reject exercising power as well in situations in which neither appropriate compensation can be expected nor the
possibility that the "natural counterpart" will do an about turn and retaliate
must be feared.[134]

29.b. Lack of Perspective on the Part of the Moral Object?

The second objection that is often raised against a universal concept of
morality, particularly by pathocentrists, is summarized by the expression
"lack of perspective." It is based on a premise generally accepted to be self-
evident that moral obligations are undoubtedly "always guided by interests" (Rescher 1980, 83). Someone who has no interests cannot be

impaired or otherwise treated immorally and therefore cannot be the object of moral consideration. While there is general consensus among many ethicists concerning this premise, a great deal of controversy still exists about *which* natural entities are thought to have interests and which do not. The spectrum of suggestions regarding those that have interests varies from all creatures capable of thinking to those capable of suffering and finally to all living creatures. Species and natural systems are very seldom thought to have interests.

If one analyzes the debate between the various different positions (the course of which has been particularly fierce in English-speaking countries), surprisingly one discovers that the controversy is not so much due to differences in the interpretation of empirical results but rather due to different opinions regarding the concept of "interest." The reason for this is the unusual ambiguity of this term. It is "so vague that everything depends upon the interpretation of it" (von der Pfordten 1996, 203). For example, if one interprets "interests" as expressible desires the way that anthropocentrists do, an interpretation that is so narrow and highly sophisticated that it requires the ability to think and speak (e.g., Frey 1980, 83), then it is obvious that only humans are capable of having interests (or at the most a few other animals in addition that are capable of thinking such as primates or whales). On the other hand, if a broader concept of the term interest is chosen, as in the case of pathocentrism, so that it also includes nonrational states of consciousness such as pleasure and pain (e.g., Nelson [1932] 1970; Feinberg 1980; Singer 1979b, 220), then even more animals, namely all those capable of conscious perception, have interests. If, finally, in keeping with biocentrism, the term interest is defined very broadly so that it also includes all unconscious tendencies to survive and maintain one's well-being, then invertebrate animals, plants, and even microorganisms undoubtedly also have interests.[135] Recent publications indicate that the discussion is currently focused on the controversy between pathocentrists and biocentrists. It appears that only very few ethicists feel that interests depend upon the ability to think and speak. Instead the current philosophical debate seems to rotate about the question of whether or not assigning interests to an entity requires that it has a particular kind of consciousness, that is, a perspective of its own associated with the ability to feel pain (see Teutsch 1985, 49).

Even if one disregards the fact that it is apparently almost impossible to reach sound conclusions about the existence and degree of consciousness in invertebrate animals by empirical means,[136] I still have serious doubts about whether settling the question of "interests" will help us to come any closer to solving the problem of the scope of moral consideration. On the contrary, I believe that this dilemma underlines and illustrates the thesis outlined earlier that when it comes right down to it, excluding parts of nature from the moral community is an arbitrary act. If the question of moral considerability is coupled onto the concept of interest, and if at the same time the definition of interests is "a matter of interpretation" (von der Pfordten 1996, 203), then this means that the scope of direct moral consideration is dependent (to a certain extent) upon a *contingent semantic construction*. Depending upon whether the term "interest" is *defined* more or less broadly, invertebrate animals and plants may find themselves within the scope of moral consideration in one case or outside of it in another. Thus the exclusion of the majority of all living things from moral consideration as in the case of pathocentrism must be regarded as the result of establishing a more or less *arbitrary axiom* and far from being absolutely conclusive on the basis of "objectively determined facts."[137]

Advocates of pathocentrism, of course, will not fancy having their axiomatic line of demarcation regarded as an arbitrary one. Thus it is not surprising that they often justify their more narrow interpretation of the term "interest" by referring to a common linguistic tradition that supposedly embraces only their own interpretation and no other one (e.g., Feinberg 1980, 169). Nevertheless there are two reasons why the exclusion of invertebrate animals and plants by definitional means and semantic justification is not convincing. First, the language we use every day suggests that at least superficially it does not seem to be completely unreasonable to attribute interests even to plants. We refer to plants as "flourishing" or "not doing very well," and we say that they "need" light, water, and nutrients for their wellbeing. When Feinberg (1980, 169) and Singer (1993, 279) call this manner of speaking purely metaphorical, then this interpretation is certainly legitimate, but it also weakens their claim for support for *their own* position on the basis of the same general linguistic tradition.

Second, particularly in ethics appeals to the everyday use of a word or to linguistic intuition are only of limited value (see Burkhardt 1983, 401, 402). While language usage is usually an expression of what *is*, that is, how the world is perceived, ethics has to do with what *ought to be*, or rather, how the world ought to be seen from a moral perspective. In view of this discrepancy, it is important to reckon with the possibility that what ought to be might not yet be evident in language use. For this reason Routley and Rout-ley (1979, 37, 38) maintain that attempts to claim that independent norma-tive propositions concerning the scope of morality are a matter of definition, something not uncommon in philosophy, are both "philosophically facile" and "methodologically unsound." According to these authors, the main rea-son why such attempts are inappropriate is that the contents of propositions are thus quietly withdrawn from critical examination. Moreover, with re-spect to the debate about the scope of moral consideration a process of this kind is comparable to "justifying discriminatory membership for a club by referring to the rules, similarly conceived as self-validating and exempt from question or need of justification." In analogy to the naturalistic fallacy one could refer to such a process as a "definitional fallacy."

Along the same line of thought, just as attempts to justify excluding cer-tain living things from moral consideration solely on the basis of *semantic* arguments fail to be convincing, attempts to demonstrate the plausibility of exclusion on the basis of certain *natural* traits must also appear to be faulty. In addition to the *logical* objection that this procedure may constitute a nat-uralistic fallacy, its *factual* basis can also be criticized since it requires pos-tulating a single, absolute discontinuity somewhere in the realm of nature, one that is also normatively relevant for the question of moral considerabil-ity. However, an absolute discontinuity of this kind is nowhere to be found. According to Skirbekk (1995, 422), the tricky thing about environmental ethics is that "biology operates with continuities whereas we are accus-tomed to thinking about morals in terms of absolute limits." Of course, by referring to continuities in biology the idea is not to deny the existence of greater differences between some groups of organisms as opposed to oth-ers. But, in the first place, there are always several such discontinuities from which to choose, and there seems to be no good reason why a partic-ular one should be selected for normative purposes rather than another. And in the second place these "discontinuities" must be seen in relative terms since they all are characterized by a transitional area and since the organisms on either side of the "discontinuity" still have a large number of traits in common. With respect to the discussion of the concept of interest this means that rational interests (of humans), conscious interest (of higher animals), and biological interests (of plants) may differ very fundamentally in their composite qualities, but there are more or less broad transitional areas between them, and they all share a common basis. The common basis is their self-referentiality, which, according to von der Pfordten (1996,

238), is characterized by autonomous generation, autonomous development, and self-maintenance.

Although I share Varner's (1990) and von der Pfordten's opinion that in view of such overlaps it is reasonable to refer to "interests" among creatures that lack consciousness, I do not consider determining which organisms have interests to be decisive for the question of whether or not invertebrates and plants are worthy of moral consideration. In the context of a holistic perspective both conceptual alternatives lead to the same conclusion. *Either* we define the term "interest" in the broadest possible manner, that is, on the basis of self-referentiality. Then it follows that all organisms without distinction have interests and are unquestionably members of the moral community, even though this does not exclude the possibility of further specifying the term "interest" secondarily on the basis of empirical data (see Williams 1980, 153). *Or* like Singer (and classical utilitarianism) the term "interest" is defined so narrowly that it requires consciousness. Then a primary distinction is made, but it cannot be considered a convincing argument for excluding nonsentient animals and plants from moral consideration. Invertebrate animals and plants may not have interests *as defined in this manner*, but this does not mean that their unconscious inclinations to survive are therefore morally irrelevant. What is most important in this context is the fact that all animals and plants—regardless of whether or not they possess consciousness—have a good of their own (Taylor 1981, 199).

Of course, this fact is strictly rejected by advocates of pathocentrism. Wolf (1987, 166), for example, finds it impossible to imagine "what might be bad about doing this or that to plants." And Feinberg (1980, 168) writes: "Trees are not the sorts of beings who have their 'own sakes,' despite the fact that they have biological propensities. Having no conscious wants or goals of their own, trees cannot know satisfaction or frustration, pleasure or pain. Hence, there is no possibility of kind or cruel treatment of trees." I, however, find Feinberg's argumentation not very convincing. In response to it Regan (1976, 490) has correctly argued that it only shows that plants are unable to experience a *particular* kind of well-being, namely that of happiness. But it may very well be that plants have a *different* kind of well-being. Accordingly, the observation that plants—as far as we know—do not experience pain only means that they cannot be tortured. It does not say that they might not be capable of being harmed in *some other* manner.

Elliot (1978, 702, 703), on the other hand, responded to Regan's rejoinder by pointing out that a "different kind of harm" must not simply be purported but has to be accounted for. Without appropriate evidence it is more reasonable to assume that consciousness is a necessary condition for having interests or inherent good. What can I say about that? First of all, it must be recalled that our analysis of the concept of morality led to the conclusion that the burden of proof must be *reversed*. The person who assumes that invertebrates and plants are direct objects of moral consideration is not the one who must convincingly demonstrate that his or her perspective is legitimate. This is the responsibility of the person who feels that the prima facie universal character of morality must be restricted to entities with consciousness. Second of all, it is difficult to discern what Elliot might conceive of as appropriate evidence that invertebrates and plants have a good of their own. It appears that the pathocentric position he favors assumes *from the very beginning* that these entities constitute no substantial reality since they have no "perspective of their own." At any rate, there can be no doubt that Feinberg (1980, 170) considers plants to be nothing but complicated machines when he writes: "An automobile needs gas and oil to function, but it is no tragedy for it if it runs out—an empty tank does not hinder or retard its interests. Similarly, to say that a tree needs sunshine and water is to say that without them it cannot grow and survive. . . . Plants may need things in order to discharge their functions, but their functions are assigned by human interests, not their own." Feinberg's comparison is clearly faulty since he overlooks the fact that a car without gas and oil remains intact while a tree without light and water *dies*. Nevertheless, what he wants to express is obvious. From a moral point of view chopping down a tree is no different from taking a car to a junkyard. Both experience no real harm.

In Chapter 26 I have already pointed out that it is, of course, always possible to regard a living thing simply as a machine and thus not as "substantial reality" (see Spaemann 1990, 133). A *methodological* example of this viewpoint is the objectivistic reductionism of natural science, while Cartesianism exemplifies it in the form of a *worldview*. While methodological reductionism is clearly not only legitimate but also absolutely indispensable for advancing knowledge in the natural sciences, all the knowledge we have accumulated so far indicates that Cartesianism must be regarded as an untenable ideology. Anyone nowadays who thinks that animals and plants are *really* machine-like entities is no longer up to date in matters of theory of science. Although this has already made the rounds in ethics with respect to vertebrates that are capable of suffering, the pathocentric position demonstrates that Cartesianism is not yet dead when it comes to invertebrates and plants. Just as Descartes denied that animals "really" experience pain, pathocentrism practically denies that invertebrates and plants can "really" die or be killed. After all, machines can't be killed. In my opinion, it is not so much the reductionism of the pathocentric position and the way it contradicts intuition that makes it untenable but rather the *inconsequence* of its reductionism. While monarch butterflies, ladybugs, clams, and Venus flytraps are radically objectified and reduced to the status of machines, pathocentrism balks at the idea of performing this kind of ontological degradation on humans, cocker spaniels, and hamsters. But what truly reasonable explanation can pathocentrism offer for such a dichotomous viewpoint? A remark by Singer (1993, 279) indicates that the answer must ultimately be sought in metaphysical beliefs. In the case of entities *without* consciousness a "purely physical explanation of life processes" is considered plausible, but not in the case of entities *with* consciousness. Birnbacher (1989, 399) seems to adhere to the same view when he writes: "Life is no longer a 'mystery.' Not only the processes of life but also its origins have been made accessible to scientific explanation. If there is anything that is still mysterious and not yet explainable by scientific (evolutionary) theory, it is the existence and origin of consciousness."

In response one could note that the mysteriousness of life is by no means diminished by gaining insights into the rules that govern it.[138] Moreover, the very fundamental question that the discussion of mystery evokes is what exactly counts as an "explanation" of emergent phenomena such as crystals, liquids, life, consciousness, mind, and the like. In Chapter 5, I already pointed out quite clearly that in my opinion there are fundamental limits to the epistemological scope of the scientific method in these areas.[139] However, if pathocentrists, like the scientistically inclined, wish to operate on the premise that a scientific explanation of the rules governing such phenomena is a *sufficient* one, then they must also concede that there is no longer any valid justification for claiming that consciousness is the *only* phenomenon capable of infinitely eluding the powers of ontological reductionism. As modern brain research and the discussion of the mind-brain problem accompanying it have shown, both neurobiology and materialistic philosophy of the mind claim that like the phenomenon of life the phenomenon of consciousness is also accessible to "scientific explanation" (see, for example, Dennett 1993). If one consequently adheres to the concept of "metaphysical naturalism," that is, to the belief "that both nature and our ability to reason are exclusively rooted in natural processes" (Vollmer 1986, xviii), then only two options remain. Either we regard *all* living things as complicated machines, the way ontological reductionism and behaviorism do. Then higher animals and *we too* are machines. Or we acknowledge that all living things without exception have a reality of their own beyond such mechanistic and reductionistic interpretation. But then *all* living things also have *a good of their own* (see Sprigge 1979, 142).

Just because it is possible to identify these two options doesn't, mean that they represent two equally acceptable philosophical positions. In my opinion ontological reductionism is a weaker standpoint since it can only be maintained *theoretically* but fails to hold true in *everyday life*. To me it seems to be almost impossible for someone to regard *himself* or *herself* as an electrochemical machine without continually becoming entangled in contradictions. And just as this option fails when contemplating oneself, it will also present problems when considering all those creatures that we regard as similar to ourselves and to which we have emotional bonds, in particular our fellow human beings near and (possibly also) far away and (under certain circumstances) certain vertebrates to which we are attached.

Even though the incoherence of a mechanistic standpoint seems to be obvious in view of these considerations, one could, of course, still espouse the idea that this is insignificant for the social life of humans and maintain that the psychological effect of empathy is normally strong enough to compensate for the opposing effects of ontological reductionism. However, from the standpoint of environmental ethics, pragmatism of this kind is by no means satisfactory. It would mean that by and large we would actually only recognize those parts of nature as having a reality of their own entitling them to complete moral consideration that we are able to perceive as sufficiently similar to us. In other words, other beings would have to be able to demonstrate that they share basic things with us as, for example, consciousness, if they wish to be protected for their own sake. It can hardly be denied that this condition is a highly *arbitrary* barrier. Why should only those life-forms be subject to protection that *like us* have evolved a highly sophisticated nervous system associated with having a perspective of their own and experiencing pain? The simple fact that conscious suffering is highly relevant *for us* in matters of morality is certainly not a convincing argument for making the ability to suffer a strict requirement for moral considerability among all other entities as well.

It seems to me that in view of these reflections we have good reason to reverse an objection commonly directed toward biocentrism and holism by representatives of restricted concepts of moral consideration, namely that of *anthropomorphism*. It is often claimed that granting direct moral consideration to forms of nature below the level of animals is the result of an unwarranted projection of physiological conditions familiar to humans and higher animals upon lower ones, plants and landscapes. But it is also justifiable to claim the reverse, namely that denying forms of nature below the level of animals a good of their own is an expression of a limited concept of a "good of one's own," one based on projecting experiences with other humans and pets on the rest of nature. Because *we* cannot imagine a life without pleasure and pain, or at least only with some difficulty, we assume that the life of a tree, which (as far as we know) experiences no pain or pleasure, is not a "real" life, that is, not one that can really be harmed. It shouldn't be difficult to recognize that an anthropomorphic view of this kind is hardly compatible with the universal perspective of a moral standpoint. Espousing a universal perspective does not mean that we judge the existence or the life of another entity by comparing it *with us,* but rather that we do justice to its existence and its life *as such.* In Chapter 25.b I believe I have shown that—in spite of all due modifications—this is at least in principle possible from an *anthroponomic* perspective.

29.c. Lack of Goal-Directedness?

The third and most frequent objection raised against a holistic concept of morality can be summarized by the expression "lack of goal-directedness." This means that inanimate natural objects and entire systems such as species, rivers, and ecosystems cannot be the objects of direct moral consideration since they cannot even begin to exhibit anything like intentionality, that is, aspirations in a particular direction (e.g., Cahen 1988, 202). Without such "interest" in the broadest sense of the word it is impossible to talk about either fulfillment or lack of fulfillment or about good or harm. However, the ability to apply these bipolar terms (Ricken 1987, 16) is a central and absolutely necessary part of any concept of morality and thus a "noncontingent" and compelling criterion for moral considerability (Goodpaster 1980, 281, 282).

In order to examine the legitimacy of this objection we must first analyze what exactly goal-directedness ("telos" or "natural ends") in nature involves. The starting point for such an analysis is the observation that in principle the state of each and every thing of the world is capable of changing, whereby these alterations eventually lead to a new (more or less temporary) end state. This gives the impression of goal-directedness oriented toward such an end. However, a more exact examination of the many phenomena that may at first appear to be goal-directed or teleologically oriented reveals that there are not only several different kinds of goal-directedness in nature but also that many processes that *appear* to be goal-directed really are not. If one starts off by differentiating between virtual and real goal-directedness, then according to Mayr (1982, 48, 49) *truly goal-oriented* phenomena can also be divided into two groups of processes, *teleomatic* and *teleonomic* ones.

Teleomatic processes are those through which "a certain goal is reached exclusively by means of physical laws." They are "end-orientated" in a manner that is passive, automatic, and dictated by external forces and conditions and thus occur mainly in the realm of inanimate matter. When, for example, a falling stone reaches its endpoint, the earth, "this has nothing to do with searching for goals or intentional or programmed behavior but is solely due to the laws of gravity. The same is true for a river that unwaveringly flows to the ocean." According to Mayr (1982, 49) the entire process of cosmic evolution from the big bang to the present is the result of a series of teleomatic processes upon which stochastic disturbances are superimposed. Among the laws of nature the laws of gravity and thermodynamics are the ones that most frequently determine teleomatic processes.

Teleomatic processes must be clearly distinguished from teleonomic ones. The goal-directedness of the latter can be attributed to the "operation of a program" and therefore is found in nature only in connection with living organisms (Mayr 1988a, 44, 45). Examples are organismic ontogeny and goal-oriented forms of behavior such as searching for food, courtship, reproduction, or migration. Such behavior is characterized by the existence of a final point, goal, or end encoded in the program. The program itself, which can be either closed or open for learning processes, is the result of natural selection and is constantly adjusted anew on the basis of the selective value of the endpoint that has been attained (Mayr 1988a, 43).

In addition to teleomatic and teleonomic processes, *virtual* goal-directedness also exists as shown by the dynamics of ecosystems and the evolution of species. Even if ecosystems and species give the impression that they have made adjustments that are conducive to their own good (e.g., adjustments resulting in stability or the preservation of a species), these characteristics of the system are really only byproducts of processes executed at the level of active individuals (see Sober 1986, 185, 186). The behavior of the individuals involved is teleonomic, but no programs for the entire system or species can be discerned that facilitate the attainment of a predetermined end state (telos). At the most there is a kind of open-end potential that can further develop under the influence of internal or external factors. For example, how a species evolves in the future is certainly seriously limited by the gene pool it has in the present, but within these limits its further course will primarily be determined by the ever-changing constellation of selective pressures. At any rate, it is certainly not programmed by its gene pool. If we can thus exclude the possibility of a telos for the evolution of an *individual species,* then this must also hold true for the *entire process of evolution.* Even if the course of evolution appears to be progressive, extending from unicellular organisms on up to human beings, modern evolutionary biology almost unanimously rejects the idea of goal-directedness in evolution (orthogenesis, progressionism, "cosmic teleology"). The currently accepted view is that the evolution of "higher forms" is solely the result of selective pressure due to competition among individuals and species and colonization of new zones of adaptation (Mayr 1982, 50).

What does this briefly outlined system of classifying real and virtual goal directedness mean with respect to the normative question of criteria for moral considerability? From the standpoint of the objection presented above that moral considerability requires the existence of some sort of telos, one would have to postulate that purely adaptive processes such as those characteristic of species, ecosystems, and the evolutionary process cannot be granted moral consideration while all teleomatic and teleonomic processes would indeed be entitled to it. However, it can be readily seen that proponents of restricted concepts of morality, particularly biocentrists, would most likely advocate including only natural entities that function by *teleonomic* means, that is, organisms, within the moral community. Entities such as rivers, mountains, dripstone caves, or drifting sand dunes that arose and continue to develop by *teleomatic* means, on the other hand, would not be viewed as worthy of moral consideration. Why? The argument usually offered is that these wholes originated and developed "practically in a completely heteronomous manner" and moreover that after they have attained a (temporary) final state, they exhibit no signs of attempting to maintain themselves. And if no tendencies toward "self-maintenance" can be observed, then there can be no obligations on the part of others to maintain such entities (von der Pfordten 1996, 239).

Regarding this view two things must be noted. First, it would be a misunderstanding to assume that the holistic standpoint postulates obligations on the part of others to *maintain* natural entities and systems *in a certain state*. An ethical claim of such far-reaching dimensions would be absurd since satisfying it would not only require constant intervention in nature by humans. It would also practically demand bringing all forms of natural dynamics to a standstill. Instead the holistic position requires that *spoiling* or *destroying* such entities be prohibited prima facie, and in practice this means permitting natural dynamics to occur to the greatest extent possible. Contrary to the previously described assertion that in dealing with inanimate objects the bipolarity between good and evil that constitutes the very substance of morality is lacking (Goodpaster 1980, 282), two analogous poles can indeed be identified. One of these is characterized by a pristine state and autonomous development (e.g., that of a meandering river), while destruction or massive interference by humans (e.g., by means of a dam, canalization, or draining, etc.) constitute the other. Granted, the fact that inanimate objects lack clearly defined outer limits and self-referentiality presents epistemological difficulties in objectively determining these poles (see also Kantor 1980, 167). However, I do not consider these *practical* problems to be insurmountable.

A second point relevant to the discussion developed above is whether failure to satisfy the criteria of "autonomous generation," "autonomous development," and "self-maintenance" that von der Pfordten (1996, 238) considers to be prerequisites for the existence of interests, makes exclusion from the moral community inevitable. Besides the fact that even organisms are never so autonomous that they can originate, develop, and maintain themselves completely free from the influence of external factors and forces, the objection could be raised that we are dealing here with a form of naturalistic deduction. It is neither logically conclusive nor factually compelling that moral considerability be made contingent upon the three criteria listed above. Thus to me it is not at all clear why only those parts of nature whose goal-directedness is governed by an *internal program* should be protected for their own sake, but not those whose goal-directedness occurs by *purely physical ways and means*. Couldn't one argue just as well that for the latter the need for moral protection is even greater for the very reason that they are not able to maintain themselves? To me it seems that this question can only be dismissed if one associates moral considerability with the paradigm of life and living things from the very beginning. But then one should admit that abstract criteria such as "goal-directedness," "self-identity," or "interest" are really only reconstructions of the main and fundamental, graphic, and global criterion "life" that have been presented after the fact.

However, regarding the criterion "life" Hunt (1980) has shown, that as a critical quality for determining moral considerability it is no less arbitrary than those favored by anthropocentrism and pathocentrism, which have already been examined and found to be contingent. In keeping with the thesis of reversing the burden of proof, he believes that it is at least necessary to offer rigorous justification for selecting *this particular* criterion and not the more comprehensive one of *existence*. According to the pathocentric ethicists Birnbacher and Singer a truly convincing argument in favor of this choice is not in sight. Regardless of all the criticisms of holism they otherwise have, both find it highly implausible to grant radical privileges to animate nature over inanimate nature to the extent that value in and of itself and thus also moral respect is attributed in principle to the former but not to the latter. For Singer (1993, 280), for example, it is "not obvious why we should have greater reverence for a tree than for a stalactite. . . ." And Birnbacher (1987, 65) poses the following question: "Why should beauty, wholeness, symmetry and complex organization be valuable in natural objects that are alive but not in those that are inanimate?" It is probably impossible to answer this question in a convincing manner without resorting to vitalistic assumptions (which are obsolete according to current theory of science).

Since the modern view of the world, upon which science has left its stamp, makes it difficult to justify a normative gap between life and nonlife at a *theoretical* level as biocentrism does, advocates of biocentrism (and others) like to refer to the *practical* consequences that would ensue from including inanimate parts of nature in the moral community, consequences that are supposedly contrary to intuition. Thus it would simply be too much to expect us to respect stones and therefore render the use of gravel in construction a matter of moral philosophy. However, an extreme example of this kind doesn't really provide the kind of proof we need right now. It simply shows that the question of the appropriate treatment of *certain* inanimate things (such as *individual* stones, for example) apparently is of very little (almost negligible) moral significance. However, what it cannot demonstrate is whether it would *always* be absurd to judge the treatment of inanimate objects from the standpoint of direct moral responsibility.

In order to make this point clear, all we need is an appropriate counter example such as the mental experiment that Pluhar (1983, 53) devised (even if it seems quite utopian at the moment). Let us assume that advanced space technology might someday make it possible for us to exploit the resources on the planet Mars and transport them back to earth for a reasonable price, and let us assume further that no form of recent life exists on Mars. Would it then be ethically acceptable to raze Olympus Mons, the 25-kilometer-high and thus highest mountain of the solar system, in order to produce some luxury articles?[140] Regardless of how one replies to this question, the biocentric standpoint, according to which it is *in principle* impossible to consider a matter of this kind one of direct moral responsibility, would seem to be rather apodictic. It is at least not clear how a biocentrist could convincingly argue that the teleonomy of a *single* bacteria demands more ethical respect than the teleomatic inclinations of this magnificent mountain. In this case, instead of constructing a radical moral boundary between life and nonlife it seems to be much more consistent to consider the meaning and achievements of morality in its most comprehensive sense, namely that of "protecting that which exists from destruction" (Steinvorth 1991, 886).

However, since that which exists is always something that *has come to be* historically and under closer scrutiny also is found to be something that *will come to be* in the future, protection from destruction cannot be restricted to those parts and processes of nature that (for the time being) prove to have a telos that is either governed by a program or determined by physical laws. It must also include those entities and processes of nature that are not goal-directed but have come to be through spontaneous self-organization and natural selection, namely ecosystems, species, the biosphere, and the process of biological evolution as a whole. These are the systems in which the teleonomic structures and processes of individual organisms are embedded, which have made them possible in the first place, and which will hopefully secure their survival in the future. In the context of an holistic concept of ethics intrinsic value must be attributed to these systems as well.

One objection sometimes raised against this moral postulate is that systems and processes that lack goal-directedness cannot be granted moral consideration since their formation and dynamics are determined "merely" by chance and necessity (see, for example, von der Pfordten 1996, 180,

181). The mechanisms of systemic wholes (such as ecosystems and the biosphere) are "explained as being merely the result of particular behavioral patterns of individual organisms, the changes in the environment that they generate, and external physical and chemical factors as well" (von der Pfordten 1996, 242). If the aim of this objection is to describe something that is the opposite of teleonomy of individual organisms, it is a futile one. It seems to overlook the fact that not only ecosystems and their regulatory mechanisms can be viewed from a strictly reductionistic and mechanistic standpoint (based on cybernetics and systems theory), but also all organisms (including humans) and their purposes as well. As Salthe and Salthe (1989, 360) have shown, both Darwinian evolution theory and modern theories of self-organization are based on the assumption that the goal-oriented behavior of organisms is exclusively *causal* in nature. That is, it can be explained without recourse to *teleology*. This is summed up in the term *teleonomy*. When it comes right down to it, according to this causal-analytical perspective organisms also do not really pursue any goals or purposes. It is simply for heuristic reasons that we describe them *as if* they pursue "goals" and "purposes." From this point of view their "interests" are also determined by chance and necessity or chaos and regularity, just like the regulation mechanisms of ecosystems.

Wolters (1995, 249) provides a good example of this mechanistic and reductionistic viewpoint by comparing the regulatory processes of living systems in nature with the behavior of "a central heating system that is regulated by a thermostat." As already pointed out a number of times, I find a comparison of this kind legitimate as a model for generating *methodological* solutions in science. However, in my opinion Wolters goes too far when he employs such a model as an argument to demonstrate the apparent inconsistency of a holistic concept of intrinsic value. Thus he writes, "Most of us would certainly hesitate to attribute any kind of intrinsic value requiring moral consideration to our central heating system." It is not difficult to discern exactly where the inappropriateness of this kind of argumentation lies. A model that was originally derived from engineering science, a cybernetic machine model, is projected upon nature in order to reach the "compelling" conclusion that one obviously cannot attribute intrinsic value to a pure cybernetic system. In other words, the author concludes something that was originally formulated as a premise. It is quite clear that for neither ontological reductionism nor for anthropocentrism can convincing evidence be provided in this manner, that is, by drawing upon the *methodological* reductionism of science. On the contrary, it is important to remember that at least two pieces of evidence discussed in this book very definitely contradict the idea of ontological reductionism. The first of these is the departure from strict adherence to the paradigmatic elements of mechanism and determinism that can be observed in the worldview of science nowadays (see Chapter 7), and the second is the practical inconsistency of ontological reductionism. As I tried to demonstrate using Feinberg's comparison of trees with cars, this inconsistency is revealed by the fact that the ontological reductionist readily reduces nature to the level of a machine, but is careful to exclude himself from this kind of total reduction.

If those who advocate a restricted concept of moral consideration persist in viewing ecosystems and certain nonhuman organisms as cybernetic machines in spite of this evidence, then it is obvious that they will, of course, also subject *species* to the same kind of reduction. Species too can be regarded and treated as if they had no substantial reality of their own, as if they were "no more than" a bundle of mutated, recombined, and selected genes, as if their value consisted simply and exclusively in being of use to human beings. But as in the case of organisms, a shortsighted view of this kind will probably become more and more difficult to maintain the more its adherents allow themselves to "attentively observe" their counterparts in nature. In Chapter 26 a necessary prerequisite for this kind of "enhanced perception" was shown to be at least a minimal amount of knowledge about this counterpart, in addition to (at least partial) transcendence of an egocentric perspective ("awakening to reality"). Only when someone has a relatively adequate concept of what constitutes the "nature" of a species, what a species "is," can one judge what it means when such a manifestation of life is irrevocably extinguished.

In order to emphasize the significance of the loss of species in our day and age from an ethical point of view, in the next chapter I shall attempt to summarize some of the most important characteristics of a species that are relevant for an ethical discussion and demonstrate the consequences that result from them for both practical species conservation and the theoretical concept of pluralistic holism. By specifying the term species protection once again (and for the last time) in the course of this explication, I will return to the starting point of the line of argumentation I have been pursuing, that is, to a postulate of general species protection basically rooted in intuition, even though it should be clear that sufficient justification for it has been provided in the meantime.

30. Species Protection as a Paradigm of Pluralistic Holism

Although people's everyday concept of a species was certainly adequate for the discussion so far, for further examination of the problem at hand it is necessary to define the term species more precisely. One of the most frequently cited definitions in the context of the modern biological species concept was formulated by Mayr (1942, 120) and is as follows: Species are "groups of actually or potentially interbreeding populations that are reproductively isolated from other such groups." Expressed more succinctly, a species is often defined as "the greatest potential reproductive community that exists."

If the criterion of *reproduction* is used as a starting point for characterizing a species, one might be tempted to assume that the main task of practical species protection is to save as many specimens of a particular species as are required to maintain propagation and secure the specific gene pool of the species for time to come. To attain this goal it would simply be necessary to make sure that the course of propagation is not interrupted. It follows that it would completely suffice to protect a small number of individuals in reserve, or, if this should prove to be too expensive, to at least maintain them for future generations in zoos or (in the case of plants) in gene banks. As a result of increasing progress in gene technology it might even be possible someday to revive species that have already become extinct by means of genetic engineering.

However, nothing could be more mistaken than such a technical understanding of species protection. The basic error of this way of thinking is the assumption that a species is nothing more than a collective of identical individuals and that the existence of a species can be guaranteed by keeping a few individuals of the species alive by whatever means possible. Contrary to this reductionistic view of a species and species protection, ecologists and population biologists in particular point out that individuals of an endangered species should never be regarded as isolated entities, and that in the long run they can only be successfully protected in their natural surroundings and in relation to other members of the species.

In order to illustrate the meaning of the particular context in which an individual organism or species lives, Pianka (1985, 685) compared an animal in a zoo to a single word out of context. If one removes a word from a paragraph, a significant part of its meaning and informational content is lost. According to Pianka something similar takes place when an organism is removed from its natural surroundings. Just as a word can assume the grammatical function of a subject, an object, or an adjective and is embedded in complex relationships to other words within a paragraph, an organism can also play the part of a producer, a consumer, or a decomposer within an ecosystem. It is confronted with members of other species in the form of enemies, predators, and potential competitors and may be a predator itself. In addition to such interspecific relationships individual organisms maintain numerous relationships to members of their own population, for example, to offspring, relatives, potential partners, or neighbors in adjacent territories. If these relationships are prevented or curtailed, as is almost inevitably the case in small reserves, zoos, or botanical gardens, then an organism is deprived of a significant part of its ecological and ethological identity. In the end, it is reduced from a complex and intricately interconnected natural entity to, biologically speaking, a more or less amputated cultural entity.

However, it would be a mistake to think that the restricted living conditions that prevail in small reserves or zoos are basically a problem of animal welfare but insignificant for species protection. As Slobodkin (1986, 239f.) has shown, these conditions are no less disastrous for species protection since they prevent the adaptations to a continually changing environment that are necessary for the survival of a species. Species have evolved in a process of interaction with their natural surroundings and thus can only continue to evolve by further interaction of this kind. If they are no longer given sufficient opportunity to react to ecological and climatic changes (e.g., unexpected epidemics, new enemies, etc.), they will eventually "lose touch" with evolution. A few species that are not particularly demanding can be kept alive in the care of humans for relatively long periods of time. However, without the permanent challenge posed by natural selection, their chances of surviving on their own in the wilderness are reduced.

This discussion, of course, raises the question as to how the "reaction" of a species to such challenges is to be envisioned and what factors determine whether it succeeds in meeting the challenge. In order to understand this mechanism it is important to realize that the gene pool of a species it not a uniform and homogeneous thing. On the contrary, it is composed of many different kinds of genotypes. As a matter of fact, no two organisms of a population are completely identical with respect to their genomes, a phenomenon biologists refer to as *genetic variability*. The particular significance of the varying genetic makeup of individuals for the survival of the species to which they belong is that they represent a kind of potential "insurance" in the event of unexpected developments. The more diverse the gene pool of a population is and the greater the number of subpopulations a species encompasses, the greater the probability that among all the members of the species by chance a few exist whose genotypes will be able to withstand the pressure of a sudden change in the environment. It is the rare genetic constellations of these "outsiders" that give them a selective advantage and make it possible for the species as a whole to continue the relay race of life on into the next generation in the face of selective pressure. It is thus clear that genetic variability is "of decisive importance for both the development of individual species and the evolution of life in general" (Weber et al. 1995, 187).

What conclusions can be drawn from these findings for further specifying a rather general concept of species protection and for turning this knowledge into action? The first thing that can be concluded is that to exercise effective species protection it is not enough to concentrate on *global* extinction processes. Since the probability of global extinction increases as genetic variability decreases, it is also important to protect subspecies and *local* populations in order to facilitate species protection. It follows that *regional* lists of endangered species are also legitimate since they call attention to losses and a corresponding impoverishment of genetic variability at a higher level (see Blab 1985, 616). However, it is always necessary to interpret decreases in numbers at a local level in a greater context of time and space. As the chapters about "generalization" (4.e) and "ecological stability" (11.b) clearly demonstrated, a reduction at a local level is not always grounds for concern, as it may sometimes reflect the natural dynamics of the ecosystem (see Norton 1986, 113; 1987, 33).

The second conclusion to be drawn has already been implied and is closely connected to the first. Just as a species cannot be properly protected without sufficiently taking *intraspecific* (genetic) relationships into account (see Matthies et al. 1995), species protection will also be ineffective if it fails to take *interspecific* (ecological) relationships into consideration, that is, if it does not succeed in protecting the natural surroundings of the species (Fritz 1983, 301; Soulé 1985, 728). Therefore most biologists and conservationists agree nowadays that species protection can usually best be implemented by *protecting the habitats* in which species occur. According to Leitzell (1986, 253) habitat protection provides "the best possibility of increasing, or at least protecting, the diversity of species on the earth." And Roweck (1993, 18) thinks that we should abandon the idea all together that we could solve the fate of wild animals and plants at the species level: "The very number of all the species involved makes this impossible."

If the most appropriate and promising goal of species protection thus seems to be protecting the habitat rather than the species itself, it is nevertheless important to keep in mind that heading for this goal by a direct route may raise rather serious epistemological problems. While the requirements of individual species for survival can be determined (whereby a species here is considered to be the sum of all its individuals), it is not possible to establish a *definite* state of well-being for larger wholes such as biotopes, ecosystems, and landscapes that a conservationist might be able to secure or enhance through intervention. As discussed in Chapter 12 it is not possible to define a "healthy" state of an ecosystem that would correspond to what we mean by the health or well-being of organisms. For ecosystems we can postulate at the most a kind of *dynamic* state of "well-being" stemming from maximal autonomy such that the processes of self-organization and natural selection take place with as little interference as possible. Thus to do justice to an ecosystem means to minimize interference with it or at least interference of an irreversible kind.

In the course of thinking about ecosystem protection we arrive at a third specification of species protection that goes beyond what we might envision intuitively. In the end, to protect a species means to protect the integrity of *natural processes* (see Soulé 1985, 731; Scherzinger 1991; Smith et al. 1993b). To the extent that nature is itself a process, nature protection means process protection. In the context of a modern scientific worldview nature is no longer considered to be a mere collection of *things*, as many people still tend to think. Instead it is commonly viewed nowadays as a set of interconnected *occurrences*. As a result, the most comprehensive set of occurrences imaginable gradually comes into focus as the ultimate goal of holistically oriented nature and species protection, namely evolution itself. The very process by which matter has developed and which has brought forth such an enormous diversity of life forms as well as properties of systems such as consciousness and the human mind must be regarded in general as *valuable in itself*, even if many problems remain puzzling and questionable when examined in greater detail. As the most fundamental manifestation of nature on our planet, this process must not be curtailed without some kind of existential reason. Instead, it must be given the opportunity to develop further on the basis of the complexity and diversity that has accrued so far.

Obviously, an understanding of species protection that is extended in this manner to include the evolutionary dimension requires that our basic intuition to maintain *all* species be modified. No moral obligation can be postulated to the effect that humans might be required to maintain species that would die out *of their own accord*. Paleontological results indicate that even though there may be no inherent mechanism of extinction that results in the death of species, it is apparently still the "ultimate fate of all species" (Eldredge 1991, 58). In view of these findings, any attempt to try to prevent a natural extinction process would seem to be a case of *protecting nature against itself*. However, compared to species extinction induced by humans the particular case of *natural extinction* is quantitatively so insignificant that we can readily ignore it in the context of this discussion. The relationship between the natural and the anthropogenical rate of extinction is usually estimated nowadays to be at least 1:1000 (E. U. von Weizsäcker 1992, 128). Moreover, one can only refer to *natural* species extinction with sufficient certainty when this occurs in areas in which the course of ecological processes is for the most part indeed autonomous. However, this is probably only true in the few real wilderness areas left on earth and perhaps also in some of the larger national parks. In smaller protected areas complete succession cycles usually do not take place so that "letting nature take its course" in these areas may sometimes result in systemic conditions that correspond to neither the original goal of protecting the natural dynamics of the system nor that of maintaining species diversity (Remmert 1990, 164). If we take this situation into account, at the present time we must always assume that humans are more or less responsible for extinction processes. Of course, this is especially true for extinction processes that occur in cultural and industrial landscapes that are marked by human activity, because on a larger scale no autonomous processes take place in these environments any more at all.

It is not difficult to see that these circumstances pose a problem for the third specification of species protection discussed above. If species protection really means process protection and requires as little interference with natural systems as possible, and if process protection of this kind can only take place in a very few, particularly spacious wilderness areas, or rather, if it is only *there* that the basic goal of maintaining undiminished evolutionary potential is really guaranteed, then in a large number of protected areas the "principle of noninterference" or rather "letting processes run their course" cannot be the sole guiding principle. In order to make sure that the widely spread conditions in which human influence predominates do not confound the original aim of this principle, the first-order rule of process protection must be supplemented by a second-order rule. The latter would call for measures that compensate for the consequences of human influence and thus permit the original goal of process protection to be attained at least approximately.[141] Scherzinger (1991, 27) summarized the relationship between the two antagonizing procedures as follows: "As much dynamics as possible, as much management as necessary." He specifically recommends that as many and as large areas as possible be left to develop on their own, but also that these "wilderness islands" be connected with one another by means of smaller protected areas in which biotope management is restricted to only the most necessary interventions. According to Scherzinger (1991, 28), a combined program such as this would be most likely to result in implementing a kind of nature protection that does not require "constantly struggling against natural development" but rather permits "as great a degree of natural dynamics as possible."[142] Of course, the question that then automatically arises is what exactly might serve as a measure for determining "the greatest degree of natural dynamics." Is it possible to find a criterion for operationalizing moral respect for the rather abstract "whole" of the evolutionary process in a useful manner?

The answer to this question is provided by a fourth specification of species protection, which leads us back to the classical understanding of the term by taking genetic variability, ecosystem protection, and process protection into consideration. *Total diversity*—a concept defined by MacArthur (1965, 528) as "the total number of species in a fairly wide geographic area" and now extended to include the genetic, ecological, and evolutionary perspective—appears to be the most appropriate and practical criterion for judging human dealings with wild nature. Since species are the basic units of the evolutionary process (Mayr 1982, 296), the developmental state of total diversity represents a direct measure for determining the extent to which natural processes are being impaired by humans. According to estimates by Altner (1985, 568), "if we are successful in maintaining species diversity—with respect to a biotope, a landscape and the local infrastructure— . . . this is the best indication that deficits in the relationship between humans and nature have been overcome."

This is not to say that the various aspects of biological diversity to which we aspire (habitat diversity, species diversity, genetic variability, etc.) are all somehow "subsumable" under one unifying principle at the systemic level of the species (see Weber et al. 1995, 188, 189; Hengeveld 1994, 1). I also do not wish to suggest that the conflicts discussed previously that exist between the individual, the population, the species, or the ecosystem should always be resolved in favor of the species. But of all the entities of nature there are, it seems to me that the species occupies a position that makes it particularly well suited to be a major focus of environmental ethics. With the exception of the biosphere the species represents the most highly organized unit of life processes that can be grasped objectively in time and space, that is, without having to resort to making arbitrary distinctions. It is therefore a unit that can be readily operationalized for the purposes of nature conservation. The temporal and spatial boundaries of a species, as opposed to those of an ecosystem, are not dependent upon the judgment of an observer but determined "by nature itself" (Willmann 1985, 5).

It must be noted, however, that an *objective* status of a species such as that described above is explicitly contested by various representatives of nonholistic concepts of ethics. Thus Hampicke and colleagues (1991, 20), for example, maintain that a species is "not a natural entity but the result of human concept formation and therefore an abstraction." They further claim that "the boundaries of a species are often contested." Gethmann (1993, 248) expresses a similar view when he refers to a species as a "secondary abstractum." And von der Pfordten (1996, 165) believes that "the category species is purely the result of classification by an observer and in reality not involved in ecological interactions." Obviously, if estimations of this kind were accurate, they would not only be significant for theory of science but also have normative consequences. If species in a conventional sense were only *classes* devised by humans, then it would be difficult to explain why humans should hold any particular kind of responsibility toward them. To demand that we should exercise moral respect toward an arbitrary construct of human concept formation would be quite a futile undertaking. Thus at the end of the quotation cited above Gethmann (1993, 248) does indeed come to the following conclusion: "In a non-metaphorical manner of speaking there is no such thing as dignity of a species or genus."

However, the premise from theory of science on which this normative conclusion is based is no longer tenable nowadays. A species *is not* a "secondary abstractum." As the evolutionary biologist Mayr (1988a, 315) has emphatically pointed out, a species is "not an invention of taxonomists or philosophers but it has reality in nature." According to Mayr (1988a, 331), "Modern biologists are almost unanimously agreed that there are real discontinuities in organic nature, which delimit natural entities that are designated as species." Granted, the concept of a species as a class formulated by Gethmann, von der Pfordten, and Hampicke is one that many biologists, paleontologists, and philosophers used to hold, but this does not invalidate the fact that the *typological or morphological concept of a species* that it reflects is nowadays rejected by most taxonomists as unscientific (Mayr 1988a, 338). According to the modern *biological concept of a species*, membership in a species is no longer assigned on the basis of *subjective* similarity, that is, because of exhibiting or not exhibiting certain properties that at best could serve as circumstantial evidence. Instead, *objective criteria* are employed, namely the ability to reproduce (fertility) and reproductive isolation. In order to take the results of evolutionary biology into proper account, which show that species *continue to develop*, a species currently is understood to be a continually changing reproductive community rather than a type with a constant form. As Willmann (1985, 129) emphasizes, a species can have a very different appearance at different times in the course of evolution (species transformation or modification). However, as long as no reproductive isolation mechanisms occur between two subpopulations leading to phylogenetic separation, the entire line that has come to be in the course of evolution is considered to belong to one and the same species. According to Willmann (1985, 118), a new species can only be generated by a speciation event. "It begins when the mother species splits up into two subpopulations, and it ends as soon as the subpopulation that represents a new species has itself separated into two daughter species" or when it ceases to produce offspring. Both the changeability and the mortality of a species indicate that the "nature" of a species is not compatible with the concept of a species as a class. A class, which by definition has a constant essence, can neither develop nor can it as an abstract (and thus infinite) unit simply die off (Mayr 1988a, 347, 348; Willmann 1985, 58).

If species are indeed not classes, what then are they? One answer that was proposed by Ghiselin (1974) and has gained more and more support in discussions on the philosophy of biology in the past twenty years is as follows: Species are *historical individuals* and the name of a species is a proper name.[143] A species really does satisfy all the four conditions that Mishler and Brandon (1987, 399) describe as necessary conditions for individuality. These include "(1) spatial boundaries, (2) temporal boundaries, (3) cohesion, and (4) integration." Of these the most important criterion is certainly that of *cohesion*. Through their gene pool, organisms that form a species exhibit interconnections with one another and a continuity that the members of a pure class of objects would never have. It should be noted, however, that internal cohesion does not mean that the individuals must be connected by some kind of *physical* continuity or be somehow "inseparable" from one another. According to Ghiselin (1981, 274), even our earth and our solar system are individuals, although at the same time each can be regarded as a composite whole comprising separate individuals. He maintains that this shows that the term "individual" can indeed be applied in the same manner at different levels of organization. As far as bioscience is concerned, it is clear that the usual concept of an "individual" is mainly that of a single creature or organism. "But the cells that make up an organism are also individuals, and on the other hand, individual organisms are parts of an even more comprehensive individual, namely that of a biological species " (Willmann 1985, 57).[144] Concerning the relationship between a single organism and a species this means that the former is not simply a *member* or *an example* of a species. This would be the case if species were classes. If, however, a species is itself an individual, then each of the organisms that belong to it must be regarded as an integral *part* of it. Each is an integrated component of a super-ordinate unit of life, a manifestation of a dynamic lifeline with a unique biological and historical identity.

After having taken a look at some of the findings from biology and theory of science concerning the "nature" of a species that are relevant for the topic of this book, it is now perhaps somewhat easier to estimate what it means *from an ethical standpoint* to eliminate a species. Now we can clearly see that the extinction of a species does not just entail the disappearance of a "secondary abstractum" or, so to speak, the annulment of a construct of the human mind. To extirpate a species means to irreversibly destroy a *real* and *central* unit of the process of life. A biologically informed ethic must take the results discussed above into account and recognize that life processes are driven at several different systemic levels that are complementary to one another—at the level of the genes, in the form of individual organisms, and in the form of species. An ethic of this kind will come to the conclusion that the moral significance of destroying a species (usually) is much greater and more serious than that of killing an individual organism. Even though a unique kind of individuality is extinguished in both cases, the death of the *historical* individual that a species represents means the death of *a whole line of life* in addition to the death of the organismic parts of which it is composed. From a biological perspective it is not only an individual genotype that is lost, a more or less replaceable exemplar of this line of life. What is destroyed is an entire gene pool that has developed over millions of years. As Wilson (1995, 33) explains, "The DNA of each of almost all the species there are, whether amoebas or humans, [consists] of . . . one to ten billion basepairs or letters of life. This information would fill all the editions of the *Encyclopaedia Britannica* that ever appeared. Each species achieved this complex set of rules through innumerable mutations and acts of natural selection which caused this species to adapt optimally to a particular environment and led it to become integrated among many other different organisms to form an ecosystem" [Translation P.N.]. From this perspective to destroy a species means to suddenly knock out a process of genetic "discovery" and "learning" that has successfully asserted itself over millions of years amid innumerable challenges and dangers. "To kill a species is to shut down a unique story" (Rolston 1985, 723).

But wait! Wasn't it *nature itself* that continually interrupted such life histories in the course of evolution? If it is true that 98 percent of all the species that ever existed became extinct *naturally*, what is so particularly offensive about the extinction of species brought about by humans? This argument crops up repeatedly in ecological debates and in literature on environmental ethics.[145] Nevertheless, there are two good reasons why it is still not a very convincing one for mitigating the moral significance of anthropogenically induced species extinction. First of all, it overlooks the fact that many cases of natural extinction differ from those induced by humans in that they occurred through separation of a single former species into two new ones (speciation), except, of course, for the instances of mass species extinction. Since speciation results in the extinction of one species and its replacement by two successor ones, its disappearance does not mean that its history suddenly stops but rather that it continues on in the lifelines of the daughter species. Willmann (1985, 120) refers to this process as one of "resolution" in order to underline the difference between this process and that of death and extinction *without successor species*.

Another thing that is overlooked when people point to natural extinction is that it is not only logically contestable but also completely unacceptable from an ethical standpoint to justify human *action* by simply referring to it as a natural *occurrence* (naturalistic fallacy, see Chapter 8).

Just as the observation that people have always died as a result of natural catastrophes cannot serve as adequate justification for humans murdering other human beings, it is also not legitimate to refer to past examples of mass extinction or normal "background extinction" in order to mitigate the ethical significance of the current incidence of species extinction for which humans are responsible. *Natural* species extinction is to *human-induced* species extinction as the natural death of an organism is to killing it. Natural death is fate, while the act of killing, at least in the first instance, is one of moral injustice. Of course, in spite of striking parallels it is important to point out that there is a very obvious difference between killing an individual and destroying a species. In the first case we are dealing "only" with the destruction of *organismic* individuality, while the second involves the destruction of *historical* individuality in addition. In order to draw attention to the special quality of this moral iniquity, Rolston (1985, 723) referred to the extinction of a species as a case of "superkilling." "It kills forms (species), beyond individuals. It kills 'essences' beyond 'existences,' the 'soul' as well as the 'body.' It kills collectively, not just distributively." According to Rolston it is not only the loss of potentially valuable information for humans that constitutes the tragedy of this killing. It is the irreversible loss of biological information, regardless of whether or not it is useful for humans. Or, to use more graphic imagery, it means the loss of a unique masterpiece of nature.

But what if I am neither a sensitive aesthete nor in any way interested in natural history or biology? Why should I bother about the loss of another form of life or the biological information contained in it? After what has been said in the past chapters in an attempt to provide theoretical justification for preserving species, it should be clear that questions of this kind are irrelevant once I have reached a decision in favor of a moral standpoint after having been faced with the basic choice between egoism and altruism. Then respecting a species as a "masterpiece of nature" is not *only* a question of aesthetics or scientific interest. It is ultimately a matter of ethical self-determination. Of course, the *aesthetic* dimension can certainly help us to grasp the *ethical* dimension of species preservation both emotionally and motivationally, especially if it is also enriched by knowledge. But the objective claims of ethics go beyond this. If we view the world from a holistic perspective, then we very simply have "just as little right to destroy a species as an individual manifestation of life as we do to destroy human individuals, regardless of whether we like or dislike them" (Reichholf

1996, 63). Of course, advocates of a restricted concept of morality will not accept parallel argumentation of this kind, since they can at best conceive of duties toward organisms but not toward species. Remarkably, however, many of them do seem prepared to make *one* exception that leads them to recognize moral duties toward a historical individual instead of just toward organismic individuals, and that is with respect to the species *Homo sapiens*. At any rate, Rolston (1985, 722) feels justified in maintaining that "all ethicists say that in *Homo sapiens* one species has appeared that not only exists but ought to exist."[146]

Why then, Rolston (1985, 722, 723) rightfully asks, do advocates of a restricted concept of morality postulate this only for our own species? Why not postulate a more general concept of moral respect for a historical individual and extend it to other species? Certainly there is no doubt about it that "moral actors," creatures exhibiting the phenomenon of a moral conscience and the capacity for reflection, can only be found among the highly developed species of *Homo sapiens*. But it is this very observation that makes it seem paradoxical when *the only species capable of morality* in its dealings with other species is only capable of acting in its own (collective) self-interest. Rolston (1988, 157) puts his finger on the heart of this paradox with the following statement in which he remarks on the current situation of biological diversity and the almost exclusively instrumental importance that is usually attached to it: "Several billion years' worth of creative toil, several million species of teeming life, have been handed over to the care of this latecoming species in which mind has flowered and morals have emerged. Ought not this sole moral species do something less self-interested than count all the products of an evolutionary ecosystem as rivets in their spaceship, resources in their larder, laboratory materials, recreation for their ride? Such an attitude hardly seems biologically informed, much less ethically adequate. It is too provincial for superior humanity. Or, in a biologist's term, it is ridiculously territorial. If true to their specific epithet, ought not *Homo sapiens* value this host of species as something with a claim to their care in its own right?"

31. Balancing Interests and Dealing with Conflicting Duties

"Theoretically that's all very well," a proponent of an anthropocentric position in environmental ethics might reply to the argumentation in favor of the intrinsic value of nature and species presented here. "But as far as practical consequences go, the pluralistic-holistic ethics on which your position is based is completely unacceptable. If you place *everything in nature*, not just all species but also every blade of grass on the edge of a path, every insect, every tree and every pond under moral surveillance and thus declare it to be basically inviolable, that means no less than condemning humans to complete inactivity. People would then neither be allowed to mow a meadow nor farm the land nor make paths through the woods, let alone kill animals. Were we to pursue an ethics that requires moral consideration for everything that exists right down to the letter, this would not only lead to the end of human culture, it would very simply mean that humans would be condemned to starve.[147] How, of all things, can an ethics that claims to be ecologically informed overlook the fact that humans are situated at the top of the food pyramid and have no other choice than to constantly intervene in the processes of nature and live at the expense of other organisms?"

Before delving into the problem of balancing interests in the context of the holistic and pluralistic concept of morality addressed above, it is necessary to clarify one important misunderstanding that is at the root of this frequently raised objection, namely the idea that duties that exist *prima facie* are always completely equivalent to *actual* ones (see Ricken 1989, 186). The fact that a prima facie duty to respect the good of other creatures and natural systems exists does not mean that there is always a corresponding actual duty to abstain from any kind of intervention in nature. According to Frankena (1963, 24) "something is a prima facie duty if it is a duty other things being equal, that is, if it would be an actual duty if other moral considerations did not intervene." The essence of a prima facie rule is that it is not subject to any qualifications, because it is as such in keeping with the goals of all other members of the moral community (see Ricken 1987, 18). The rule that one should keep a promise, for example, is valid prima facie, without further qualification. In any case, this rule expresses a duty that we must try to fulfill. But the unrestricted validity of this prima facie rule does not mean that no other prima facie obligations exist that might conflict with it under certain circumstances, and that sometimes may have priority over it. The prima facie obligation to keep a promise can, for example, be overridden by another prima facie rule that says that we should try to save an innocent person from execution by a totalitarian government.

The conclusions drawn from this example involving interpersonal relationships can be applied in the same manner to the case of how humans deal with nature. In this case and in the context of holistic and pluralistic ethics a prima facie rule forbids us categorically to harm the physical and ecological integrity of other natural entities and systems. However, this rule can, for example, be superceded by the prima facie duty to protect one's own life or that of other humans. Holistic ethics does not require that a person sacrifice his own life for nonhuman nature. In fact, in view of the duty to respect oneself that morality also demands, it might even be considered immoral to sacrifice oneself on principle (see Günzler 1990a, 97). What I hoped to demonstrate with the example of the innocent victim of persecution is that conflicts between different prima facie rules are not just a problem of *holistic* ethics but can also crop up in the ethics of *interpersonal* relationships. However, I have to admit that the *number* of potential conflicts within a moral community that adheres to holistic ethics is decidedly greater than in a community that advocates restricted moral consideration. There are two reasons why this is not surprising. First, the number of members of the community that are subject to moral consideration is simply much greater. Second, as a "biological consumer" a human being cannot avoid living at the expense of other organisms in order to survive. If one views human beings' dealings with nature from a holistic perspective, one is confronted almost everywhere (at least with respect to other organisms) with existential conflicts. Since it is often necessary for a moral actor to violate the prima facie absolutely valid rule against killing nonhuman individuals, the thought, of course, occurs that a rule of this kind is useless and that the ethics upon which it is based is unrealistic and incoherent. In other words, what good is a moral rule that evokes more exceptions than cases of compliance? The answer to this is that in spite of all the compromises that must be made when balancing interests, and in spite of unavoidable concessions to the self-interest of human beings, a rule of this kind sounds out the maximum number of possibilities available for protecting nature since interventions in nature are placed in principle under the burden of proof. Contrary to anthropocentric ethics, holistic ethics requires that damage to or impairment of nonhuman life be justified *in principle*. And contrary to the claims of many critics of extended concepts of morality that when we reach the point of balancing interests anthropocentric and nonanthropocentric concepts of morality eventually come to the same conclusions (convergence hypothesis), it is the very process of balancing interests that shows that anthropocentric and holistic approaches are fundamentally different. Whereas anthropocentrism requires that specific cases involving *restrictions* on free exploitation of nature be justified in principle, holism demands that specific cases involving *exploitation* of nature that encroach upon its basic inviolability be justified. The crucial point of nonanthropocentric concepts of morality is that they *reverse the burden of proof*.

Of course, it still remains to be seen whether reversing the burden of proof really has any *practical* consequences. At this point in the discussion I must return to an objection I already mentioned, the objection that even at the theoretical level, when it comes to balancing interests, the difference between anthropocentrism and holism I have proposed is an illusion. According to this objection, even in the context of a holistic concept of morality *we humans* are always the ones who decide whether or not sufficient "proof" has been presented to justify damage to natural entities or systems in favor of our own interests. Strictly speaking, so goes this objection, the process of balancing interests in the context of holism bears the mark of hidden anthropocentrism. I have already discussed this objection in Chapter 25.b (Is Anthropocentrism Inescapable?) and demonstrated that whether or not it is justified depends upon how the supposedly unavoidable violation of a particular prima facie rule is regarded by the nonanthropocentric concept in question. With respect to the ethical status of such violations two theoretical options were put up for debate: (1) a relative or moderate version of nonanthropocentrism, and (2) an absolute or radical one. While *moderate* versions of nonanthropocentric ethics are based on an a priori hierarchical value system ("scala naturae") that on principle permits ethical justification of sacrificing "lower" forms of life for "higher" human interests (e.g., Attfield 1983, 176; Rolston 1988, 223f.; Hösle 1990, 73), advocates of a radical version of nonanthropocentric ethics very definitely reject the idea of forcing the concept of intrinsic value into a hierarchical system (e.g., Schweitzer [1923] 1974; Taylor 1986). For the latter the conflict between the moral claims of nature and the interests of humans represents a moral dilemma that basically cannot be completely resolved.

Although I cannot deny that our unreflected everyday intuition tends to quite naturally assume a value hierarchy of all organisms with humans at the top and protozoans or inanimate matter at the bottom, in the following discussion I will argue in favor of a modified form of the second, radical, or absolute concept of ethics. My skepticism about scaling the concept of intrinsic value from the very beginning is based on both theoretical and practical considerations.

From a *theoretical* perspective it must be pointed out that there is no truly convincing evidence for such a hierarchy in nature. As Schweitzer (1991, 157) correctly noted, "in the long run, any attempt to establish general differences in value among organisms . . . involves making judgments on the basis of feelings about whether we consider them to be closer to us or farther away, a highly subjective measure of value. Who knows what meaning another organism might have in and of itself and in the context of the world?" As far as the position of an organism within an *ecological* whole is concerned, it is obvious that the subjective measure of our intuition is not supported by ecological evidence. On the contrary, the logical problem of a naturalistic fallacy notwithstanding, more empirical evidence contradicts such a measure than supports it. When we rely on subjective feelings, we not only seriously underrate the most "ecologically significant" species, for example, "ecosystem engineers" (earthworms, trees, etc.) or microorganisms, whose contribution to energetic, trophic, or symbiotic relationships is indispensable. We also fail to account for the temporal and spatial variability of ecological relationships when we rely on a strict value hierarchy based on intuitive assumptions. As the discussion about the concept of "keystone species" has shown (see Chapter 22.b), it is often impossible to assign fixed "ecological value" to many species because whether or not a species functions as a keystone species may vary significantly in time and space.

If attempts to establish a value hierarchy of organisms by referring to *their role* in a (hypothetical) whole are destined to produce highly questionable and contradictory results, how much more dubious must be the various attempts to evaluate organisms *as such*, that is, on the basis of their properties or functional capacities. Of course, our intuition tends to tell us that a value hierarchy in which humans have the greatest intrinsic value compared to all other animals and plants is self-evident. And we justify this intuition by pointing out that only humans have brought forth such unique things as rationality, aesthetic creativity, self-determination, morality, and culture. Nevertheless, if one casts an unprejudiced eye on this intuition, she will wonder why *these* particular properties (and not some other unique ones) lend their holder greater intrinsic value (see Taylor 1984). As Taylor (1981, 212) has pointed out, "it is true that a human may be a better mathematician than a monkey, but the monkey may be a better tree climber than human beings. If we human beings value mathematics more than tree climbing, that is because our conception of civilized life makes the development of mathematical ability to be more desirable than the ability to climb trees." However, this has nothing to do with greater *intrinsic* value.

Moreover, we must be aware that it is highly ambivalent to make the intrinsic value of an organism dependent upon the presence, scope, or degree of certain abilities and achievements. Because if we were really consistent in our thinking, we would also have to scale the intrinsic value of individuals *within* the human race, which ultimately would lead to the completely unacceptable conclusion that healthy, intelligent, and mature people are in and of themselves more valuable than, for example, sick or mentally disabled people or infants. One argument that is often brought forth in contradiction to this idea is that it is not the immediately existing abilities that are decisive but rather the potential for such abilities (Skirbekk 1995, 425). It seems to me that this auxiliary argument is not very convincing. It would be more authentic to admit that when it comes right down to it, we attach greater value to belonging to the species *Homo sapiens*.

But how and why should this genetic fact alone lend greater value to an organism? In my opinion no convincing answer can be given to this question without going back to the metaphysical premises of the anthropocentric worldview that have already been criticized. Only if we assume that the evolution of life forms is a progressive process that leads to higher forms, of which humans represent the current or perhaps even final culmination, is the concept of human superiority plausible. However, nowadays the assumption of "progress" in evolution is rejected by most evolutionary biologists as an example of ideological projection (Wuketits 1995). What objective measure might there be for determining such "progress"? Certainly it cannot be denied that, for example, *complexity* has increased since the beginning of life, for example, that the organization of relatively recent mammals is much more complex than that of any billion year old unicellular organisms. But whether and why exactly an increase in complexity should be considered "progress" is a moot point. "We expect a progressive system to be highly efficient and robust. But complex systems in particular are susceptible to disturbances. As efficient as they may be in solving their problems, they are very ineffective when confronted with more or less dramatic changes in their environment (Wuketits 1995, 35).

Instead of complexity we might just as well refer to *reproductive success* as a measure of evolutionary progress. In this case many an insect would have to be considered more highly evolved than mammals, even though this might strongly contradict the intuitions of most people. But according to Wuketits, when we speak in a naïve and intuitive manner of "primitive" and " more highly developed" organisms, we miss the point about what is at stake in evolution and the lives of organisms. "Each organism has to solve problems, so to speak. It must manage to survive in an environment that is by no means friendly. But how an organism solves these problems is unimportant as long as the solutions guarantee its survival, or, more precisely, its reproductive success. A mole that blindly burrows through the earth has not solved its major survival problem any less successfully than a primate with its capacity for spatial and color vision. Otherwise there would be no moles."

If we really did declare *the ability to solve major problems of life* to be the decisive criterion for evaluating a species, then humans would have quite a chore ahead of them to prove their (self-claimed) superior worth. In view of the great experiment with the global climate (greenhouse effect) and the destruction of biological diversity (extinction) we have recently entered upon, it is very possible that the not too distant planetary future will show that we have sawed off the branch on which we were perched. Thus it is quite fitting when Wuketits (1988, 172) poses the following question: "What 'developmental level' can be attributed to a species that is able to reflect on evolution but at the same time does a great deal to destroy the biosphere!?" Regardless of what reply we give to this question, it seems at least to illustrate that neither an unequivocal measure of evolutionary "progress" nor an *objective* criterion for a value hierarchy of organisms exists.

Having shown that any attempt to establish a value scale for organisms is destined to fail at a theoretical level, *practical* objections to moderate versions of nonanthropocentric ethics then become evident. If there is no conceptual basis for a hierarchy, then the risk is great that operating as if there were such a thing will result in arbitrary, thoughtless, and self-righteous practice. Thus Teutsch (1990, 102) has good reason to suspect that when we differentiate between higher, intermediate, and lower levels of value with respect to organisms, this can be interpreted to mean that "humans naturally occupy the highest level and therefore have the right to regard all inferior forms of life as means placed at their disposal for attaining their own ends." Human behavior that leads to the loss of plant or animal life would only be considered reproachable if this life were thought to be of relatively great value (Vossenkuhl 1993b, 10). Instead of granting all non-human life as great a degree of protection as possible, a position of moderate and relative nonanthropocentric ethics would have exactly the opposite effect. With its value hierarchy it would in principle justify harm to a large part of nature. Thus what might at first appear to be a "realistic" and "viable" way of dealing with the problem of balancing interests turns out to be a highly dubious matter when examined more closely. Ways of dealing with nature that are thoroughly subjective and basically guided by self-interest are raised to a level of ethical "correctness" and finalized by developing criteria that are appropriate to this behavior (see Günzler 1996, 165). However, it is obvious that a strategy of easy and general moral self-exoneration such as this is not what ethics is all about. As Günzler (1990a, 98) points out with reference to Schweitzer ([1923] 1974), "It may satisfy our will for self-assertion when we successfully manage to defend ourselves against pain or destruction, but this does not give us the right to justify having had to sacrifice the lives of other beings to achieve this end as a moral act by referring to a value hierarchy." The outcome of this way of thinking would be moral self-complacency and relinquishment of moral reflection. If when trying to reach a decision by balancing interests we start off with wholesale partiality in favor of our own interests, then there's really no reason to consider whether or not harm to other beings is necessary for asserting ourselves. Before we even perceive a conflict of interest as such, we neutralize it with a value hierarchy that operates to the disadvantage of nature. Vossenkuhl (1993b, 10) is thus right when he maintains that "morality of this kind does not provide any kind of orientation to help us improve our behavior but instead helps us to excuse the negative consequences of our behavior."

But what then is the theoretical alternative to such a moderate and relative version of nonanthropocentric environmental ethics? In the context of an absolute concept of ethics for which I will argue in the following section, the idea of pluralistic holism is that we are obligated prima facie to exercise the same moral consideration toward all species. This obligation is not weakened when interests are balanced against one another and will automatically lead to a *moral dilemma* in the case of a conflict with other prima facie duties.[148] If conflicting obligations or self-interest make it necessary to resolve such a dilemma to the disadvantage of nature, then the harm to nature that ensues must be conceived of as *guilt*, the dimensions of which depend upon the degree of necessity involved.[149] In the words of Schweitzer (1991, 42): "In the case of a conflict between saving my life and destroying or harming the life of another being, I can never reconcile the ethical and the necessary by a relative concept of ethics but must choose between the ethical and the necessary. And if I choose the latter, then I must accept guilt for having harmed the life of another."

Three objections are usually raised against this concept of ethics. The first criticism refers to the accumulation of moral dilemmas that it implies; the second is directed toward what is perceived as an inappropriate extension of the concept of guilt; and the third points to the risk of subjectivism and relativism. Concerning the first problem it is clear that any accumulation of moral dilemmas is unacceptable for ethical theories that fundamentally deny the possibility of conflicts of duties. In scholastic ethics (after Aristotle and Plato) but also in Kantian ethics, conflicts of duties are regarded as merely fictitious. Since these ethical theories are based on the idea of a unified system of morality with a hierarchical order of goods and duties, they assume that any such apparent conflicts can be resolved objectively by referring to a greater obligation. However, there are reasonable doubts that an abstract concept of a scaled order of duties is really adequate for dealing with the problems connected with specific situations involving vital necessity. In particular these concepts do not seem to take the significance of the context of moral claims sufficiently into account. After his rejection of *theoretical* objections to the possibility of moral dilemmas Vossenkuhl (1992c; 1993a, 145) draws attention to the *practical* dangers of a decontextualized, formalistic concept of ethics. "A view of the world in which a hierarchical structure of laws and obligations predominates not only restricts personal moral judgement. It also generates false certainty with respect to judgement." Thus pragmatic considerations lend greater plausibility to a concept of ethics that allows for a plurality of nonreducible moral norms as well as values and obligations not ordered in a hierarchical fashion, even if the unpleasant price that has to be paid when this concept of ethics is applied is often that of moral dilemmas (see Forschner 1992, 212).

If one accepts this theoretical view of ethics, then there is no reason to deplore the accumulation of dilemmas that arise in the context of pluralistic holism that I advocate. As Vossenkuhl (1992b, 188) remarks, in interpersonal ethics as well moral dilemmas are not at all restricted to extreme situations (as, for example, abortion or other matters of life and death). In fact they are "commonplace, when, for example, professional duties conflict with duties to the family and when these both conflict with duties to oneself." If we usually do not tend to perceive these situations as dilemmas, it is because the norms connected with them, which are mostly determined by the society in which we live, are more or less strongly represented in our consciousness. Thus we will rarely consciously perceive the dilemmatic nature of the following situation in the area of environmental ethics that Vossenkuhl (1992c, 164) describes: "I am, for example, supposed to avoid harming nature with waste or pollutants, but I am also supposed to feed myself, heat my home, etc.; I must therefore prevent both waste and pollution, but I must also produce it since I cannot nourish myself without waste and pollution." This example not only demonstrates how commonplace moral dilemmas are. It also shows that attempts at resolution are bound to fail. It is not enough to refer to the deontic principle that "can" is a prerequisite of every "should" and maintain that it is therefore absurd to try to moralize the question of releasing pollutants by heating and eating. The inability to master everything in life is not a legitimate criterion for what I ought to do. The fact that I cannot live *without* producing waste and pollutants does not help me to decide *how much* and *under which circumstances* I am allowed to produce these things. The structural difficulty connected with responding to such questions is what Vossenkuhl (1992c, 163) refers to as the problem of "normative over-determination of the moral choice." Not only are we lacking an unequivocal hierarchy of obligations that might facilitate reaching a decision in a specific situation. We also have no generally valid criteria for estimating the relationship between our obligations and the particular context at stake. But it is this very context that to a large extent determines what is morally right. Whether or not I should turn the heat on during the season between summer and winter; whether I should heat all of my rooms or only one; whether I should set the thermostat to 59 degrees F or 77 degrees F—from a standpoint of environmental ethics these things not only depend upon how seriously I judge the danger of the greenhouse effect and my contribution to it to be. My decision will also depend upon how sensitive I am to the cold; whether or not I have acute or chronic bronchitis; whether or not I consider it aesthetically acceptable to put on a sweater; and whether or not I am prepared to argue with my roommates about the room temperature. If you consider the entire conglomerate of conditions associated with this decision, it seems to be a complete illusion to expect ethics to provide *general* or *objective* directives for defining how to act in a particular situation. Ethics simply cannot be that "useful." And if a subjective choice is made in favor of the apparently lesser evil (e.g., heating a single room to a temperature of 68 degrees F), ethics is also unable to show that the infringement upon a competing prima facie duty (the duty not to exacerbate the greenhouse effect more than necessary) that accompanies this decision is morally justified. The obligation that would have to be ignored in this case still continues to be an obligation. "An obligation doesn't just disappear because contingent matters make it impossible to uphold it" (Vossenkuhl 1992c, 166).

The question is, of course, how we are to interpret the act of disregarding a prima facie duty. Are we dealing with a justifiable restriction, or is it really a matter of *guilt*, as suggested above? In the literature two arguments are usually presented against using the concept of guilt in the manner that characterizes Schweitzer's ([1923] 1974) ethics. First of all, it is argued, the concept of guilt should be reserved for instances in which freedom and culpability really exist. If we broaden its application to include unalterable facts of life such as the role of humans as consumers, then we overextend the scope of ethics and blur the distinction between deliberate and compulsory or existentially necessary transgression of a moral rule. Strey (1989, 126), for example, finds it "inappropriate to talk about a state of guilt if and when we are forced to live at the expense of other organisms." To apply the concept of guilt in this manner would ultimately mean "that humans are guilty simply by virtue of being humans." According to Strey a position of this kind could easily provide a way of "excusing one's behavior because of being unable to either make amends for guilt or avoid it. If there is no innocence in these very basic matters of existence, then it is a waste of time to reflect at any length on other forms of culpability."

I do not deny that there is indeed a certain risk involved with such broad application of the concept of guilt. But one can counter Strey's argumentation with the observation that his own position is no less subject to interpretative risk. If there is *in principle* no guilt connected with the role of humans as biological consumers, then it is difficult to criticize the *excessive* consumption of nature they exercise at present. Strey (1989) would reply that his exoneration of humans from the onus of guilt only holds true for those instances in which we use parts of nature "which we need in order to live" (p. 131), but elsewhere he admits that there is "no *factual* argument" that would allow us to define an "objective" line of demarcation along a scale extending from merely surviving to living luxuriously in order to distinguish between what is legitimate and what is not. As was made clear in the discussion about the problem of moral dilemmas, here too there is no *ethical* argument that can help us out of the predicament. Ethics cannot tell me, for example, how I can eat in an "ethically irreproachable" manner, that is, for instance, whether I should eat meat at all, under which circumstances, and to what extent. It can provide prima facie rules that are relevant to this matter (e.g., against killing, cruelty to animals, or irreversible damage to natural systems), but as far as transgressions of these rules are concerned, it cannot define an objective borderline between what is permitted and what is forbidden. If we thus accept that a simple alternative between black and white does in fact not exist, but that there is instead a continuous spectrum of various shades of gray between black and white, then it doesn't seem to be completely unfounded to correlate this spectrum with a graduated concept of guilt. In this case innocence ("white") is impossible because of the constant presence of moral dilemmas, but the "degree of grayness" can at least be minimized if prima facie duties are only circumvented under the pressure of existential necessity. Obviously this only makes sense if we as actors are really interested in reducing the "grayness" of our behavior, which means taking the concept of guilt seriously and not passing over it lightly.

But this is the very point at which the second argument against employing a concept of guilt in the context of absolute ethics crops up. "If harming life or obstructing it is bad, but, on the other hand, I cannot live without harming other life, then those who take this ethics completely seriously could fall victim to neurotic guilt feelings that paralyze them in their actions instead of inspiring them" (Günzler 1990b, 120). This could indeed be risky, particularly in the context of environmental education for children and adolescents, and it must be carefully considered from case to case. "Environmental education should not be implemented at the expense of causing people to be heavily burdened with feelings of guilt and deprived of their ability to enjoy life" (Günzler 1990b, 120). However, it is difficult to see that *exaggerated* feelings of guilt currently pose a serious danger in areas other than the very sensitive one of children and adolescents.[150] Symptoms such as the loss of species, industrial animal husbandry, or processes causing more and more of the surface of the earth to be sealed off (296 acres per year just in Germany) seem to indicate that at least in Western industrial nations the problem is not so much large numbers of humans plagued by a guilty conscience but rather that the very existence of a large part of nonhuman nature is threatened. It is not *unusually strong* feelings of guilt regarding nature that give us reason to be concerned but rather an almost total *inability to see that guilt might be involved at all*, a blind spot generated by an anthropocentric worldview. If we assume that *these* conditions are generally valid, and if we also take the power of human self-interest into account that probably predominates anyway, then the absolute concept of ethics that I propose here and the use of the term guilt associated with it do not seem to be terribly problematic.

On the contrary. In my opinion this concept offers at least four opportunities for better ways of dealing with nature. First, it probably mobilizes individual feelings of responsibility and capacities for moral judgment to a greater extent than a relative concept of ethics. Since it does not provide individuals with a ready-made hierarchy of values or clear lines of demarcation between "allowed" and "forbidden" that have been approved by expert ethicists, it challenges the individual to decide for himself from case to case exactly where his personal limits to maintaining and supporting other life lie. In each case he is forced to try to exercise as much humanitarian behavior as possible toward his environment (see Schweitzer 1991, 40, 44). Second, the experience of "guilt" that surfaces once a moral dilemma has been resolved serves as an important guideline for future decisions. As Marcus (1987, 200) has shown, it causes us to try to avoid dilemmatic situations wherever possible. People will automatically endeavor "to act such that if they are supposed to do x and y, they will also be able to do x and y."[151] Third, the experience of guilt encourages us to think about something like compensation, atonement, or simply gratitude toward nature (see Schüz 1990, 148, 149). Anyone who feels guilty about doing harm to nature will most likely seek to reduce this guilt at least to some extent by helping other living things wherever possible or by providing support for them. This could be achieved in connection with animal or species protection or biotope conservation. Fourth, an absolute concept of ethics refines the attention and perception of a moral actor by constantly revealing the innumerable and basically irreconcilable conflicts that exist between humans and nature. Instead of dulling her senses with prefabricated compromises and ultimate ethical solutions it enables her to experience conflicts more profoundly from time to time, to sensitize her conscience, and thus to gradually internalize an attitude toward humankind and nature that is characterized by as much consideration and sympathy as possible.

Nevertheless, we must still consider the third and final objection to this kind of ethics, the objection that it is tainted by subjectivism and unrealistic rigorism. Isn't it expecting too much of an individual if we leave it up to his or her judgment alone to find a way to master the often very complex ecological dilemma situations that occur when dealing with nature? Isn't it unrealistic and even quite simply irresponsible to foster a concept of ethics so narrow that it is limited to propagating *ideal* prima facie rules that cannot always be maintained in everyday life, a concept whose authority ceases to function as soon as *practical* decisions have to be made that force us to violate certain prima facie rules? I regard this as a legitimate objection, but for reasons discussed above I still consider an absolute version of pluralistic ethics to be more convincing and productive. Therefore I would like to draw upon a suggestion made by Burkhardt (1981, 324f.; 1983, 426f.) and propose maintaining an absolute concept in the first instance but complementing it with a relative concept of ethics at a second level. According to Burkhardt's dual concept of ethics, *absolute* ethics forms the foundation upon which prima facie duties are based. "It tells us why some things simply *have to be* and why other things *are not allowed*" (1983, 426). If, however, circumstances force us to violate these prima facie duties, then *relative* ethics must come into play. At such a secondary level violations of prima facie duties cannot be ethically justified after the fact, but at least it is possible to provide trans-subjective reasons for such violations. Furthermore the function of relative ethics is to convey principles that help an individual to minimize his guilt in the course of weighing the probable consequences of his actions. I envision two advantages to such a dual concept of ethics that includes both absolute and relative ethics. First of all, it prevents decisions that are of necessity subjective from completely escaping the domain of ethics and the obligation of ethical justification. And second of all, it permits us to incorporate both deontological and consequentialist aspects in ethical discourse. The latter seems to be all the more important since a combination of both approaches has proven to be absolutely essential for an effective "ethics of a technological era" (cf. Chapter 17; Zimmerli 1991, 404).

However, in the context of the discussion presented here it would be too great a task to try to develop something like a catalog of principles for the relative level of pluralistic holism. Just how voluminous an undertaking of this kind might be is suggested by Norton's (1987, 180) estimation that it would be a "monumental task not yet begun" to try to develop practical rules in the context of pluralistic holism. The main difficulty is that pluralistic holism does not just involve weighing the interests of individual organisms (that is, entities belonging to the same level of systemic organization) against one another. It must also deal with conflicts that span different systemic levels (e.g., between individuals, populations, species, and ecosystems). And how exactly we might ethically cope with conflicts of this kind is still quite uncertain (Norton 1987, 179). But in spite of these methodological problems I don't want to end the discussion with abstract theoretical considerations. Instead, allow me to discuss at least five "priority principles" that Taylor (1986, 263) proposed for fair solutions of conflicts of interest in the context of his *biocentric* concept of morality. These are the Principles of (1) Self-Defense, (2) Proportionality, (3) Minimum Wrong, (4) Distributive Justice, and (5) Restitutive Justice. Let me very cautiously advance a few ideas about whether and how these principles could be extended to accommodate the broader scope of holism, or rather, how they might possibly be modified to achieve this.

The Principle of Self-Defense. As in the case of interpersonal relationships so also in conflicts with nature can killing a dangerous aggressor be considered excusable if it is the only way there is to save one's own life. In a holistic context this principle can be extended to include contests with entire species or even their extinction, as, for example, when dealing with the threat posed by lethal viruses or bacteria. However, the controversial discussion about what to do with the last laboratory specimens of the smallpox virus (Dixon 1976) showed that a decision in favor of complete extinction of a *species* is much more difficult to reach than in a situation involving self-defense among individuals, even from a purely anthropocentric standpoint. Neither the future risk of a pathogenic species for humans nor its potential usefulness for medical research can be calculated with certainty. Regardless of what one thinks about self-defense in conflicts with entire species, it is important to note that the concept of intrinsic value of a species does not imply that the species remains unviolable *in every possible situation*. Contrary to what advocates of restricted concepts of morality sometimes maintain, a person who supports the intrinsic value of species is not obligated to save the bacteria that cause tuberculosis. Just as killing a criminal kidnapper in self-defense does not disprove the universality of human dignity, recognizing the dangers or undesirableness of TB bacteria does not render the idea of assigning intrinsic value to other species absurd (see Taylor 1983, 241).

The Principle of Proportionality. If *basic* interests of (harmless) animals and plants are in conflict with *nonbasic* ones of humans, the basic interests should in principle be given priority over the others, regardless of their origin. According to Taylor (1986, 273), nonbasic human interests can be recognized by the fact that they vary from person to person, whereas basic interests, which are an essential component of personal existence, are shared by all humans. Taylor distinguishes between nonbasic human interests that are *intrinsically* incompatible with respect for nature and those that are *extrinsically* incompatible, that is, as a result of their consequences. As an example of intrinsically incompatible interests, the pursuit of which, in his opinion, is the least excusable, he mentions killing elephants in order to sell ivory to tourists, picking rare wildflowers for private collections, or hunting and fishing *solely* for pleasure. What is characteristic of all these examples is that other creatures are treated in a *completely* instrumental manner in the course of a conscious act involving no basic needs. Cases that Taylor (1986, 276) considers to be examples of pursuing nonbasic interests that are not intrinsically incompatible with respect for nature but may still have serious consequences include, among other things, constructing airports, highways, public buildings, or parks in areas that were formerly not used or rarely used; damming free-flowing rivers to produce electricity; or clearcutting a virgin forest to plant a commercial tree plantation. Obviously this distinction between intrinsic and extrinsic incompatibility makes no sense when we try to apply it to a holistic perspective that grants intrinsic value to entire systems as well as individual organisms. Flooding a valley by building a dam might be merely extrinsically incompatible with respect for nature if only the individual organisms involved are considered, but if we take the valley as a whole into consideration, then the incompatibility of this act is clearly intrinsic. With respect to the valley, the act of flooding it is one of complete instrumentality. Thus if the principle of proportionality is applied from a holistic standpoint, this would mean that not only the basic interests of nonhuman organisms but also the "right" of inanimate entities (rivers, mountains, etc.) and collective wholes (ecosystems, species, etc.) to exist would have to be given priority over nonbasic human interests. Massive damage to superordinate wholes for the purpose of satisfying nonbasic human interests would obviously be the least excusable kind of intervention.

The Principle of Minimum Wrong. Taylor (1986, 282) summarizes this principle approximately as follows: If rational, informed, and autonomous people who have already quite clearly assumed an attitude of respect for nature feel that they are unable to give up certain of their nonbasic interests even though they are extrinsically incompatible with the basic interests of other species, then they should try to pursue these interests in a manner that would cause as little damage to nature as possible. I have already listed a few examples of nonbasic interests that Taylor considers to be extrinsically incompatible with respect for nature. In these and in similar instances the principle of minimum wrong requires that before human interests are asserted, it is necessary to check whether the goals associated with such undertakings can be realized in a less destructive manner. In keeping with his biocentric approach, Taylor (1986, 284) refers exclusively to the number of creatures harmed as a measure of destruction. But, as already noted, with a pluralistic and holistic ethic the effects on superordinate wholes such as populations, species, and ecosystems would also have to be considered. Minimizing wrong in the context of holism would usually mean that superordinate wholes' "right to exist" would have priority over the good of the individual parts. At any rate, the irreversible extinction of a species for the sake of nonbasic interests would have to be prevented.

The Principle of Distributive Justice. If basic interests of humans are in competition with those of other species for space to live and natural resources, then all parties should be given the same amount or at least a fair share of these. In essence this means that an appropriate percentage of land and water surface should be maintained as free from utilization by humans as possible (nature reserves, national parks), and that if humans are permitted to use these areas, that this be executed in a manner that is as amenable to nature as possible.[152] According to Taylor (1986, 293) the principle of distributive justice requires us in general to "devise ways of transforming situations of confrontation into situations of mutual accommodation whenever it is possible to do so." Taylor (1986, 296) points out that if we adhere to the principle of distributive justice in a logically consistent manner, we would at least have to change our eating habits. We could drastically reduce the amount of agriculturally cultivated land currently used for producing feed, which usually excludes inhabitance by many other species, if we would reduce the excessive amount of meat we consume. The use of fertilizers and pesticides, the main sources of species extinction in middle Europe, would also drop. However, even though the moral dilemma that results from a conflict between the basic interests of humans and those of other species could certainly be greatly alleviated by such a move, it cannot be eliminated completely. While some cultures (as, for example, Eskimos) are forced to eat mostly meat, no one can avoid eating at least plants and intervening more or less severely in the dynamics of naturally evolved ecosystems in order to do this. In order to alleviate the effects of this dilemma, Taylor (1986, 304) proposes the following final principle.

The Principle of Restitutive Justice. This principle comes into play when species are subjected to harm even though the principles of minimum wrong (principle 3) and distributive justice (principle 4) have been observed. In cases such as these we should try to compensate for or make restitution for the damage incurred through measures such as species or biotope conservation. However, let me point out a misunderstanding that might arise in this connection. "Restitutive justice" does not mean that massive interventions in nature can be justified by claiming that a "surrogate biotope" will be set aside someplace else in place of the area that has been destroyed. In the long run this kind of "bargaining with indulgences" that nature protection agencies often grudgingly have to accept ("compensatory land purchases"; see Gerdes 1996), would be fatal. Instead of preventing a complete sell-out of nature, a concept of this kind would provide justification for such acts, not only after the fact but strangely enough even beforehand. In view of the possibility of such abuse of the idea of "restitution," it seems clear to me that it is very important not to interpret the rules of *relative* ethics listed above merely as isolated instances but always to consider them in the context of superordinate prima facie rules generated by *absolute* ethics. Then it becomes obvious that these rules are not there to absolve humans of the guilt they accrue with interventions in nature that often might have been avoided.

Instead they should help us to keep our "guilt account" as small as possible. Of course, this requires that we be really willing to minimize our individual or collective guilt, which also means seriously considering so-called "zero options."

How are we to evaluate Taylor's five priority principles *all in all* in the context of the two-tiered concept of morality I have proposed? On the one hand they are certainly useful for a couple of reasons. First of all, they demonstrate that biocentric or holistic-pluralistic ethics is not destined to fall apart as soon as it comes to a showdown in the course of balancing conflicting claims. And second, they show that the kind of nature protection they imply is usually more far-reaching than that of strict anthropocentric ethics, even if this cannot be proven by rigorous deduction in each and every case. On the other hand, however, it cannot be denied that there are severe methodological limits to the possibility of applying these rules. Neither are these principles so comprehensive that they cover all imaginable conflicts between humans and nature (not to speak of the conflicts between nature and nature induced by humans). Nor are they so specific that they permit all the conflicts they address to be unequivocally and neatly resolved. As Taylor (1986, 263) himself has emphasized, these five principles should not be understood as if they could function as premises of a deductive argument. "We cannot deduce from them, along with the facts of the case, a true conclusion expressible in a normative statement about what ought to be done, all things considered." Instead an undefined amount of uncertainty always remains in making decisions, sometimes even a definite inability to reach any decision at all.

If you recall the previous discussion about the problem of moral dilemmas, then this is not surprising. Our attempts to describe reality show that it is too complex and multilayered to be grasped by simple and general solutions of a normative nature. And in view of the fact that an "objective" analysis of the relevant context is a difficult task even for practical ethics concerning *interpersonal* relationships, this task appears to be almost completely insurmountable in the field of fundamentally complex, often nonlinear *ecological* relationships that can be generalized only to a limited extent. This must be taken into account if and when competent environmental ethics are worked out in the future (see Marietta 1995, 167). As far as pluralistic holism goes, this means that hopes for a *detailed* catalog of obligations probably have to be abandoned. Expectations of undeterred feasibility in ethics comparable to that found in applied science can certainly not be satisfied by this concept of ethics.

However, as widespread as such expectations may be, they reflect a serious misunderstanding, namely the idea that ethics is a systematic discipline for providing individual directives for action by ultimate and conclusive deduction all the way to a justified value hierarchy (see Günzler and Lenk 1990, 47). The opposite view, which Schweitzer ([1923] 1974) proposed, seems to be more realistic. Schweitzer frequently employed the image of life as a jungle in which ethics can only provide general orientation similar to that of a compass. However, a compass does not relieve the individual of the necessity of using his machete to tediously make his way through the undergrowth, constantly making new decisions and assuming responsibility for them (Günzler and Lenk 1990, 49). The consequences for the problem of balancing conflicting claims and weighing conflicting duties are obvious. Ethics cannot control everything that, according to its principles, should be controlled. Ethical thinking is simply a valuable tool that provides general orientation but cannot free the individual actor from responsibility in reaching a decision. *His* or *her* personal knowledge, attitudes, judgment, and competence in using the "ethical compass" are indispensable.

32. Conclusions and Prospects

Having completed our long journey through ecology, theory of science, philosophy of nature, and ethics it is now time for a final assessment. What, in a nutshell, are the results of this treatise? The starting point for my thoughts was the phenomenon of worldwide species extinction, which I regard not only as a major event in planetary history but also as a disconcerting symptom of a human-made crisis. Two solutions for this so-called ecological crisis were presented, which are often closely connected with one another albeit not necessarily: (1) the conviction that this crisis can be resolved by *science and technology*, and (2) the conviction that if an environmental ethic is necessary for resolving this crisis, it should be an *anthropocentric* one, in keeping with ethical tradition. In my treatise I have rejected both of these. Having shown that the first solution is incompatible with the epistemological and normative limits of ecology, the second foundered on the touchstone of species protection. Starting with the premise that general species protection (i.e., one that in principle includes *all* species) is desirable, a premise initially rooted in intuition, the anthropocentric arguments based on utility that are commonly presented (e.g., economic, ecological, and aesthetic ones) were shown to be insufficient and unstable. General species protection can only be justified in a factually convincing and psychologically coherent manner in the context of holistic ethics. Having pointed out these *practical* consequences I then attempted to demonstrate that holistic ethics, and with it general species protection, are also justifiable on a *theoretical* basis. I did this by criticizing the anthropocentric worldview and presenting a formal analysis of the concept of morality. This leads us to an answer to the question posed at the beginning of the book, the question why we should be concerned with human-induced species extinction. The answer is as follows: *Species—like individual humans—have intrinsic value (i.e., they are valuable in and of themselves). Their extinction should be prevented, not only for instrumental reasons but first and foremost for their own sake.*

These briefly formulated results may evoke skepticism. What have we gained with the proposition of intrinsic value? Is it really of any use to species? Isn't this abstract and theoretical body of arguments too far away from the everyday reality of most humans, even if it is logically conclusive? Obviously objections of this kind and others are to be expected (see, for example, Ernst 1996). Nevertheless, as far as this book is concerned, I think they are only partially justified. They are *not* justified when their aim is to express the wholesale idea that ethical argumentation "isn't very useful anyway," that ethics usually takes second place anyway compared to other everyday views, including the logics of self-interest or economic and political "necessities." I consider this objection to be negligible for two reasons. In the first place, we are unable to test the option. No one knows what reality would be like if we had no philosophical ethics. It may very well be that its indirect effect on the perceptions and attitudes of many people is consistently underestimated. In the second place this objection stems from the misunderstanding that ethics is only justified if it is useful. However, when ethics is equated in this manner with other, purely instrumental disciplines such as economics, medicine, or engineering science, we overlook the fact that the *primary* task of ethics does not involve attaining defined goals. On the contrary, it is the discipline whose foremost job it is to evaluate goals and justify these evaluations. It is, of course, true that ethics must also be interested in having its considerations be put into practice and communicating a corresponding ethos. But responsibility for realizing these endeavors is by no means restricted to ethics. The challenge is just as great for pedagogy, journalism, politics, and law.

Even if a criticism of ethics *in general* is not applicable to the discussion presented here, the objections mentioned above are still legitimate for consideration *within* the field of ethics. They draw our attention to the fact that many important questions and problem areas regarding species protection have been left unsolved in this study. In the context of this book it was not possible to offer a complete system of environmental ethics, nor could I even begin to examine all the relevant areas of overlap with empirical disciplines or devise detailed solutions to the problem of species extinction. As outlined in the introduction, this was not the intention of the book. My main aim was rather a more modest one, that is, to open up a perspective that acknowledges that the death of our planet's species does indeed have a *directly ethical* dimension (not just an instrumental one related to future generations of humans). It is based on the conviction that the problem of species extinction cannot really be solved *without* an altered perspective such as the one presented here.

Of course, even if we accept that it is *necessary* to acknowledge an ethical dimension to species extinction in order to master it, it goes without saying that this is certainly not *sufficient*. If the concept of pluralistic holism is to be put into practice—something that this ethics like any other one should also be interested in—further ethical and empirical investigations must be undertaken. Let me mention just three aspects of environmental ethics of the many that were not pursued further in this book that definitely warrant further consideration for the putting the concept of pluralistic holism to work.

The first of these concerns the relationship between species extinction and *everyday behavior*. The problem here is that the moral conflict between human self-interest and the "right" of other species to exist only rarely presents itself in a clear and unequivocal fashion, at least not in modern industrial countries. In everyday life we are almost never confronted with the problem of whether or not to eliminate a specimen of an endangered species for the sake of private interests. Species are usually threatened by the unintentional but nonetheless accepted side effects of everyday activities, which, taken by themselves, appear to be relatively harmless. One example of such an everyday activity with numerous disastrous effects on nature is driving a car (see E. U. von Weizsäcker 1992, 82f.). Let me mention only two of these effects: (1) atmospheric warming as a result of the carbon dioxide released ("greenhouse effect") with all the consequences for global ecosystems that ensue (see Peters and Darling 1985; McKibben 1990), and (2) widespread nitrogen input from the exhaust of automobile engines, which threatens the existence of many species in our cultivated surroundings (see Reichholf 1993, 181f.). According to Reichholf (1993, 205) with every car trip we take we fertilize protected areas far away from the source of this fertilization. Fertilizing nitrogen compounds are generated in particular at high speeds when atmospheric nitrogen is oxidized in the motor. "Thus motor vehicle traffic must be given a large portion of the blame for the loss of species." It is quite obvious that complex relationships between actions and side effects of this kind represent a novel and difficult problem for ethics. Not only does the traditional distinction between the purpose of an action and its unintended "side" effects begin to founder. Just as confusing is the fact that the greater significance of a side effect compared to the primary effect may only become visible when the sum of all the side effects is taken into account. Usually it is only the sum of all the side effects that leads to consequences. According to Spaemann (1990, 190, 191) this phenomenon threatens the way we see our actions in a manner thus far unknown in history. "As far as having a significance of its own is concerned, the individual action appears to disappear altogether. It seems to be characteristic of such an action that it is part of an ensemble which the actor doesn't see. And that reduces the significance of the action once again."

Questioning individual responsibility in this very fundamental manner addresses the second aspect of environmental ethics that deserves further consideration, namely the relationship between species extinction and the *socioeconomic context* in which it occurs. If appealing to individual, personal responsibility is to have more impact than a simple Sunday sermon, political and economic structures must be established that support any individual feelings of responsibility that arise rather than constantly obstructing them. Let us return to the example of motorized vehicles. It seems ridiculous to ask people to leave their cars at home as often as possible in deference to other species and global ecosystems when at the same time this request is undermined by a political and economic reality that not only continually subsidizes motor vehicle traffic (if you consider the balance of deficits for the national economy that results from it), but also does everything it can to bring total motorization to billions of people in other parts of the world as well. In spite of all due skepticism about whether ethics can achieve *anything at all* in this case, these considerations mean that for environmental ethics concerned with species conservation the traditional but apparently inadequate concept of *individual* responsibility must be supplemented by a concept of *institutional* or *collective* responsibility. Collective action, the world of institutions, must also be subject to moral rules of the game if an individual who is prepared to take on responsibility is not to be destined to failure from the word go (Günzler 1996, 163; see also Lenk, 1993). However, acknowledging the necessity for collective, political strategies for solving problems doesn't alter the fact that all ethical considerations ultimately always lead back to the individual and the extent to which he or she is individually prepared to assume responsibility. Not only must institutional responsibility ultimately be borne by individual people. Individual people are the ones who must go to the polls and express their will to create the socioeconomic structures that would support the responsibility they are prepared to assume as individuals. But, some might ask, what exactly might motivate these people to follow a moral impulse and opt for altruism instead of egoism in their relationships with nature?

This question leads us to a third aspect that is relevant for exercising a holistic concept of environmental ethics in practice, namely the question of the motivational connection between ethical behavior and *enlightened individual self-interest*. Since I consider this aspect to be the most important of the three discussed in this chapter, allow me to elaborate on it in greater detail before completing my presentation. The starting point for my thoughts is defined by two empirically founded premises. The first is that the ecological crisis and the worldwide extinction of species associated with it are basically the result of the wasteful lifestyle of a small part of humanity.[153] The second is that the relationship between prosperity in industrial countries on the one hand and worldwide species death on the other is not merely contingent but to a certain extent even compulsory. Of course, advocates of technical optimism described in Chapter 2 will object to the second premise. They will admit that economic growth led to serious damage to the environment *in the past*, but at the same time they will claim that *in the future* "intelligent" technical solutions will allow us to increase material wealth without depleting nature. Economic growth is not necessarily incompatible with the demands of "ecology" but rather the very prerequisite for efficient environmental protection and ecological research. Nevertheless, this economic version of technical optimism must also be abandoned as illusionary wishful thinking. Empirical facts contradict the possibility of resolving differences between contemporary human interests and the well-being of future generations and other species *solely* by means of advanced technology. Consider the basic fact that land cannot be multiplied or the almost inevitable side-effects of energy production or the second law of thermodynamics. As the discussion of this fundamental law of nature in Chapter 11.d has shown, recycling is also unable to cause the dilemmas associated with growth to simply vanish into thin air.[154] If we are willing to take these facts seriously, we cannot avoid the conclusion that the ecological crisis and species extinction will only cease or at least be alleviated if people in rich industrial nations very clearly limit their material wants and their disproportionately high degree of consumption. Both moral considerations and long-term human self-interest indicate that material self-restraint is an inevitable consequence. In this sense E. U. von Weizsäcker (1992, 258) has written the following: "It is absolutely compulsory that we inhabitants of northern countries adjust our lifestyles to less consumption. The current form of prosperity that we believe to have attained and that we consider to be the foundation for future golden ages is *untenable*. If we repress this very trivial fact, we are preparing the grounds for a political and ecological wildfire that will incite the world."

Nevertheless, it is clear that even if we recognize the empirical and moral necessity of material self-restraint, this insight alone will alter almost

nothing. Implications and claims of this kind are too unpopular and undesirable. Therefore, in order to motivate people to assume a moral standpoint, it would be very advantageous to be able to show that lowering our
level of material consumption does not necessarily mean that we will have
to settle for a poorer quality of life but rather just the opposite, that it will
generate more internal and ultimately also external opportunities than
those that appear to have been lost. In other words, it is well worth it to
show that a moral lifestyle is a better one, even in ecological respects. In
view of the fact that for more than two thousand years philosophical ethics
has been concerned with what constitutes a morally good life and what
happiness is, it seems to me that this discipline is particularly well suited to
demonstrate the fundamental relationships and experiences involved. In
this connection it would be worthwhile examining the following aspects:
(1) the question of personal freedom, (2) the real possibilities of self-denial,
and (3) the moral quality of needs.

Let us first take a look at *personal freedom*. With respect to our dealings
with nature, personal freedom could be related to two different things: external nature, that is plants and animals, and internal nature, which means
one's own person. Regarding freedom with respect to *external* nature it is
quite clear that this concept is often used in a manner that is both confusing and abusive. Although it seems to be quite well accepted nowadays that
absolute freedom (in the sense of "doing what you want to") is not possible
in the area of interpersonal and social relationships between humans, when
it comes to nonhuman nature it is still quite common to apply a concept of
freedom that allows us to do anything we want and are able to do (Maurer
1982, 21). As in the case of slaves and aliens in former times, we still tend
to regard other organisms, species, and landscapes as pure resources that
we are free to use whatever way we like (as long as the rights of other humans are not indirectly violated). But just as the scope of moral consideration and responsibility has been extended farther and farther in the course
of ethical history (from Aristotle to medieval Christianity and modern social ethics) in order to account for an increased level of integration in interpersonal and social systems, it seems to me that our present knowledge
about the position of humans in nature makes it imperative to extend the
radius one more (and final) time to encompass *all* of nature within the
scope of direct moral responsibility. Not only our social environment but
nature too must come to be understood as a fundamental part of human

life, without which a morally good life is impossible. Damaging or destroying part of one's own life cannot be indicative of real freedom.

That brings us to the *internal* aspect of the concept of freedom, that is, our relationship to our own nature, to ourselves. In this case philosophical analysis can refer to a number of classical ethical studies that have all come to similar conclusions. Internal freedom is never associated with excessiveness but rather with acknowledging personal limits. Acknowledging internal and external limits means being free to determine these limits oneself. Wise self-determination in turn will result in voluntary self-limitation. In view of the global environmental problems that are becoming more and more evident, there seem to be only two alternatives to setting such limits: forced sacrifice for purposes of prevention in the context of totalitarian planning or forced sacrifice as a result of catastrophes. It is clear that both alternatives are ethically unacceptable since they would not only result in the loss of freedom but also in the loss of humanitarian principles (see Furger 1976, 82).

However, a very fundamental problem of ethics seems to stand in the way of *denying oneself something* for ecological reasons, namely the fact that acting according to moral principles offers no guarantee that someone who acts in a moral manner will personally profit from what can be morally gained by such actions. More precisely, if only a few people exercise voluntary self-denial in order to prevent damage to nature or an ecological catastrophe, it is quite probable that in spite of (and in addition to) the burden of self-denial, they will also have to bear the consequences of the catastrophe that results. For this reason it would serve as a kind of "starting capital" for a corresponding change in attitude if it could be shown that a certain degree of self-restraint is accompanied by internal gains in addition to the external gains we initially aimed for, that external and internal good coincide. This brings us to an idea that was common among many philosophers from antiquity to modern times: the conviction that moral greatness and fulfilled human existence are inconceivable without exercise in self-restraint ("temperentia"). Even if external goods offer possibilities for human self-realization, these goods become really valuable when they are used with moderation, that is, when we are able to avail ourselves of them without internal compulsion (Furger 1976, 81).[155]

In our day and age the idea that continually enhanced consumption doesn't necessarily lead to a comparable increase in happiness (i.e., subjective contentment) is supported by empirical evidence as well. As modern consumer research has shown, people's wishes often grow faster than the possibilities for satisfying them. This means that with respect to happiness, the gross gains that accompany increased consumption are just about zero (see Birnbacher 1979b, 47). Moreover, in the context of upward spiraling consumption, our wants are always a step ahead of the means we have for satisfying them as well as our incomes so that the subjective feeling of happiness we experience when one want has been satisfied is immediately in danger of being nullified by new wants. In view of the fact that "productive dissatisfaction" of this kind is economically desirable and even encouraged by advertising, the current economic system with its orientation toward quantitative growth seems to be doubly absurd. It obviously conflicts not only with external, ecological requirements but also with the internal, moral nature of humans. In my opinion, it is the twofold character of this conflict that makes an *ethical* solution to the ecological crisis, one that goes beyond simply alleviating symptoms, so very difficult. In order to circumvent the *internal* aspect of the problem, that is, the struggle with one's own needs and the attendant motives and worldviews, the *external,* ecological aspect of the problem is declared to be a purely technical one. If environmental ethics would work on exposing these often unconscious but frequently also structurally well-supported "immunization strategies," it could contribute more to solving the ecological crisis than many a "hardcore" research project in technical environmental conservation.

Closely connected with the problem of quantitative growth is the question of the *moral quality of needs* (see Meyer-Abich 1979; Maurer 1984). Is it possible to distinguish between good and bad, authentic and compensatory needs? Even if a person might dispute the possibility of evaluating needs in an *intrinsic* sense, he or she would have to admit that some human needs have insignificant consequences for nature while others have very serious ones. Playing an instrument or sailing, for example, are less disturbing for nature than car racing or heated swimming pools. As needs of the latter kind increase, their satisfaction will not only reduce the possibilities of future generations and other creatures for a good life. They will cause species to become extinct as well. If we do not wish to simply take this for granted, and if the current status of technology does not allow us to reduce undesired consequences, then the only option left for making changes is to work on altering our needs. Environmental ethics will not be able to avoid pointing out that some ways of satisfying needs (e.g., what kind of recreation we pursue) can no longer be regarded as a morally neutral matter of taste (Singer 1993, 285). Subtly differentiated, *extrinsic* judgment of such actions seems to be unavoidable. To achieve this we must consider not only the frequency of such needs and the current state of technology but also a large number of ecological parameters that allow us to determine the degree to which nature has been utilized and, in particular, the amount of irreversible depletion involved. Here too the goal of protecting species poses the greatest moral challenge. Using species extinction as a highly sensitive instrument of measure it will be possible to see to what extent the ecological crisis has been perceived as an *ethical* one and consequently also as a crisis of the culture of our needs. However, visions of intellectual, bodily, and spiritual happiness that can be attained without the use of a great deal of technology are probably more effective than criticism. As experts in the art of living in all eras of history have continually confirmed, experiences of this kind are the most intensive and most productive.

33. NOTES

Introduction: The Basic Problem and Possible Solutions

1. In view of the varying assumptions that exist about the total number of species on earth, it is not surprising that the numbers regarding the current rate of extinction published in disciplinary literature also differ. Thus some experts claim that one species dies every day while others talk about one hundred per day. The extinction rate of 27,000 species per year cited here, which Wilson (1992, 280) himself calls a well-founded estimate, is based on a total species estimate of 10 million and thus is about half way in between the range of 5 to 15 million total species and 17,500 to 35,000 extinctions per year quoted by Stork (1993, 217).

2. For example, Ziswiler (1965), Werner (1978), Ehrlich and Ehrlich (1981), Bauer (1985), Vermeij (1986), Wehnert (1988).

3. For example, Trepl (1991), Plachter and Foeckler (1991), Kiemstedt (1991), Beierkuhnlein (1994).

4. According to Duden's German dictionary (Duden 1969, 86) the term anthropocentric refers to "a way of thinking which regards human beings as being the center and final purpose of the world." In philosophical discussions, however, the term is used in at least three different ways: (1) *Ontological* (or metaphysical) anthropocentrism refers to a worldview according to which humans are by nature "really" the center and final purpose of the world. I shall deal with this position in Chapter 27. (2) *Epistemological* anthropocentrism is the idea that we can basically only recognize and judge the things of the world from a specifically human perspective. In keeping with a suggestion made by Teutsch (1988) this concept will be referred to as "anthroponomy" (see Chapter 25.b). (3) *Ethical* anthropocentrism finally postulates that *only* humans have intrinsic value and therefore that only humans can be objects of *direct* moral consideration (see Chapter 18). This is the version of anthropocentrism with which this book will primarily be dealing.

A. Hopes for an "Ecological Solution"

I. Ecology as the Epitome of Controlling Nature?

5. In this context ecology is defined according to Ernst Haeckel (1886) as the science of the relationships among organisms as well as between organisms and their environment. It is primarily a subdiscipline of biology, but because of its involvement with relationships with the abiotic world, it extends into other scientific disciplines as well. By referring to the *science of ecology* rather than simply to ecology, I wish to emphasize that I do not employ the term ecology in the popular sense of a worldview, a program of nature conservation, or a political orientation.

6. See Popper (1959), Albert (1968), Stegmüller (1969a), Vollmer (1975, 1989).

7. Thus in his foreword to the leaflet titled *Ecosystem Research in the Waddensee* (Borchardt et al. 1989) Bernd Neumann, a former parliamentary adjunct secretary in Germany's Ministry of Research and Technology, writes the following: "At the moment we cannot estimate to what extent the ecosystem is capable of absorbing these burdens without damage and just when irreparable damage might occur. Although the Waddensee has been studied intensively by researchers from different scientific disciplines for many years, we still are lacking a scientifically well-founded *total analysis* of this complex ecosystem" (my emphasis).

8. A position of this kind was, for example, expressed by Armin Grünewald, a former spokesman for the German government on economic policy (cited in Schütze 1989, 52). In response to a survey by the magazine *Natur* in which he was asked what he knows about the principle of entropy and what he thinks about it, he replied that the laws of thermodynamics are theoretical ones and that he never put much stock in theory. He further claimed that his profession's way of thinking is not very much influenced by science and that he finds science too "mechanistic and too logical."

9. It would be a contradiction if the technical optimist, who is very confident in science on the one hand, would so blithely skip over its evidence on the other hand.

10. The general formula for calculating the number of possible relationships B between n building blocks is $B = 2^{n(n-1)/2}$. The number of possible relationships can be nicely illustrated by drawing four points on a piece of paper and then connecting them with lines. The following possibilities can be imagined: no lines at all, six single lines, fifteen double lines, twenty cases of three lines at a time, fifteen cases of four lines, six cases of five lines, and one case of six lines at a time. That adds up to sixty-four possibilities of connecting four points with one another by straight lines (Kafka 1989, 23).

11. As Luhmann (1990, 33) points out, reducing complexity is, of course, a common procedure outside of science as well, since "everything that appears to be certain is a case of reduced complexity." "Every system must reduce the complexity of its surroundings, in particular by perceiving the environment in a limited and categorically preformed manner."

12. See, for example, Wodzicki (1950), Kowarik and Sukopp (1986), Sukopp and Sukopp (1993).

13. See Dwyer and Perez (1983, 320), West and Goldberger (1987, 354).

14. A further misunderstanding, which Vossenkuhl (1992a, 98, 99) has pointed out, consists of the assumption that risk assessment permits us to control the probability of possible damage.

15. According to the position of the Scientific Advisory Committee (Wissen-schaftsrat 1994, 13) on environmental research in Germany, ecosystem research does indeed aim for such a noble goal. Thus they write, "Ecosystem research . . . intends to go beyond understanding individual processes or process chains and to grasp the complex interactions that occur between *as many parts of a habitat as possible* as well as those that occur between these parts and the environment" (my emphasis).

16. Causal analysis of fluctuations in population size in the field has, however, shown that regulation of population density in predator/prey relationships is often not reciprocal. "In reality it is not the bobcats which regulate the hares but rather the hares which determine how many bobcats can exist" (Kurt 1977, 139).

17. Autecology studies the relationships between an individual organism and its surroundings.

18. According to Vollmer (1986c, 122, 123) the relationship between reality and scientific knowledge (but not only scientific knowledge) can be illustrated by a model of graphical projection. In this case the object that is projected corresponds to "reality;" the projection method is comparable to the signals that our sensory organs are capable of registering; the screen is the same as our mind while the picture elicited corresponds to perception or simple experience. Therefore the structure of a picture generated by projection depends upon the structure of the object, the manner of projection, and the structure of the screen that receives the projection.

19. The term is not always used in the same manner. For example, Lorenz (1986, 87) defines it as the "belief that the only thing that can be regarded as reality is that which can be expressed in terms of the exact sciences and which can be proven by quantitative methods." Popper (1957, 105), on the other hand, refers to scientism as "a name for the imitation of *what certain people mistake* for the method and language of science." I myself refer to scientism in the sense of Garaudy (1991) as a combination of epistemological positivism and technical optimism.

20. "It is superstition to regard the absolute as an object or to envision an object as something absolute. Therefore I call it scientific superstition when on the basis of scientific evidence something is regarded as being real as described by this evidence or when science is expected to be able to solve all the problems of humanity" (Jaspers 1968, 65).

II. The Science of Ecology as a Normative Authority?

21. For example, Searle (1967), Tranøy (1972), Kadlec (1976), Nordenstam (1982).

22. Beck (1986) provides a detailed discussion of this problem in his book titled *Die Risikogesellschaft*. Gethmann and Mittelstraß (1992, 21) criticize equating acceptance (validity in practice) and acceptableness (normative validity) as "relativism" or "sociologism."

23. Thus, for example, Bauer (1985, 572) writes the following: "Every ecosystem has a different purpose in the total economy of nature. Species protection must take this into account. Species protection requires no further justification of any kind. Species protection is a safeguard for reaching the ultimate goal of well functioning ecosystems."

24. Thus for Schönherr (1987, 318, 332) ecology is "the systematic site at which Heidegger's ontological philosophy and schizoanalysis meet and where the technical challenges to humans and nature are illuminated. . . . Ecology is weak hermeneutics."

25. It follows that in a strictly descriptive sense there is also no such thing as an "ecological crisis." I have chosen to use this term in my book anyway, because I feel that it is sufficiently well established and not readily subject to misunderstanding.

26. After all attempts to establish an equilibrium between predators and prey under laboratory conditions failed to succeed for longer periods of time, in spite of highly sophisticated technical perfection, serious doubts have arisen about whether such an equilibrium really functions under the stochastic conditions that exist in the field. According to the mosaic-cycle hypothesis (see Chapter 11.b), a long-term equilibrium between predators and prey wouldn't be necessary anyway (Remmert 1990, 68).

27. This seems to explain the current tendency to automatically attribute every natural catastrophe or incidence of extreme weather to anthropogenic influences. Of course, there may indeed be some connection, but for reasons described in Chapter 4.e this connection can in principle not be demonstrated in enough time to avert the results that ensue.

28. Since evolution constantly generates new things, the fact that an ecological phenomenon is new and therefore unique is not a good argument against this phenomenon. There is a first time for everything.

29. See definitions by Tischler (1976, 116), Zwölfer (1978, 15), Markl (1983, 74), Remmert (1984, 260), Pimm (1984, 322), Krebs (1985, 581), Grimm and Wissel (1997).

30. According to Peters (1976, 7) the climax theory is only of limited scientific value anyway. Since its fundamental arguments are formulated such that they apply to every possible case, it cannot be refuted and is therefore strictly speaking not a scientific theory.

31. See, for example, Weissert (1994), Margulis and Hinkle (1991), and a critical discussion by Kirchner (1991).

32. In keeping with many philosophers I assume here that it makes sense to attribute interests to all living things (see Teutsch 1985, 49). In Chapter 29.b I shall pursue this rather controversial topic in greater detail.

33. In order to better grasp the rather vague concept of "many species," modern ecology often uses the Shannon-Weaver diversity index. This term covers both the number of species and the relative frequency of their members. "A system with a large number of species, in which, however, 99 percent of the individuals belong to only one species, exhibits very low diversity according to this method of calculation. On the other hand, an ecosystem with relatively few species, each with approximately the same number of individuals, has a relatively high diversity index" (Remmert 1984, 203f.).

34. Right now worldwide 50 acres of rain forest are being irrevocably lost every minute or 110,000 square kilometers per year.

35. Thus the biotopes that the German Advisory Committee for Landscape Protection (Deutscher Rat für Landespflege 1985, 547) suggests should be protected are ones that "either enhance the ecological equilibrium ('stability') of nature's economy and a diversity of landscapes, or which exhibit a *great deal of species diversity* [my emphasis] or harbor rare or endangered species of plants and animals." At least a few other criteria for protection are mentioned in addition to species diversity, which serve to further qualify the criterion of species diversity. But that doesn't in any way improve the usefulness of this term as a general guiding principle, at least not as long as it is not clear how it compares with other (often contradictory) criteria for protection.

36. In Part B, Section II, which deals with ethics, I shall clarify whether or not "greater naturalness" is in and of itself normatively relevant, or in other words independent of what this means with respect to ecological stability or species diversity.

37. This resembles Arrow's (1973) impossibility theorem in social philosophy. Arrow has shown that the aggregation of preferences associated with social choice leads to paradoxes.

38. See, for example, von Haaren (1988), Trepl (1991), Kiemstedt (1991), Schweppe-Kraft (1992), Kaule and Henle (1992), Beierkuhnlein (1994).

39. It seems to me that this interpretation is supported by the fact that many suggestions about how to resolve deficits in conservation research once again take recourse in the idea of better empirical science. Once it can no longer be denied that nature conservation problems are first and foremost *normative* ones that can only be resolved with the help of the humanities and social sciences, the function of these disciplines is then often seen as consisting only of providing *descriptive* analyses. Thus in the summary of their article titled *Research deficits in the area of species and biotope protection* Kaule and Henle (1992, 134) define the specific functions of the humanities and social sciences as "dealing

with questions of how changing values affect attitudes to the environment, how to promote the acceptance of conservation measures, and determining the reasons for the structurally underprivileged status of nature conservation in administration and government." The fundamental problem of *justifying* norms is mentioned only briefly in this article. Just as little attention is given to the need for normative reflection by Beierkuhnlein (1994, 17), who correctly notes that the "major models" of nature conservation "are influenced by social processes" and can "therefore vary," but who otherwise apparently sees no need to discuss these models with respect to their *conceptual legitimacy*.

40. If this criterion for the definition of "waste" is rejected on the grounds that the substances in question do eventually return to some kind of cycle following a fundamental change in the system some time or other, then the term becomes almost completely meaningless. After all, it is possible that even the waste that humans produce might someday also be recycled (e.g., as part of the flow of material in primitive ecosystems consisting mostly of bacteria that might arise after humans have become extinct).

41. Clapham (1973, 229) writes the following: "Resiliency can be defined as the ability of an ecosystem to maintain itself in a healthy state in the face of outside perturbation; . . . health refers to the resemblance of the ecosystem to the normal steady state condition for the area."

42. At this point I shall not deal with other (in my opinion not very convincing) criteria for the health of an ecosystem proposed by others as, for example, "naturalness" or "species diversity." For this the reader is referred to Bayertz (1986, 94f.).

43. "We *know* an object when we are able to *construct* it" (Habermas 1973, 32). C. F. von Weizsäcker (1960, 172) expresses similar views: "Scientific thinking is really only validated by action, by a successful experiment. To experiment means to exercise power over nature. In the long run having such power is the ultimate proof that scientific ideas are true."

44. Regarding the problem of "mental health" see Engelhardt and Spicker (1978). According to Bleuler (1983, 120) the term "mental disease" cannot be grasped objectively and can neither be defined nor described by scientific means. "The term is not a theoretical concept; instead it is strongly influenced by personal experience with oneself and with others."

45. See Richardson (1980), O'Neil et al. (1986), Trepl (1988; 1994, 139f.).

46. Since the differences between ecosystems and organisms listed here also apply to the greatest of all ecosystems, the biosphere (at least with respect to points 1, 3, and 4), in summary they can also be regarded as an objection to concepts of "planetary health" such as those proposed in the context of Lovelock's Gaia Theory (1988, 177) or in Meyer-Abich's natural philosophy of holism (1991, 164). In both cases it is not at all clear how the term "health of the whole"

(Meyer-Abich 1991, 165) can be operationalized beyond the very simple idea of maintaining life in and of itself.

47. I shall deal with *theoretical* objections to ascribing interests to "lower" animals and plants in Chapter 29.b.

48. For example, in his inaugural lecture in 1869 Haeckel (1924, 49), the founder of the term "ecology," referred to this discipline as the "science of economy, the economy of animal organisms." Kreeb (1979, 71) also characterizes ecology as an "economic science."

49. It is this ambiguity that hampers so-called "ecological balancing," an instrument of environmental technology with which proponents attempt to evaluate damage to the environment both quantitatively and qualitatively. According to Lang and colleagues (1994, 115) an "objective and scientifically well-founded measure of environmental damage . . . is not in sight."

III. What Ecology Has to Offer

50. In his book *Grundlegung zur Metaphysik der Sitten* (Fundamental Principles of the Metaphysic of Morals) Kant ([1785] 1965, 67) writes as follows: "*Empirical principles* are of no use anywhere for justifying moral laws. The universality with which they are expected to be valid for every reasonable creature without exception, the unconditioned practical necessity with which they are thus imbued, ceases to apply when reasons for moral laws are sought in *particular aspects of human nature* or in the chance circumstances in which it is placed."

51. Examples of such literature are Plachter (1990), Dierssen (1994), Blab and Völkl (1994), Müller and Müller (1992), Arndt et al. (1987), Mader (1985), Heinrich and Hergt (1990, 265f.).

52. In this case the term "attitude" refers to a tendency to behave or act in a similar manner in certain situations because of a cognitively and emotionally founded value orientation. "It does not mean inflexible conditioning but rather a schema of thinking and behavior which is constantly subject to adaptation" and enables us to do the right thing in everyday situations quickly and without detailed thought (Teutsch 1985, 29).

53. Consider the amazing ability of some "primitive" cultures to maintain their population density at a relatively stable level (see, for example, Norberg-Hodge 1991, 56).

54. Since it is also possible to criticize this observation as an example of scientism and as a contradiction of itself, I wish to point out that such a reproach is also subject to epistemological limits and thus ultimately incapable of being proven. Regarding the related problem of the supposed inconsistency of hypothetical realism see Vollmer (1985, 251, 252).

55. A combination of both kinds of "error unfriendliness" (irreversibility and pro-
hibitive dimensions) characterizes the continued uncontrolled consumption
of fossil fuels in industrialized countries, which ultimately is something like
an enormous experiment with the earth's climate. According to Kafka (1989,
103) "the mere suggestion that an action might seriously alter the biosphere"
ought to be enough to cause us to refrain from employing it. At this point I
cannot go into detail about ethical justification for giving priority to negative
predictions rather than positive ones as Kafka suggests. Instead I wish to refer
the reader to Jonas (1984, 70–83).

56. According to Fischbeck (1976) only twelve plant species make up almost the
entire food supply of humans nowadays—from a total number of 10,000
available plant species. If only one of these few high-input species would suddenly fall victim to a disease (a possibility that can never be excluded completely), the only realistic chance of retaining this species is to introduce
"resistance factors" from uncultivated relatives, provided these forms have not
been eliminated and are still available in sufficient diversity (see E. U. von
Weizsäcker 1992, 132).

57. Of course, not all ethicists share this view. For example, von der Pfordten
(1996, 101) considers Jonas' thesis to be "too simplistic and too strong." But
he does admit that in the modern period "greater focussing and a shift in the
spectrum of discussion topics in the direction of anthropocentrism" has taken
place.

B. The Debate about an Ethical Solution

58. Teutsch (1985, 22) introduces a fifth basic type called an "egocentric" form of
environmental ethics that is exclusively oriented toward the personal interests
of the individual (or a group of individuals with the same interests). Since this
type of ethics can be readily criticized as a form of "ethical egoism" (see
Frankena 1963, 16), and since it doesn't play a very significant part in the debate about environmental ethics, I have chosen not to consider it in this
discussion.

59. In this passage Kant ([1797] 1990, 83) distinguishes between duties "*toward*"
something and duties "*in view of*" something. "On the basis of pure reason a
human being has no duties other than those towards humans (himself or another human being); . . . and his supposed obligation towards other creatures
is merely an obligation to himself, a misunderstanding which comes about
when he mistakes his duty *in view of* other creatures with a duty *towards* these
creatures."

60. Bentham ([1789] 1970, 283), the founder of utilitarianism and one of the first
representatives of pathocentric ethics in the modern period of the Western
world, expressed the critical reason for regarding animals as direct objects of

moral behavior in the following statement: "The question is neither can they *think* nor can they *speak* but can they *suffer?*"

61. I wish to point out here that the term "holism" as I shall use it in the following ethical discussion must be clearly distinguished from the term "holism" as it is used in theory of science discussed in Chapter 12.

62. For example, Hartkopf and Bohne (1983, 68f.), Norton (1984), Meyer (1986, 155), Pearce (1987, 9), Hampicke et al. (1991, 24).

I. A Pragmatic Approach: Is Anthropocentrism Sufficient?

63. "Petitio principii" means a logical fallacy in which a premise is assumed to be true without warrant or in which what is to be proved is implicitly taken for granted.

64. This idea was expressed by no one less than Schopenhauer ([1840] 1922, 587) when he stated the following in his paper titled *Preisschrift über die Grundlage der Moral* (Treatise on the Foundations of Morality): "One might perhaps want to counter that ethics has nothing to do with how human beings really act but rather that it is science which determines how they *should* act. However, this is the very principle which I reject. . . . I propose instead that the purpose of ethics is to interpret the highly diverse kinds of actions of human beings with respect to morality, to explain them and to trace them back to their fundamental sources."

65. Quotation cited in the German magazine *Der Spiegel* 10/90, p. 248.

66. The objection that it is *practically* impossible to save *all* currently endangered species from extinction (Gibbons 1992, 1386) does not apply here. This piece of contingent evidence has no effect on the fundamental obligation to try to come as close as possible to attaining the goal of general species protection.

67. In general a (positive) time preference means that "what is closer to us in time is assigned greater value while what is farther away in time is devaluated" (Birnbacher 1988, 30). In economic science this devaluation of future utility or harm is represented by constant decrease in value of an object in the course of time, comparable to a kind of reverse interest rate.

68. Since many "pests" occur frequently and have also succeeded in withstanding all human attempts to eradicate them, one might assume that they are not a matter for the species protection movement. However, the example of the black rat (*Rattus rattus*), which is on the Red List of endangered species in Germany, shows that this estimate is premature.

69. The biologist Ehrenfeld (1976, 648) shares a similarly skeptical position regarding the economic value of most species. He refers to species that lack utility in the usual sense as "nonresources."

70. See van Dersal (1972, 7), Gunn (1980, 24), Norton (1987, 124).

71. For example, Kurt (1982, 130), Amery (1982, 128), Wehnert (1988, 140).

72. See Jackson and Kaufmann (1987) and Palumbi and Freed (1988) as well as Gautier-Hion and Michaloud (1989).

73. "Classical" biologists, in other words, zoologists and botanists with broad knowledge in systematics and field experience, are themselves members of what might be called an endangered species.

74. Of course this doesn't mean that the *causes* of a reduction in the numbers of these species (e.g., pollution) are also irrelevant from the standpoint of system ecology.

75. In the literature a variety of different definitions of the term *biodiversity* (diversity of life) can be found, each of which emphasizes only a particular aspect of biological diversity (e.g., species diversity, habitat diversity, or genetic diversity) (Whittaker 1972; Cousins 1991; Platnick 1992; Walker 1992, 19). According to Hengeveld (1994, 1) these definitions have neither a uniform approach nor a common conceptual framework. Regarding the practical difficulties for species protection that this leads to, see for example Westman (1990). Norton (1986, 112) refers to the term "total diversity" in his argumentation, in keeping with the definition of MacArthur (1965, 528f.). This term means "the total number of species . . . in a fairly wide geographic area composed of several habitats."

76. A tabular summary of universal system functions for terrestrial ecosystems and endogenous system functions deduced from them can be found in Woodward (1993, 272). See also the schematic representation in Schulze and Zwölfer (1987, 417).

77. According to the paleontologist Stanley (1981, 197), it would be a "gross exaggeration" to suggest that "all species in any habitat are interdependent to such a degree that we must fear the domino effect—loss of species after species when one is removed by extinction." See also the paleontological evidence of Sepkoski (1992, 87f.).

78. This is, of course, not a watertight argument against the possibility that the massive extinction of species induced by humans that we are *currently* experiencing at a *speed* that is probably unmatched in the history of evolution might indeed involve uncontrolled positive feedback loops. However, if this were true, it would represent a real *novelty*, something for which no support can be found in the general rules of ecology or evolutionary biology.

79. In order to illustrate the basic tenets of the rivet and redundant-species hypotheses, I have presented extreme examples of these positions. The graphic illustrations in Lawton (1994) and Vitousek and Hooper (1993) show that many different transitional versions are conceivable.

80. According to Ehrlich and Ehrlich (1981, 48) there is evidence that the unconscious ties to nature that many people seem to have are the result of humans' coevolutionary adaptation to their natural surroundings. In this sense "it is evident that nature in our daily life should be thought of as a part of the biolog-

ical need." Schemel (1984, 191) also considers the need for aesthetic experience in nature to be more than an elitist goal, as the following statement indicates: "Humans' emotional relationship to nature is ingrained in their very nature, part of being 'totally human' and a very basic need. . . . " Kellert and Wilson (1993) provide a detailed discussion of this relationship in their book titled *The Biophilia Hypothesis.*

81. Hegel ([1832] 1986, 175, 176) provides an example of aesthetic judgment that claims to be universally valid but is certainly not based on profound knowledge, the example of his antipathy toward particular animal species. In his book titled *Vorlesungen über die Ästhetik* (Lectures on Aesthetics) one can, for example, read that a sloth, "which drags itself from one place to the next with great effort and whose general appearance expresses its inability to act and move quickly, arouses displeasure because of its slow motions, because it is activity and mobility which represent higher ideals of life. Similarly we fail to find amphibians, certain species of fish, crocodiles, toads, many insect species etc. attractive; but in particular chimeras, forms which represent a transition from one form to another and in which different forms are mixed, catch the eye and impress us as being ugly, as for example, a duckbill platypus, which represents a hybrid between a bird and a tetrapod."

82. In philosophical discourse on the aesthetics of nature this position is discussed in connection with "positive aesthetics." As Norton (1987, 111) has shown, representatives of positive aesthetics (e.g., Carlson 1984; Callicott 1983, 353) usually refer more or less explicitly to scientific knowledge provided by ecology or evolutionary biology. According to Carlson (1984, 33), for example, the Darwinian perspective of undirected evolution provides no grounds for "regarding some [forms of life] as aesthetically inferior to others." Gould (1977, 13) expresses a similar view when he writes that in light of the view that evolution "does not inevitably lead to higher things, the 'degeneracy' of a parasite is just as perfect as the gait of a gazelle." Others such as Gunn (1984, 318) and Russow (1981, 109) express the opposite position that there are some species such as the snail darter (*Percina tanasi*) or the Houston toad (*Bufo houstonensis*) "which, by no stretch of the imagination, are aesthetically significant." In cases like these, Russow (1981, 110) claims that "lacking any alternative, we may be forced to the conclusion that such species are not worth preserving." Mannison (1980) represents a position quite different from either of the above. He maintains that nature *can never be* the object of aesthetic judgment since contrary to a piece of art, no artist is involved with intentions to which one could refer in passing judgment.

83. Certainly a lot of nonbiologists would argue that if they heard about the death of this bug they too would regret its loss. In my opinion, however, their regrets would not result from recognizing the *aesthetic and intellectual* loss that

future generations or they themselves might have experienced but rather because they intuitively attribute *intrinsic value* to this species.

84. See Krieger (1973), Tribe (1976), Norton (1987, 127f.), Gerdes (1996).

85. It is often said that there is no more nature on earth that has not been influenced by human beings. Since the advent of global warming at the latest this is certainly the case (see McKibben 1990; Peters and Darling 1985). Nevertheless, I still find it worthwhile to differentiate between, for example, natural processes that are "strongly influenced by humans" and ones that "run their course almost autonomously," and to refer to the latter as "wilderness" (in a limited sense, of course).

86. It is often claimed that if you try to weigh values derived from different value categories, you can't get around a cost-benefit analysis. However, this objection fails to take a very significant difference in the decision process into consideration. In the case of a cost-benefit analysis the reference value (e.g., the monetary value) is assigned *ahead of time*. The results of the analysis are merely a matter of arithmetic. In the process of weighing values, however, there is no general measure of this kind to facilitate decision making. Equivalent goods can only be reconstructed *after the fact* (see Kelman 1981, 40).

87. A flier of the German nature conservancy organization BUND in the series titled *Aktiver Naturschutz* (Active Nature Protection) (Thielcke 1978) demonstrates this hierarchy quite well. There you can find the following argumentation: "Species protection is necessary because plants and animals contribute to (1) our food supply, (2) the production of valuable compounds (such as medicinal products), (3) our supply of raw materials, (4) our awareness of ecological dangers (in their role as bio-indicators), and (5) the stability of ecosystems." The quintessence of topics (1)–(5) is that "to protect nature means to actively protect human beings." At the tail end of this summary consisting of twenty-five lines of brief argumentation, topic (6), "ethical tasks," is presented in two lines as a kind of appendix (cited according to Amberg 1980, 76).

88. In view of these consequences, the strategy currently employed by advocates of species protection that consists of listing *as many reasons as possible* for protecting species (Heydemann 1985, 581) must be regarded with great skepticism. Since the two categories of justification, "material utility" and "immaterial utility or ethics," seem to contradict each other with respect to their basic structure rather than complement one another (Trepl 1991, 429), it often seems wise to drop the second-order arguments (utility) in order to avoid weakening those of the first order (ethics).

89. See Barry (1977), Baier (1980), Callahan (1981), Birnbacher (1988).

90. In light of the fact that the burden of proof currently rests on the part of the skeptic (see Chapter 20), it seems to be more plausible to interpret lack of such elementary moral intuition (at least partially) as a lack of moral sense instead of assuming that existing intuitions are merely a matter of insufficient understanding.

91. See, for example, Stern (1976, 87f.), Tribe (1976, 73), Bierhals (1984, 119), Meyer-Abich (1984, 51), and Stone (1988, 43).

92. One indication that Stone's estimate is correct is a short paragraph in the previously mentioned book *Rettet die Vögel—wir brauchen sie* (Save the Birds—We Need Them) (Schreiber 1978, 189), in which the journalist and nature conservation advocate Horst Stern remarks, "We've really come a long way. Anyone who wants to protect animals nowadays has to convince people that protecting them is important for human well-being. 'Save the birds, we need them!' It's not enough simply to save the birds. Only what is useful is entitled to live."

93. Examples of this can be found in publications by Thielcke (1978), Auhagen and Sukopp (1983, 9), Deutscher Rat für Landespflege (German Advisory Council for Landscape Protection) (1985, 538), and Trepl (1991, 429).

94. Nevertheless, it is still a moot point whether a purely species-oriented approach of this kind is the best one.

95. This has been proposed by Regan (1986, 1993) and Wolf (1987) as well as by theoreticians of a utilitarian tradition (e.g., Elliot 1980; Feinberg 1980; Birnbacher 1988, 222f; and Singer 1993).

96. See, for example, Altner (1979, 124), Johnson (1991, 207f.), and Heffernan (1993, 402).

97. See, for example, Feinberg (1980, 161–167), Rescher (1980, 83), Attfield (1983, 150), Rolston (1985, 723), and Norton (1987, 170f.).

98. A case that illustrates that it apparently is not just an intellectual exercise to grant greater value to a species than to the lives of individual organisms but something exercised in real life is that of the plant *Dudleya traskiae*, which occurs in Santa Barbara, California. In order to save a few specimens of this endangered species, the U.S. National Park Service killed several hundred rabbits, that is, higher vertebrates (Primack 1993, 241).

99. This concept must be clearly differentiated from the concept of "*monistic holism*" as once proposed by Callicott (e.g., 1980, 327). According to monistic holism only the system as a whole has intrinsic value, and the value of the individual components of the system is exclusively a function of their particular relationship to the whole. Obviously an extreme concept of holism of this kind would result in unacceptable, totalitarian consequences for the individual (see Marietta 1993). Therefore Callicott seems to have moved away from monistic holism in later publications (e.g., 1993, 360).

100. This expression corresponds to the German title of Ehrlich and Ehrlich's book *Extinction* (1981).

101. In Teutsch's (1985) *Lexikon der Umweltethik* (Encyclopedia of Environmental Ethics), for example, the topic "species protection" is not listed at all.

102. In disciplinary literature in German "environmental ethics" (Umweltethik) is often referred to as "ecological ethics" (Ökologische Ethik).

II. A Theoretical Approach: Can Holism Be Justified?

103. This objection often occurs more implicitly than explicitly. Thus, for example, Wolf (1987, 166) asks what "might be bad [for plants] if one did this or that with them" and replies in the following manner: "The fact that this question cannot be answered shows that taking these organisms into consideration is not a question of morals, even if there may be other reasons, because the objects of morality are creatures capable of suffering." Since for Wolf this definition of morality has apparently been established *a priori*, it is not surprising that consequently in her view any attempt at extension, for example, one that considers direct responsibility for plants, is bound to fail under theoretical examination. If established theory in principle always has priority over reflected intuition, then *there can be no justification* capable of extending the scope of responsibility to include plants.

104. Thus in his *Nicomachean Ethics* Aristotle (384–322 B.C.; 1996) sees no reason to question slavery. It appears as natural to him as it does to all his countrymen.

105. In this connection the continued existence of war, torture, and oppression is no more an argument against this progress than the inability to eradicate criminality is an argument against the existence of morality all together.

106. The process of correcting ethical theory through reflected intuition presented here corresponds to a great extent with what Rawls (1973, 46) has described as the concept of a "considered judgement in reflective equilibrium." Ethical theory and well-thought-out judgments constitute and correct one another in a process of reciprocal feedback.

107. The term "anthroponomism" can be found as early as in Kant's (1797, A47) *Metaphysische Anfangsgründe der Tugendlehre* (Metaphysical Principles of Virtue), although it is employed for a different purpose in that treatise. There it is intended to express the capacity of humans "for self-determination realized in the form of self-imposed laws" as opposed to "empirically derived anthropology" (Wenzel 1992, 5).

108. Consider, for example, racist court decisions in the United States in the 1960s or the problem of appropriate legal evaluation of abortion or rape in the context of a mostly male-dominated judicial system.

109. Anyone who thinks that this is a construed and highly improbable example should recall the problem of dealing with refugees.

110. See in this connection Krämer (1984), Frankena (1963, 16, 67), Spaemann (1990, 45f.).

111. The mistake of the subjectivist is that he looks at only what is *coincidental* in different kinds of morality. In doing this he fails to see what is *not coincidental*, which is a basic prerequisite for recognizing anything that is coincidental. As Tugendhat (1989, 927) demonstrates, subjectivism is therefore hardly more than a verbal posture. "In real life we constantly make objective moral judgements. For example, we condemn torture or breaking a promise. When we do this, we are not expressing subjective feelings, but rather we demand the opposite kind of behavior of one another. If we were really subjectivists, we would have to change our entire intersubjective behavior in a manner that is hardly imaginable." For a criticism of subjectivism see also Spaemann (1986, 11f.).

112. Of course, at this point a noncognitivist would object that the concept of plausibility only makes sense in connection with factual statements since only these can lay claim to being objective. However, it is difficult to see why contrary to our everyday understanding the term objectivity should be limited to factual statements.

113. For this reason Schnädelbach (1987, 82) refers to the "historicity of reason." Reason is an "open concept" that is subject to cultural and historical change. Rationality cannot be defined (without circular argument); at the most it can be explicated within a certain context. According to Schnädelbach, a rationalist who believes that rationality can be systematized in an a priori and definitive manner will not accept this. "This kind of openness must drive him to desperation, and he is probably the real irrationalist . . . when it comes to reason, even though the fact that rationality has a history opens up new possibilities." Of course, Schnädelbach points out that there has to be something continuous in the course of this history. "Otherwise we wouldn't be able to understand it as a history of rationality." However, if one wishes to describe the nature of this continuity, one will not get much further than "minimal characterizations."

114. Analogous results have been attained from an analysis of two disciplines that stand for reason and rationality even more so than ethics, namely logic and mathematics. Their philosophical foundations are also not unconditional. Thus the theory of science specialist Stegmüller (1969b, 307) maintains that "regardless of the field involved there is no 'inherent guarantee' for human thinking. One cannot attain positive results completely free of any premises. One must always believe in something in order to be able to justify something else."

115. As a historical example, the religious custom of human sacrifice can be examined. Based on the premise that *this is the only way* that the gods can be appeased and that the entire welfare and woe of the tribe depends upon humoring these gods, it is rather difficult to criticize this activity. Nowadays we criticize it primarily because we can no longer accept the premises on which it is based. On the other hand, the profound respect for animals, plants, and landscapes associated with certain primeval religions indicate that basic premises that are obsolete from our present point of view can sometimes also lead to worthwhile attitudes and activities.

116. See Godfrey-Smith (1980, 46), Meyer-Abich (1984), Taylor (1986), Strey (1989, 76), Marietta (1995, 102).

117. Regan (1980, 366) maintains that this premise in itself represents a preliminary decision of sorts, namely a preliminary decision in favor of worldviews in which rationality plays an important part. Of course, this objection is correct. In this case as well Stegmüller's (1969b, 307) statement cited above is valid, namely that one first has to believe in something in order to justify something else. However, in the context of the ideas presented in this chapter I find a preliminary decision in favor of rationality to be unproblematic, because in discussions about anthropocentrism and physiocentrism in philosophy both sides claim to be rational.

118. Among philosophers it is still a moot point whether or not such evidence truly exists. According to Stegmüller (1969a, 168) this problem is one that is "absolutely impossible to solve." Anyone who argues *in favor* of such evidence becomes caught up in circularity because his argumentation must rest on evidence from the very beginning. A person who *is opposed* to such evidence automatically contradicts himself, because he too is dependent upon his argumentation being evident.

119. Of course one can argue about whether we are really dealing with whole worldviews or simply with individual elements of worldviews when considering the ontological premises connected with anthropocentrism, pathocentrism, biocentrism, and holism. However, this does not alter the fact that even if only elements are involved, these are of a worldview-like nature. Therefore for reasons of simplicity I chose to talk here about worldviews.

120. Thus in his paper titled *Geschichte und Naturbeschreibung des Erdbebens, welches 1755 einen Teil der Erde erschüttert hat* (The History and Scientific Description of the Earthquake That Shook Part of the Earth in 1755) ([1756] 1985) he writes, "Man is so taken with himself that he regards himself as the only purpose of God's dealings, just as if the latter had no other thing in mind than mankind in establishing rules for governing the world. We know that everything that nature comprises is a worthy object of heavenly wisdom and transactions. We are part of nature and want to be all of it. The rules of

the perfection of nature in a greater context should not be taken into consideration, and all that is important is to simply place everything in the proper perspective to us. Everything in the world that contributes to comfort and pleasure, so man thinks, is only there for our sake, and nature incites no alterations [such as, for example, earthquakes] except for the purpose of making things uncomfortable for us, for disciplining, threatening or taking revenge upon us."

121. The worldview that has the greatest influence nowadays is certainly the currently dominating worldview of economics, that is, viewing the world as a giant department store in which man the consumer ("homo oeconomicus") assumes a position of central importance. However, this conception of the world also seems to have its roots in one (or more) of the traditions mentioned here.

122. In order to grasp the magnitude of this number more easily, Ferris (1981, 2) devised the following mental scenario: " . . . if we were to launch expeditionary forces at such a fantastic rate that an expedition reached a new star in our galaxy every hour of the day and night, and we kept up this rate of exploration year after year, we would have visited a little fewer than half the stars in the Milky Way Galaxy in six million years, a period of time considerably longer than the present tenure of our species."

123. Mathematical analyses have shown that the tree shape or bush form is not a human artifact (in the sense of a tautology) but can be reproduced when various criteria are applied. Of course different taxa can assume different positions from case to case.

124. The well-known theologian Bultmann (1962, quoted in Liedke 1981, 73) provides an example of this kind of thinking. For him there is no question "that the history of mankind is basically different from the processes of nature, that in human history in the course of time the same does not recur in an eternal cycle but rather that novel and decisive things occur constantly. This is because history is the history of humans. However, a human being is not . . . a segment of the cosmos but rather basically distinct from the rest of the world." In Schmidt's (1975, 5) opinion views of this kind are not rare in contemporary philosophy: "The common thesis of hermeneutic philosophers from Dilthey to Gadamer and Habermas is that science deals with 'dead' or rather ahistorical objects, processes and facts of nature. The humanities, on the other hand, have to do with humankind itself and its products, which are always historically contingent."

125. In the literature the temporal data of different authors who employ such methods of projection vary to a certain extent. Nevertheless, this does not affect the quality of the statement in any way.

126. Since the creation of the atomic bomb and in view of the ecological crisis one, of course, must consider the possibility of self-incurred extinction of human civilization. Considering this threat that will also continue to exist in the future, it would be a "major accomplishment" of human beings if they would manage to survive even a tenth of the time on earth that the dinosaurs existed. After all, this group of animals, which has often been unjustly derided as an example of "bad design," still managed to survive on earth for 145 million years (Eldredge 1991, 102).

127. When the cosmos is taken into consideration as the greatest whole that exists, infinite regression concerning the question of meaning can only cease when an absolute being (God) is postulated, the meaning of whom by definition can no longer be questioned. Since the existence of this "original reason for all reality" (Küng 1978, 622) cannot be proven by pure reason, Nagel (1987, 101) concludes that life "may not only be meaningless but also *absurd*."

128. According to the ecologist Zwölfer (1989, 27), the ecological dilemma associated with population growth can only be solved if "human beings apply the potential that accounts for their *unique status*, which encompasses reason (in the sense of being able to perceive relationships in their entirety), altruism, and solidarity."

129. Representative of the almost unfathomable number of publications are the following: Scheler ([1927] 1995), Plessner ([1928] 1975), Gehlen ([1940] 1966), Müller (1974), Landmann (1976), as well as the seven-volume collected works of Gadamer and Vogler (1972).

130. Pascal (1623–1662; 1993, 189) wrote the following: "When I think about the brief duration of my life, consumed by eternity before and afterwards; when I think about the small amount of space I occupy and even that which I can see, expended in the infinite expanse of the spaces of which I know nothing and which know nothing of me, then I shudder and am amazed that I am here and not there; there is no reason why I am right here and not there, why now and not then. Who placed me here?"

131. Without claiming to be exhaustive the following, partially overlapping criteria are mentioned in literature on environmental ethics: the human condition, personhood, potential personhood, rationality, linguistic competence, moral competence, ability to cooperate, autonomy of preference, being the subject of a life, consciousness, ability to suffer, interests, biological interests, life, intentionality ("telos"), self-identity, cybernetic self-regulation, being an integral part of an ecosystem, existence.

132. The following brief example that Singer has presented (quoted in Lombardi 1983, 265) should suffice to illustrate that the principle of equality by no means requires wholesale equal treatment: " . . . concern for the well-being

of a child growing up in America would require that we teach him to read; caring for the well-being of a pig may require no more than that we leave him alone with other pigs in a place where there is adequate food and room to run freely."

133. See McCloskey (1979), Passmore (1974), Kirschenmann (1978, 368, 369), Godfrey-Smith (1980, 40), Lombardi (1983, 267f.), Taylor (1986, 219f.), Ricken (1987, 8). Taylor (1986, 251), a prominent representative of biocentrism, points out that "what are normally taken as implications of having rights, indeed, part of the very significance (meaning and importance) of being a rightsholder, would become nonsense if moral rights were ascribed to animals or plants." Even though it might be theoretically possible to modify the concept of moral rights with reference to animals and plants, these "rights" would then have very little in common with what is normally associated with rights in connection with *humans*. Since the concept of rights in ethics is usually closely connected with the concept of what constitutes a person, the usefulness of a modification of this kind might be outweighed by the confusion it causes. Furthermore, such a modification is not really necessary since it would not contribute much to extending the concept of an "ethic of respect for nature" (Taylor 1986, 254, 255). I might add that the concept of moral rights is for the most part irrelevant for ecological problems anyway. Since most ethicists feel that a prerequisite for having moral rights is the ability to have interests (Weber 1990, 119), only individual organisms could profit from the protection such rights provide, not ecosystems or species. Even if the question of *moral* rights remains unresolved, a completely different matter is whether or not *legal* rights might be assigned to nature. Here von der Pfordten (1996, 291f.), Weber (1990), Stone (1988), and Varner (1987) have shown that this is very well possible from the standpoint of legal logics and practice and also desirable from the perspective of nature conservation.

134. From this perspective it sounds rather strange when Patzig (1983, 339) criticizes Albert Schweitzer's principle of "reverence for life" with the following argument: "Animals simply are not eligible partners for the hypothetical contract of reciprocity upon which human morality is based. They also do not behave according to moral principles among one another, and they wouldn't spare us if they were superior to us in the same manner as we usually are to them." Is this supposed to mean that as creatures capable of morality our behavior toward those incapable of morality should be guided by how they would deal with us if they were superior toward us? As Teutsch (1985, 50) has rightly pointed out, the Golden Rule requires exactly the opposite, "that we behave towards those who are inferior to us in a manner in which we would want to be treated by them if the positions were reversed. And it for-

bids mistreating our subordinates in the manner we might have to expect from them if the tables were turned."

135. See, for example, Goodpaster (1978, 319), Kantor (1980, 169), Attfield (1983, 145), Ricken (1987), Varner (1990), von der Pfordten (1996).

136. The results of neurophysiology and evolutionary biology suggest that at least among higher animals (mammals and birds) it is quite reasonable to assume that consciousness exists (Eccles 1991, 83). But "how early its antecedents arise, and whether there are somewhat similar states in plants, seem . . . questions which, while interesting, are perhaps forever unanswerable" (Popper and Eccles 1977, 29). In Popper's opinion (p. 30) we can only "speculate on the conditions of the emergence of consciousness."

137. The nature of such an axiom is illustrated by an answer Birnbacher (1981, 312) gave to the following question that he himself posed: "But who has interests?" His reply: "While Barry assumes that 'wants' can only be attributed to those who can articulate them, I prefer to define less restrictive conditions for the concept of interest. A creature to which we attribute interests should first of all be conscious and secondly have needs, the denial of which will cause suffering. Similar to Leonard Nelson's concept of interest, mine draws higher animals into the circle of subjects with interests by virtue of their ability to suffer."

138. Thus the biochemist Chargaff (1978, 172), for example, refers to the "helplessness of science before life" and maintains that it is not by chance "that of all sciences, biology is the only one that is unable to define its object: we have no scientific definition of life. The most exact studies are, in fact, performed on dead cells and tissues." Chargaff continues as follows: "I say it with all due diffidence, but it is not impossible that we are encountering here a form of exclusion principle: our inability to comprehend life in its reality is due to the very fact that we are alive. If this were so, only the dead could understand life; but they publish in other journals."

139. The discussion presented in that chapter can be illustrated by the following example that Riedl (1985, 107) proposed: "If we say that the tendency for an object to fall can be explained by the law of gravity, we must admit that we are certainly quite familiar with objects falling but that we have not come even close to proving the existence of the gravity waves which the theory claims to exist. . . . We don't know why gravity exists." In view of such obvious limits to explanation Chargaff (1970, 815) rightly asks, "Do we really understand the world? We call that which we understand the world. Humans have an enormous capacity to form abstractions about things they do not understand."

140. In fact a similar problem has arisen in connection with a topic that also seems quite utopian at the moment but is nevertheless seriously debated by

planet researchers of NASA (National Aeronautics and Space Administration), the topic of *terraforming* (Sagan 1994, 329). This term refers to the process of converting a celestial body that is unpopulated at the moment (e.g., Mars or some moon of Jupiter or Saturn) to a place where life could develop by natural means. Would such massive intervention in a foreign world be legitimate? Or in the words of Sagan (1994, 348): "Can we, who have made such a mess of *this* world, be trusted with others?"

141. Let me illustrate this with an example. Although human interventions in the natural dynamics of the Waddensee in the northern German state of Schleswig-Holstein, a national park, really should not be permitted ("first-order rule"), the national park agencies still tolerate measures for coastal protection designed to maintain the bird island Norderoog, which is constantly threatened by erosion ("second-order measure"). This exception can be justified by the fact that the sandwich terns that breed there would nowadays find no comparable biotope as a substitute if the island disappeared. Although in former times before the entire coastal area was occupied by humans sandwich terns were able to respond to geological changes by moving to a different place, today *natural* dynamics have suddenly become an existential threat.

142. However, such systems of interconnected biotopes are destined to fail in the long run if they remain only small islands in an otherwise hostile sea of civilization. According to Remmert (1990, 165) it is extremely important "that the sea surrounding the islands not be too hostile." Therefore overfertilization, heavy pesticide use, and highway construction must be avoided.

143. See Hull (1976), Wiley (1980, 78), Ghiselin (1981), Willmann (1985, 56f.), Kluge (1990).

144. Of course, it must be noted that by calling a species a "historical individual" the "species problem" (that is, the question of the ontological status of a species) cannot be regarded as being resolved in all respects. It is somewhat unsatisfactory to apply the term "individual" to a species insofar as this tends to blur the difference that exists between a "singular individual" represented by the organism and a "collective individual" such as a species. Furthermore, the relationship between an organism and its cells is undoubtedly clearly different from that between a species and its individual members if you consider the type of organization and the tightness of interconnections involved. To avoid misunderstandings, Mayr (1988a, 348–350) suggests replacing the term "individual" with the term "population" or coining a new term. However, these are mostly hassles about *terminology*. They have no effect on the results that are decisive for my discussion, namely that species are *not semantic classes* but hypothetically real entities.

145. See, for example, Patzig (1983, 340), Schäfer (1987, 21); see also Ehrlich and Ehrlich (1981, 7f.).

146. Of course, there is no agreement among ethicists about how well this postulate can be supported by arguments. While Patzig (1983, 341) feels that no rational argument in support of it is in sight, Akerma (1995) maintains that justification can only be found within the context of strong metaphysical convictions. Wendnagel (1990, 32) thinks that the desire "to maintain the existence and nature of human beings" is so "universal" that it is superfluous to find it necessary to justify it.

147. This is the manner in which, for example, Brock (1924; 1925, 266) responded to Schweitzer's ([1923] 1974) biocentric ethics.

148. A review of the discussion about moral dilemmas can be found in Vossenkuhl's encyclopedic contribution (1992b, 188) as well as in Gowans' (1987) anthology titled *Moral Dilemmas*.

149. Instead of the term *guilt* the bioethicists Beauchamp and Childress (2001, 406) refer to a "moral residue" that remains and may be associated with "attitudes of regret, contrition, sorrow, and the like" when a prima facie obligation is overridden.

150. This may have been different in former times when the concept of guilt was used instrumentally in a counterproductive manner (in particular by the church). Much of the contemporary resistance toward the use of the concept of guilt in ethics can be explained by this "traumatic" experience. However, the term guilt should not be rashly tossed onto the junk pile of history. "Perhaps the capacity for guilt and atonement, which are essential aspects of responsibility, are a decisive part of the basic human condition which distinguishes human beings from other creatures" (Schüz 1990, 149).

151. Let me illustrate this with an example. A vacationer looking for adventure who had to kill a polar bear in order to save his own life will in the future avoid places where the probability of getting into the same kind of dilemma is great.

152. Aiken (1984, 277f.) discusses what this would mean for agriculture, while Joosten and Clarke (2002) provide a helpful guide for the wise use of mires and peatlands.

Conclusions and Prospects

153. According to E. U. von Weizsäcker (1992, 258) the industrialized nations of the north use about ten times more energy, water, land, and raw materials per capita than developing countries. Twenty-five percent of the world's population is responsible for 80 percent of the global carbon dioxide emissions (Leisinger 1994, 138). "The developing countries, whom we are currently exhorting to protect their rain forests, are aware of this relationship and do

not see why they should deny themselves anything that could be advantageous for their in some cases desperate economic situation" (E. U. von Weizsäcker 1992, 202).

154. In saying this I by no means intend to deny that efficiency of technology can be significantly enhanced by the use of intelligent technology. But it is improbable that this can be achieved to the extent needed to permit all of the growing human population to attain the lifestyle currently enjoyed by a wealthy minority. If the world population doubles in the predicted manner, and if the income level of the poor is to be raised to that of the wealthy without an additional burden to the environment, then, according to investigations by Vorholz (1995, 27), technological efficiency would have to increase forty-six-fold within a few decades. "That means that the amount of nature consumed per unit of gross national product would have to shrivel to two percent of the current level." Even the highly publicized three liter car would then be "beyond good and evil."

155. As a representative of many philosophers in antiquity who were convinced of the value of voluntary self-denial, let me mention Seneca (4 B.C. to A.D. 65; 2002), while Thoreau ([1854] 1992) is a classic example of modern thinkers who advocate (and have lived) a simple life. It is unjustified that he is overlooked in almost all works on the history of philosophy (see Wiedmann 1990, 19, 109f.).

34. BIBLIOGRAPHY

Adis, J. 1990. "Thirty million arthropod species—too many or too few?" *Journal of Tropical Ecology* 6: 115–118.

Aiken, W. 1984. "Ethical issues in agriculture." In T. Regan, ed., *Earthbound: New introductory essays in environmental ethics.* Temple University Press, Philadelphia, 247–288.

Akerma, K. 1995. *Soll eine Menschheit sein? Eine fundamentalethische Frage.* Junghans, Cuxhaven, Germany.

Albert, H. 1961. "Ethik und Meta-Ethik: Das Dilemma der analytischen Moralphilosophie." *Archiv für Philosophie* 11: 28–63.

———. 1968. *Traktat über kritische Vernunft.* Mohr, Tübingen.

Altner, G. 1979. "Wahrnehmung der Interessen der Natur." In K. M. Meyer-Abich, ed., *Frieden mit der Natur.* Herder, Freiburg, pp. 112–130.

———. 1982. "Alternative Wissenschaft und Mystik." In G. Altner, *Physik, Philosophie und Politik: Festschrift für C. F. von Weizsäcker zum 70. Geburtstag.* Carl Hanser, Munich, pp. 430–439.

———. 1984. "Umweltethik—Grundsätze und Perspektiven." *Scheidewege* 14: 36–43.

———. 1985. "Ethische Begründung des Artenschutzes." *Schriftenreihe des Deutschen Rats für Landespflege* 46: 566–568.

Alvarez, L. W., W. Alvarez, F. Asaro, and H. V. Michel. 1980. "Extraterrestrial cause for the Cretaceous Tertiary extinction." *Science* 208: 1095–1108.

Amberg, M. 1980. *Naturschutz—die große Lüge.* Kilda, Greven, Germany.

Amery, C. 1974. *Das Ende der Vorsehung: Die gnadenlosen Folgen des Christentums.* Rowohlt, Reinbek, Germany.

———. 1982. *Natur als Politik: Die ökologische Chance des Menschen.* Rowohlt, Reinbek, Germany.

Apel, K.-O. 1973. "Das Apriori der Kommunikationsgemeinschaft und die Grundlagen der Ethik. Zum Problem einer rationalen Begründung der Ethik im Zeitalter der Wissenschaft." In K.-O. Apel, *Transformation der Philosophie*, vol. 2. Suhrkamp, Frankfurt am Main, pp. 359–435.

———. 1976. "Das Problem der philosophischen Letztbegründung im Lichte einer transzendentalen Sprachpragmatik (Versuch einer Metakritik des 'Kritischen Rationalismus')." In B. Kanitschneider, ed., *Sprache und Erkenntnis.* Innsbruck, pp. 55–82.

———. 1987. "Fallibilismus, Konsenstheorie der Wahrheit und Letztbegründung." In Forum für Philosophie Bad Homburg, ed., *Philosophie und Begründung.* Suhrkamp, Frankfurt am Main, pp. 116–211.

Aquinus, Thomas. 1953. *Summa Theologica* II-II, Qu.57-79, (vol. 18: Recht und

Gerechtigkeit). German-Latin edition by Albertus-Magnus-Akademie Walberberg bei Köln. Gemeinschaftsverlag Kerle & Pustet, Heidelberg.

Aristotle (1996). *The Nicomachean ethics.* Wordsworth, Hertfordshire, UK.

Armstrong, S. J., and R. G. Botzler, eds. 1993. *Environmental ethics: Divergence and convergence.* McGraw-Hill, New York.

Arndt, U., W. Nobel, and B. Schweizer. 1987. *Bioindikatoren—Möglichkeiten, Grenzen und neue Erkenntnisse.* Stuttgart.

Arrow, K. R. 1963. *Social choice and individual values.* Yale University Press, London.

Attfield, R. 1983. *The ethics of environmental concern.* Basil Blackwell, Oxford.

Auer, A. 1985. "Im Konfliktfall gilt der Vorrang der Ökologie vor der Ökonomie. Annäherungen an eine Umweltethik." *Bürger im Staat* 35(3): 174–179.

———. 1988. "Anthropozentrik oder Physiozentrik? Vom Wert eines Interpretaments." In K. Bayertz, ed., *Ökologische Ethik.* Schnell & Steiner, Munich, pp. 31–54.

Auer, M. 1992. "Ökonomische Bewertung der Biologischen Vielfalt." *Natur und Landschaft* 67(9): 439–440.

Auhagen, A., and H. Sukopp. 1983. "Ziel, Begründungen und Methoden des Naturschutzes im Rahmen der Stadtentwicklungspolitik von Berlin." *Natur und Landschaft* 1: 9–15.

Ayer, A. J. 1936. *Language, truth and logic.* Gollancz, London.

Bachmann, K. 1990. "Wenn Räuber Opfer ihrer Beute werden." *Chaos und Kreativität (GEO-Wissen No.2).* Gruner & Jahr, Hamburg, pp. 88–96.

Bacon, F. [1619] 1968. *The Wisedome of the Ancients* (De Sapientia Veterum). Da-Capo Press, New York.

Baier, A. 1980. "The rights of past and future persons." In E. Partridge, ed., *Responsibilities to future generations: Environmental ethics.* Prometheus Books, Buffalo, N.Y., pp. 171–183.

Barrow, J. D., and J. Silk. 1983. *The left hand of creation.* Basic Books, New York.

Barrowclough, G. F. 1992. "Systematics, biodiversity, and conservation biology." In N. Eldredge, ed., *Systematics, ecology, and the biodiversity crisis.* Columbia University Press, New York, pp. 121–143.

Barry, B. 1977. "Justice between generations." In P. M. S. Hacker and J. Raz, eds., *Law, morality, and society: Essays in honour of H. L. A. Hart.* Oxford University Press, Oxford, pp. 268–284.

Baskin, Y. 1997. *The work of nature: How the diversity of life sustains us.* Island Press, Washington, D.C.

Bauer, H.-J. 1985. "Welche Ursachen führten zu Gefährdung und Ausrottung von Arten?" *Schriftenreihe des Deutschen Rats für Landespflege* 46: 572–580.

Bayertz, K. 1986. "Technik, Ökologie und Ethik. Fünf Dialoge über die moralischen Grenzen der Technik und über die Schwierigkeiten einer nicht-anthropozentrischen Ethik." In G. Bechmann and W. Rammert, eds., *Technik und Gesellschaft*, vol. 4. Frankfurt am Main.

———. 1987. "Naturphilosophie als Ethik. Zur Vereinigung von Natur- und Moralphilosophie im Zeichen der ökologischen Krise." *Philosophia Naturalis* 24: 157–185.

———. 1988. "Ökologie als Medizin der Umwelt? Überlegungen zum Theorie-Praxis-Problem in der Ökologie." In K. Bayertz, ed., *Ökologische Ethik*. Schnell & Steiner, Munich, pp. 86–101.

Beauchamp, T. L., and J. F. Childress. 2001. *Principles of biomedical ethics*. Oxford University Press, Oxford.

Beck, U. 1986. *Risikogesellschaft: Auf dem Weg in eine andere Moderne*. Suhrkamp, Frankfurt am Main.

———. 1988. *Gegengifte: Die organisierte Unverantwortlichkeit*. Suhrkamp, Frankfurt am Main.

Beierkuhnlein, C. 1994. "Methodische Defizite in der Naturschutzforschung— aufgezeigt an Beispielen aus der Vegetationskunde." *Naturschutzzentrum Wasserschloß Mitwitz—Materialien* 1/94: 17–18.

Bentham, J. [1789] 1970. *An introduction to the principles of morals and legislation*. Edited by J. H. Burns and H. L. A. Hart. Athlone Press, London.

Benton, M. J. 1985. "Mass extinctions among non-marine tetrapods." *Nature* 316: 811–814.

———. 1986. "The evolutionary significance of mass extinctions." *Trends in Ecology and Evolution* 1(5): 127–130.

Berndt, R., and M. Henß. 1967. "Die Kohlmeise, Parus major, als Invasionsvogel." *Die Vogelwarte* 24: 17–37.

Bertalanffy, L. von. 1973. *General system theory: Foundations, development, applications*. Penguin University Books, Harmondsworth, UK.

Bezzel, E., and J. Reichholf. 1974. "Die Diversität als Kriterium zur Bewertung der Reichhaltigkeit von Wasservogel-Lebensräumen." *Journal für Ornithologie* 115: 50–61.

Bierhals, E. 1984. "Die falschen Argumente? Naturschutz-Argumente und Naturbeziehung." *Landschaft + Stadt* 16(1/2): 117–126.

Birch, T. H. 1993. "Moral considerability and universal consideration." *Environmental Ethics* 15: 313–332.

Birnbacher, D. 1979a. "Plädoyer für eine Ethik der Zukunft." *Zeitschrift für Didaktik der Philosophie* 1: 119–123.

———. 1979b. "Was wir wollen, was wir brauchen und was wir wollen dürfen." In K. M. Meyer-Abich and D. Birnbacher, eds., *Was braucht der Mensch, um glücklich zu sein?* Beck, Munich, pp. 30–57.

———. 1980. "Sind wir für die Natur verantwortlich?" In D. Birnbacher, ed., *Ökologie und Ethik*. Reclam, Stuttgart, pp. 103–139.

_____. 1981. "Sind die Normen der ökologischen Ethik universalisierbar?" In: E. Morscher and R. Stranzinger, eds., *Ethik: Grundlagen, Probleme und Anwendungen. Akten des 5. internationalen Wittgenstein-Symposiums 1980 in Kirchberg am Wechsel.* Hölder-Pichler-Tempsky, Vienna, pp. 312–314.

_____. 1982. "A priority rule for environmental ethics." *Environmental Ethics* 4: 3–16.

_____. 1987. "Ethical principles versus guiding principles in environmental ethics." *Philosophica* (Gent) 39(1): 59–76.

_____. 1988. *Verantwortung für zukünftige Generationen.* Reclam, Stuttgart.

_____. 1989. "Ökologie, Ethik und neues Handeln." In H. Stachowiak, ed., *Pragmatik: Handbuch pragmatischen Denkens*, vol. 3. Meiner, Hamburg, pp. 393–417.

_____. 1991. "'Natur' als Maßstab menschlichen Handelns." *Zeitschrift für philosophische Forschung* 45: 60–76.

Bischoff, M. 1993. "Über einige Schwierigkeiten, ein neues Natur-Bewußtsein zu entwickeln." In R. Schäfer, ed., *Was heißt denn schon Natur?* Callwey, Munich, pp. 49–60.

Blab, J. 1985. "Sind die Roten Listen der gefährdeten Arten geeignet, den Artenschutz zu fördern?" *Schriftenreihe des Deutschen Rats für Landespflege* 46: 612–617.

Blab, J., and W. Völkl. 1994. "Voraussetzungen und Möglichkeiten für eine wirksame Effizienzkontrolle im Naturschutz." *Schriftenreihe für Landschaftspflege und Naturschutz* 40: 291–300.

Bleuler, E. 1983. *Lehrbuch der Psychiatrie.* Springer, Berlin.

Bond, W. J. 1993. "Keystone species." In E.-D. Schulze and H. A. Mooney, eds., *Biodiversity and ecosystem function.* Springer, Berlin, pp. 237–253.

Bookchin, M. 1971. *Post-scarcity-anarchism.* Harper & Row, New York.

Borchardt, T., B. Scherer, and E. Schrey, eds. 1989. *Ökosystemforschung Wattenmeer. Das Projekt im Überblick*, Part B. Nationalpark Agency of the Waddensea in Schleswig-Holstein, Tönning.

Bossel, H. 1982. "Ansätze einer ökologisch orientierten Wissenschaft." In G. Michelsen, U. Rühling, and F. Kalberlah, eds., *Der Fischer Öko-Almanach II. Daten, Fakten, Trends der Umweltdiskussion.* Fischer, Frankfurt am Main, pp. 36–42.

Breckling, B., K. Ekschmitt, K. Mathes, H.-J. Poethke, A. Seitz, and G. Weidemann. 1992. "Gedanken zur Theorie in der Ökologie." *Verhandlungen der Gesellschaft für Ökologie* 21: 1–8.

Brennan, A. 1984. "The moral standing of natural objects." *Environmental Ethics* 6(1): 35–56.

Briggs, J., and F. D. Peat. 1989. *Turbulent mirror: An illustrated guide to chaos theory and the science of wholeness.* Harper & Row, New York.

Brock, E. 1924, 1925. "Rezension zu: Albert Schweitzer (1923): Kultur und Ethik." *Logos* 13: 264–269.

Brown, J. H., D. W. Davidson, J. C. Munger, and R. C. Inouye. 1986. "Experimental community ecology: The desert granivore system." In J. L. Diamond and T. J. Case, eds., *Community ecology*. Harper & Row, New York, pp. 41–61.

Brown, J. H., and E. J. Heske. 1990. "Control of a desert-grassland transition by a keystone rodent guild." *Science* 250: 1705–1707.

Brumbaugh, R. S. 1978. "Of man, animals and morals: a brief history." In R. K. Morris and M. Fox, eds., *On the fifth day: Animal rights and human ethics*. Acropolis Books, Washington, D.C., pp. 5–25.

Bundesminister des Innern. 1985. *Umweltpolitik: Bilanz und Perspektiven*. Kohlhammer, Stuttgart.

Burkhardt, A. 1981. "Kant und das Verhältnis der relativen Ethik zur absoluten. Zur Begründung einer ökologischen Ethik." In E. Morscher and R. Stranzinger, eds., *Ethik: Grundlagen, Probleme und Anwendungen*. Akten des 5. internationalen Wittgenstein-Symposiums 1980 in Kirchberg am Wechsel. Hölder-Pichler-Tempsky, Vienna, pp. 321–324.

———. 1983. "Kant, Wittgenstein und das Verhältnis der relativen Ethik zur absoluten: Zur Begründung einer ökologischen Ethik." *Zeitschrift für Evangelische Ethik* 27(4): 391–431.

Cahen, H. 1988. "Against the moral considerability of ecosystems." *Environmental Ethics* 10: 195–216.

Callahan, D. 1981. "What obligations do we have to future generations?" In E. Partridge, ed., *Responsibilities to future generations: Environmental ethics*. Prometheus Books, Buffalo, N.Y., pp. 73–85.

Callicott, J. B. 1980. "Animal liberation: A triangular affair. *Environmental Ethics* 2: 311–338.

———. 1983. "The land aesthetic." *Environmental Review* 7: 345–358.

———. 1993. "The search for an environmental ethic." In T. Regan, ed., *Matters of life and death: New introductory essays in moral philosophy*. McGraw-Hill, New York, pp. 322–382.

Campbell, D. 1974. "'Downward causation' in hierarchically organized biological systems." In F. Ayala and T. Dobzhansky, eds., *Studies in the philosophy of biology: Reduction and related problems*. Macmillan, New York.

Canguilhem, G. 1974. *Das Normale und das Pathologische*. Carl Hanser, Munich.

Capra, F. 1983. *Wendezeit: Bausteine für ein neues Weltbild*. Scherz, Bern.

Carlson, A. 1984. "Nature and positive aesthetics." *Environmental Ethics* 6(1): 5–34.

Caughley, G. 1976. "The elephant problem—an alternative hypothesis." *East African Wildlife Journal* 14: 265–283.

Caughley, G., and J. H. Lawton. 1981. "Plant-herbivore systems." In R. M. May, ed., *Theoretical ecology: Principles and applications*. Blackwell, Oxford, pp. 132–166.

Chaitin, G. J. 1975. "Randomness and mathematical proof." *Scientific American* 232: 47–52.

Chapman, J. S. 1974. "Environmental health, environmental deterioration." In F. Sargent II, ed., *Human Ecology*. North-Holland Publishing, Amsterdam.

Chargaff, E. 1970. "Vorwort zu einer Grammatik der Biologie. Hundert Jahre Nukleinsäureforschung." *Experientia* 26(7): 810–816.

———. 1978. *Heraclitean fire: Sketches from a life before nature*. Rockefeller University Press, New York.

———. 1991. "Erforschung der Natur und Denaturierung des Menschen." In H.-P. Dürr and W. C. Zimmerli, eds., *Geist und Natur: Über den Widerspruch zwischen naturwissenschaftlicher Erkenntnis und philosophischer Welterfahrung*. Scherz, Bern, pp. 355–368.

Chesson, P. L., and T. J. Case. 1986. "Nonequilibrium community theories: Chance, variability, history, and coexistence." In J. Diamond and T. J. Case, eds., *Community ecology*. Harper & Row, New York, pp. 229–239.

Cicero, M. T. 1997. *The nature of the gods* (De natura deorum). Translated with introduction and explanatory notes by P. G. Walsh. Clarendon Press, Oxford.

Clapham, W. B., Jr. 1973. "Resiliency and fitness of ecosystems." In W. B. Clapham Jr., *Natural ecosystems*. Macmillan, New York, pp. 229–238.

Clark, C. W. 1973. "Profit maximization and the extinction of animal species." *Journal of Political Economy* 81: 950–961.

Clements, F. E. 1936. "Nature and the structure of climax." *Journal of Ecology* 24: 252–284.

Cobb, J. B., Jr. 1972. *Is it too late?* Bruce Publishing, New York.

Cody, M. L., and J. M. Diamond, eds. 1975. *Ecology and evolution of communities*. Harvard University Press, Cambridge, Mass.

Commoner, B. 1972. *The closing circle: Nature, man, and technology*. Alfred A. Knopf, New York.

Cousins, S. H. 1991. "Species diversity measurement: choosing the right index." *Trends in Ecology and Evolution* 6(6): 190–192.

Cramer, F. 1979. "Fundamental complexity, a concept in biological science and beyond." *Interdisciplinary Science Reviews* 4: 132–139.

———. 1986. "Die Evolution frißt ihre Kinder—der Unterschied zwischen Newtonschen Bahnen und lebenden Wesen." *Universitas* 41(2): 1149–1156.

Cramer, J., and W. van den Daele. 1985. "Is ecology an 'alternative' natural science?" *Synthese* 65: 347–375.

Dahl, J. 1989a. *Der unbegreifliche Garten und seine Verwüstung: Über Ökologie und über Ökologie hinaus*. Klett-Cotta, Stuttgart.

———. 1989b. *Die Verwegenheit der Ahnungslosen: Über Genetik, Chemie und andere Schwarze Löcher des Fortschritts*. Klett-Cotta, Stuttgart.

Daily, G. C., ed. 1997. *Nature's services: Societal dependence on natural ecosystems*. Island Press, Washington, D.C.

Darwin, C. 1859. *On the origin of species by means of natural selection*. John Murray, London.

DeAngelis, D. L., W. M. Post, and C. C. Travis. 1986. *Positive feedback in natural systems*. Springer, Berlin.

Dennett, D. C. 1993. *Consciousness explained*. Penguin Books, London.

Dersal, W. R. van. 1972. "Why living organisms should not be exterminated." *Atlantic Naturalist* 27: 7–10.

DeSanto, R. S. 1978. *Concepts of applied ecology*. Springer, New York.

Descartes, R. [1637] 1956. *Discourse on methods and meditations* (Discours de la Méthode). Translated by L. J. Lafleur. Liberal Arts Press, New York.

DesJardins, J. R. 1999. *Environmental ethics: Concepts, policy, and theory*. Mayfield Publishing, Mountain View, Calif.

Deutscher Rat für Landespflege. 1985. "Warum Artenschutz?" *Schriftenreihe des Deutschen Rats für Landespflege* 46: 537–559.

Dierssen, K. 1994. "Was ist Erfolg im Naturschutz?" *Schriftenreihe für Landschaftspflege und Naturschutz* 40: 9–23.

Ditfurth, H. von. 1991. *Innenansichten eines Artgenossen: Meine Bilanz*. Deutscher Taschenbuch Verlag, Munich.

Dixon, B. 1976. "Smallpox-imminent extinction, and an unresolved dilemma." *New Scientist* 69: 430–432.

Dörner, D. 1993. "Denken und Handeln in Unbestimmtheit und Komplexität." *Gaia* 2(3): 128–138.

Duden. 1969. rororo Lexikon (Duden's German dictionary), Taschenbuchausgabe, vol. 1. Rowohlt, Reinbek, Germany.

Dürr, H.-P. 1991. "Wissenschaft und Wirklichkeit. Über die Beziehung zwischen dem Weltbild der Physik und der eigentlichen Wirklichkeit." In H.-P. Dürr and W. C. Zimmerli, eds., *Geist und Natur: Über den Widerspruch zwischen naturwissenschaftlicher Erkenntnis und philosophischer Welterfahrung*. Scherz, Bern, pp. 28–46.

Durrell, L. 1986. *Gaia—State of the ark: An atlas of conservation in action*. Bodley Head, London.

Dwyer, R. L., and K. T. Perez. 1983. "An experimental examination of ecosystem linearization." *The American Naturalist* 121(3): 305–323.

Eccles, J. C. 1970. *Facing reality: Philosophical adventures by a brain scientist*. Springer, New York.

———. 1973. *The understanding of the brain*. McGraw-Hill, New York.

———. 1991. "Der Ursprung des Geistes, des Bewußtseins und des Selbstbewußtseins im Rahmen der zerebralen Evolution." In H.-P. Dürr and W. C. Zimmerli, eds., *Geist und Natur: Über den Widerspruch zwischen naturwissenschaftlicher Erkenntnis und philosophischer Welterfahrung.* Scherz, Bern, pp. 79–89.

Eckschmitt, K., K. Mathes, and B. Breckling. 1994. "Theorie in der Ökologie: Möglichkeiten der Operationalisierung des juristischen Begriffs 'Naturhaushalt' in der Ökologie." *Verhandlungen der Gesellschaft für Ökologie* 23: 417–420.

Eddington, A. 1939. *The philosophy of physical science.* Cambridge University Press, London.

Ehrendorfer, F. 1978. "Geobotanik." In E. Strasburger, ed., *Lehrbuch der Botanik für Hochschulen,* 31st ed., chap. 4. Gustav Fischer, Stuttgart, pp. 862–987.

Ehrenfeld, D. W. 1976. "The conservation of non-resources." *American Scientist* 64: 648–656.

———. 1989. "Why put a value on biodiversity?" In E. O. Wilson, ed., *Biodiversity.* National Academy Press, Washington, D.C., pp. 212–216.

Ehrlich, P. R. 1991. "Population diversity and the future of ecosystems." *Science* 254: 175.

———. 1993. "Biodiversity and ecosystem function: Need we know more?" Foreword to E.-D. Schulze and H. A. Mooney, eds., *Biodiversity and ecosystem function.* Springer, Berlin, pp. vii–xi.

Ehrlich, P. R., and A. Ehrlich. 1981. *Extinction: The causes and consequences of the disappearance of species.* Random House, New York (German edition: P. R. Ehrlich and A. Ehrlich. 1983. *Der lautlose Tod: Das Aussterben der Pflanzen und Tiere.* Krüger, Frankfurt).

Ehrlich, P. R., A. Ehrlich, and J. P. Holdren. 1973. *Human ecology: Problems and solutions.* W. H. Freeman, San Francisco.

Ehrlich, P. R., and H. A. Mooney. 1983. "Extinction, substitution, and ecosystem services." *BioScience* 33(4): 248–254.

Eigen, M., and R. Winkler. 1975. *Das Spiel: Naturgesetze steuern den Zufall.* Piper, Munich.

Eilenberger, G. 1989. "Komplexität. Ein neues Paradigma der Naturwissenschaften." In H. von Ditfurth and E. P. Fischer, eds., *Mannheimer Forum 89/90.* Boehringer Mannheim GmbH, Mannheim, Germany, pp. 71–134.

Ekman, G., and U. Lundberg. 1971. "Emotional reaction to past and future events as a function of temporal distance." *Acta Psychologica* 35: 430–441.

Eldredge, N. 1991. *The miner's canary: Unraveling the mysteries of extinction.* Prentice-Hall, Englewood Cliffs, N.J.

Elliot, R. 1978. "Regan on the sorts of beings that can have rights." *Southern Journal of Philosophy* 16: 701–705.

———. 1980. "Why preserve species?" In D. S. Mannison, M. A. McRobbie, and R. Routley, eds., *Environmental philosophy.* Monograph Series, No. 2, Department of Philosophy, Australian National University, Canberra, pp. 8–29.

———. 1982. "Faking nature." *Inquiry* 25: 81–93.

Engelhardt, T., Jr., and S. Spicker. 1978. *Mental health: Philosophical perspectives*. D. Reidel, Dordrecht.

Engels, E.-M. 1990. *Erkenntnis als Anpassung? Eine Studie zur Evolutionären Erkenntnistheorie*. Suhrkamp, Frankfurt am Main.

———. 1993. "George Edward Moores Argument der 'naturalistic fallacy' in seiner Relevanz für das Verhältnis von philosophischer Ethik und empirischen Wissenschaften." In L. H. Eckensberger and U. Gähde, eds., *Ethische Norm und empirische Hypothese*. Suhrkamp, Frankfurt am Main, pp. 92–132.

Erbrich, P. 1990. "Natur- und Umwelterziehung als Aspekte des Religionsunterrichts—Philosophische Grundüberlegungen zum Thema." *Berichte der ANL* 14: 3–9.

Ernst, C. 1996. "Rechte der Fauna. Rezension zu: Dietmar von der Pfordten (1996): Ökologische Ethik. Zur Rechtfertigung menschlichen Verhaltens gegenüber der Natur." *Süddeutsche Zeitung* No. 242 (Oct. 19, 1996): v.

Erwin, D. H. 1989. "The end-Permian mass extinction: What really happened and did it matter?" *Trends in Ecology and Evolution* 4(8): 225–229.

Erz, W. 1984. "Zwischen Wissenschaft und Ideologie. Zur Akzeptanz eines neuen Begriffs." *Das Parlament* 34(19): 1–2.

———. 1986. "Ökologie oder Naturschutz? Überlegungen zur terminologischen Trennung und Zusammenführung." *Berichte der ANL* 10: 11–17.

Eser, A. 1983. "Ökologisches Recht." In H. Markl, ed., *Natur und Geschichte*. Oldenbourg, Munich, pp. 349–396.

Fäh, H. 1987. *Biologie und Philosophie in ihren Wechselbeziehungen: Menschenbilder—Erkenntnisweisen—Weltbilder*. Metzler, Stuttgart.

Farb, P. 1976. *Die Ökologie*. Rowohlt, Reinbek, Germany.

Feinberg, J. 1980. "The rights of animals and unborn generations." In J. Feinberg, *Rights, justice and the bounds of liberty: Essays in social philosophy*. Princeton University Press, Princeton, N.J.

Ferris, T. 1981. *Galaxies*. Sierra Club Books, San Francisco, Calif.

Fischbeck, G. 1976. "Moderne Pflanzenproduktion und Umweltbeeinflussung." *Bayerisches Landwirtschaftliches Jahrbuch* 53, Sonderheft 3: 60–67.

Fisher, A., and M. Hanemann. 1984. "Option values and the extinction of species." *Working Paper* 269. Giannini Foundation of Agricultural Economics, Berkeley, Calif.

Forschner, M. 1992. "Pflichtenkollision." In O. Höffe, M. Forschner, and W. Vossenkuhl, eds., *Lexikon der Ethik*. Beck, Munich, pp. 211–212.

Forum für Philosophie Bad Homburg, ed. 1987. *Philosophie und Begründung*. Suhrkamp, Frankfurt am Main.

Frankel, O. H. 1974. "Genetic conservation: Our evolutionary responsibility." *Genetics* 78: 53–65.

Frankena, W. K. 1979. "Ethics and the environment." In K. E. Goodpaster and K. M. Sayre, eds., *Ethics and problems of the 21st century*. University of Notre Dame Press, London, pp. 3–20.

———. 1963. *Ethics*. Prentice-Hall, Englewood Cliffs, N.J.

Frey, R. G. 1980. *Interests and rights: The case against animals*. Clarendon Press, Oxford.

Friedmann, E. I., M. Hua, and R. Ocampo-Friedmann. 1988. "Cryptoendolithic lichen and cyanobacterial communities of the Ross Desert, Antarctica." *Polarforschung* 58: 251–259.

Fritz, E. C. 1983. "Saving species is not enough." *BioScience* 33(5): 301.

Fromm, E. 1947. *Man for himself: An inquiry into the psychology of ethics*. Holt, Rinehart & Winston, New York.

Früchtl, J. 1991. "Der faule Kern der Handelsware Öko-Ethik." *Die Neue Gesellschaft/Frankfurter Hefte* 4/91: 344–348.

Furger, F. 1976. "Freiwillige Askese als Alternative." In G.-K. Kaltenbrunner, ed., *Überleben und Ethik: Die Notwendigkeit, bescheiden zu werden*. Herder, Freiburg, pp. 77–90.

Gadamer, H.-G., and P. Vogler, eds. 1975. *Neue Anthropologie* (7 vols.). Stuttgart.

Garaudy, R. 1991. "Der Sinn des Lebens und der Dialog der Kulturen." In H.-P. Dürr and W. C. Zimmerli, eds., *Geist und Natur: Über den Widerspruch zwischen naturwissenschaftlicher Erkenntnis und philosophischer Welterfahrung*. Scherz, Bern, pp. 369–380.

Gautier-Hion, A., and G. Michaloud. 1989. "Are figs always keystone resources for tropical frugivorous vertebrates? A test in Gabon." *Ecology* 70: 1826–1833.

Gehlen, A. [1940] 1966. *Der Mensch: Seine Natur und seine Stellung in der Welt*. Frankfurt am Main.

Gerdes, A. 1993c. "Grimmige Klimakapriolen." *Die Zeit* 29/93.

Gerdes, J. 1993a. "Die Verwaltung der Natur." *Kosmos* 4/93: 60–61.

———. 1993b. "Synthetische Natur?" In R. Schäfer, ed., *Was heißt denn schon Natur?* Callwey, Munich, pp. 135–146.

———. 1996. "Ist Natur ersetzbar?" *Kosmos* 4/96: 74–75.

Gerstberger, P. 1991. "Erarbeitung eines floristisch-vegetationskundlichen Verfahrens zur Bewertung der Schutzwürdigkeit von landwirtschaftlich genutzten Mähwiesen." In G. Kaule and K. Henle, eds., *Arten- und Biotopschutzforschung für Deutschland: Berichte aus der Ökologischen Forschung*, pp. 318–322.

Gethmann, C. F. 1987. "Letztbegründung versus lebensweltliche Fundierung des Wissens und Handelns." In Forum für Philosophie Bad Homburg, ed., *Philosophie und Begründung*. Suhrkamp, Frankfurt am Main, pp. 268–302.

———. 1993. "Naturgemäß handeln?" *Gaia* 2(5): 246–248.

Gethmann, C. F., and J. Mittelstraß. 1992. "Maße für die Umwelt." *Gaia* 1(1): 16–25.

Ghiselin, M. J. 1974. "A radical solution to the species problem." *Systematic Zoology* 23: 536–544.

———. 1981. "Categories, life, and thinking." *The behavioural and brain sciences* 4: 269–286.

Gibbons, A. 1992. "Mission impossible: Saving all endangered species." *Science* 256: 1386.

Gilbert, L. E., and P. H. Raven, eds. 1975. *Coevolution of plants and animals*. University of Texas Press, Austin.

Glasauer, H. 1991. "Im Einklang mit der Natur? Über das Auseinanderklaffen von Umweltbewußtsein und Umweltverhalten. Eine Polemik." *Garten + Landschaft* 7/91: 9–12.

Gleich, A. von, and E. Schramm. 1992. "Mathematische Modelle und ökologische Erfahrung." *Verhandlungen der Gesellschaft für Ökologie* 21: 15–21.

Global 2000. 1981. *The global 2000 report to the president: Entering the 21st century*. Blue Angel, Charlottesville, Va.

Godfrey-Smith, W. 1980. "The rights of non-humans and intrinsic values." In D. S. Mannison, M. A. McRobbie, and R. Routley, eds., *Environmental philosophy*. Monograph Series, No. 2, Department of Philosophy, Australian National University, Canberra, pp. 30–47.

Goodpaster, K. E. 1978. "On being morally considerable." *The Journal of Philosophy* 75: 308–325.

———. 1979. "From egoism to environmentalism." In K. E. Goodpaster and K. M. Sayre, eds., *Ethics and problems of the 21st century*. University of Notre Dame Press, London, pp. 21–35.

———. 1980. "On stopping at everything: A reply to W. M. Hunt." *Environmental Ethics* 2(3): 281–284.

Gorke, M. 1990. "Die Lachmöwe in Wattenmeer und Binnenland. Ein verhaltens-ökologischer Vergleich." *Seevögel* 11, Sonderheft 3: 1–48.

Gould, S. J. 1977. *Ever since Darwin*. W. W. Norton, New York.

Gowans, C. W. 1987. *Moral dilemmas*. Oxford University Press, New York.

Grimm, V., and Wissel, C. 1997. "Babel or the ecological stability discussions. An inventory and analysis of terminology and a guide for avoiding confusion." *Oecologia* 109(3): 323–334.

Guggenberger, B. 1986. "Für einen ökologischen Humanismus. Die Erhaltung einer fehlerfreundlichen Umwelt als zukunftsethischer Imperativ." In T. Meyer and S. Miller, eds., *Zukunftsethik 1 (Zukunftsethik und Industriegesellschaft)*. J. Schweitzer, Munich, pp. 52–58.

Gunn, A. S. 1980. "Why should we care about rare species?" *Environmental Ethics* 2: 17–37.

———. 1984. "Preserving rare species." In T. Regan, ed., *Earthbound: New introduc-*

tory essays in environmental ethics. Temple University Press, Philadelphia, pp. 289–335.

Günzler, C. 1990a. "Ehrfurchtsprinzip und Wertrangordnung. Albert Schweitzers Ethik und ihre Kritiker." In C. Günzler, E. Grässler, B. Christ, and H. H. Eggebrecht, eds., *Albert Schweitzer heute: Brennpunkte seines Denkens.* Katzmann, Tübingen, pp. 82–100.

———. 1990b. "Ehrfurchtsethik und Umwelterziehung. Zur pädagogischen Fruchtbarkeit der Schweitzerschen Ethik." In C. Günzler, E. Grässler, B. Christ, and H. H. Eggebrecht, eds., *Albert Schweitzer heute: Brennpunkte seines Denkens.* Katzmann, Tübingen, pp. 110–124.

———. 1996. *Albert Schweitzer: Einführung in sein Denken.* Beck, München.

Günzler, C., and H. Lenk. 1990. "Ethik und Weltanschauung. Zum Neuigkeitsgehalt von Albert Schweitzers 'Kulturphilosophie III'." In C. Günzler, E. Grässler, B. Christ, and H. H. Eggebrecht, eds., *Albert Schweitzer heute: Brennpunkte seines Denkens.* Katzmann, Tübingen, pp. 17–50.

Haaren, C. von. 1988. "Beitrag zu einer normativen Grundlage für praktische Zielentscheidungen im Arten- und Biotopschutz." *Landschaft + Stadt* 20(3): 97–106.

———. 1991. "Leitbilder oder Leitprinzipien?" *Garten + Landschaft* 2/91: 29–34.

Haber, W. 1984. "Über Landschaftspflege." *Landschaft + Stadt* 16(4): 193–199.

———. 1986. "Über die menschliche Nutzung von Ökosystemen—unter besonderer Berücksichtigung von Agrarökosystemen." *Verhandlungen der Gesellschaft für Ökologie* 14: 13–24.

———. 1993. "Von der ökologischen Theorie zur Umweltplanung." *Gaia* 2(2): 96–106.

Habermas, J. 1973. *Erkenntnis und Interesse.* Suhrkamp, Frankfurt am Main.

———. 1981. *Theorie des kommunikativen Handelns* (2 vols.). Suhrkamp, Frankfurt am Main.

Haeckel, E. 1866. *Generelle Morphologie der Organismen.* G. Reimer, Berlin.

———. 1924. "Über Entwicklungsgang und Aufgabe der Zoologie." In E. Haeckel, *Gemeinverständliche Werke,* vol. 5. Kröner, Leipzig.

Hampicke, U. 1992. "Kosten des Naturschutzes." *Jahrbuch für Naturschutz und Landschaftspflege* 45: 184–202.

Hampicke, U., T. Horlitz, H. Kiemstedt, K. Tampe, D. Timp, and M. Walters. 1991. *Kosten und Wertschätzung des Arten- und Biotopschutzes.* Schmidt, Berlin.

Hannah, L., D. Lohse, C. Hutchinson, J. L. Carr, and A. Lankerani. 1994. "A preliminary inventory of human disturbance of world ecosystems." *Ambio* 23(4-5): 246–250.

Hare, R. M. 1952. *The language of morals.* Clarendon Press, Oxford.

Hargrove, E. C. 1989. *Foundations of environmental ethics.* Prentice-Hall, Englewood Cliffs, N.J.

Hartkopf, G., and E. Bohne. 1983. *Umweltpolitik 1: Grundlagen, Analysen und Perspektiven.* Westdeutscher Verlag, Opladen, Germany.

Haverbeck, W. G. 1978. *Die andere Schöpfung: Technik—ein Schicksal von Mensch und Erde.* Urachhaus, Stuttgart.

Heffernan, J. D. 1993. "The land ethic: A critical appraisal." In S. J. Armstrong and R. G. Botzler, eds., *Environmental ethics: Divergence and convergence.* McGraw-Hill, New York, pp. 398–411.

Hegel, G. W. F. [1832] 1986. *Vorlesungen über die Ästhetik I* (Lectures on Aesthetics), Werke vol. 13. Suhrkamp, Frankfurt am Main.

Heidegger, M. [1927] 1977. *Sein und Zeit,* 14th ed. Klostermann, Frankfurt am Main.

Heinrich, D., and M. Hergt. 1990. *dtv-Atlas zur Ökologie.* Deutscher Taschenbuch Verlag, Munich.

Heisenberg, W. 1969. *Der Teil und das Ganze: Gespräche im Umkreis der Atomphysik.* Piper, Munich.

Helbing, C.-D. 1995. "Naturschutz als touristischer Wirtschaftsfaktor—am Beispiel des Nationalparkes 'Niedersächsisches Wattenmeer'." *Seevögel* 16(4): 97–99.

Hemminger, H. 1986. "Das Wirklichkeitsverständnis der Naturwissenschaft." *Impulse* 23(3) der Evangelischen Zentralstelle für Weltanschauungsfragen, Stuttgart.

Hengeveld, R. 1994. "Biodiversity—the diversification of life in a non-equilibrium world." *Biodiversity Letters* 2: 1–10.

Heydemann, B. 1981. "Das Ende der Menschheit—na und?" *Natur* 11/81: 24–31.

———. 1985. "Folgen des Ausfalls von Arten—am Beispiel der Fauna." *Schriftenreihe des Deutschen Rats für Landespflege* 46: 581–594.

Himmelheber, M. 1974a. "Rückschritt zum Überleben". *Scheidewege* 4: 61–92.

———. 1974b. "Die Entstehung des Krisenbewußtseins." In H. Schäfer, ed., *Folgen der Zivilisation: Therapie oder Untergang? Bericht der Studiengruppe "Zivilisationsfolgen" der Vereinigung Deutscher Wissenschaftler.* Umschau, Frankfurt am Main, pp. 97–98.

Hoekstra, T. W., F. H. A. Timothy, and C. H. Flather. 1991. "Implicit scaling in ecological research." *BioScience* 41(3): 148–154.

Höffe, O. 1981. *Sittlich-politische Diskurse.* Suhrkamp, Frankfurt am Main.

Honnefelder, L. 1993. "Welche Natur sollen wir schützen?" *Gaia* 2(5): 253–264.

Hösle, V. 1991. *Philosophie der ökologischen Krise: Moskauer Vorträge.* Beck, München.

Hsü, K. J. et al. 1982. "Mass mortality and its environmental and evolutionary consequences." *Science* 216: 249–256.

Hubbell, S. P., and R. B. Foster. 1986. "Biology, chance, and history and the structure of tropical rain forest tree communities." In J. Diamond and T. J. Case, eds., *Community ecology*. Harper & Row, New York, pp. 314–329.

Hughes, J. D. 1980. "The environmental ethics of the Pythagoreans." *Environmental Ethics* 2(3): 195–213.

Hull, D. L. 1976. "Are species really individuals?" *Systematic Zoology* 25: 174–191.

Hume, D. [1740] 1968. *A treatise of human nature*. Clarendon Press, Oxford.

———. [1748] 1999. *An enquiry concerning human understanding*. Edited by T. L. Beauchamp. Oxford University Press, Oxford.

Hunt, W. M. 1980. "Are mere things morally considerable?" *Environmental Ethics* 2(1): 59–65.

Hutchins, M., and C. Wemmer. 1987. "Wildlife conservation and animal rights: Are they compatible?" In M. W. Fox and L. D. Mickley, eds., *Advances in animal welfare science 1986/87*. Humane Society of the United States, Washington, D.C., pp. 111–137.

Irrgang, B. 1989. "Hat die Natur ein Eigenrecht auf Existenz?" *Laufener Seminarbeiträge* der Akademie für Naturschutz und Landschaftspflege (Laufen/Salzach) 4/89: 43–56.

Jablonski, D. 1991. "Extinctions: A paleontological perspective." *Science* 253: 754–757.

Jackson, J. B. C., and K. W. Kaufmann. 1987. "Diadema antillarum was not a keystone predator in cryptic reef environments." *Science* 235: 687–689.

Jaspers, K. 1968. *Nikolaus Cusanus*. Deutscher Taschenbuch Verlag, Munich.

Johnson, E. 1984. "Treating the dirt: Environmental ethics and moral theory." In T. Regan, ed., *Earthbound: New introductory essays in environmental ethics*. Temple University Press, Philadelphia, pp. 336–365.

Johnson, L. E. 1991. *A morally deep world*. Cambridge University Press, Cambridge.

Jonas, H. 1973. "Die Natur auf der moralischen Bühne. Überlegungen zur Ethik im technologischen Zeitalter." *Evangelische Kommentare* 6: 73–77.

———. 1982. *The phenomenon of life: Toward a philosophical biology*. University of Chicago Press, Chicago.

———. 1984. *Das Prinzip Verantwortung: Versuch einer Ethik für die technologische Zivilisation*. Suhrkamp, Frankfurt am Main.

———. 1988. *Materie, Geist und Schöpfung: Kosmologischer Befund und kosmogonische Vermutung*. Suhrkamp, Frankfurt am Main.

Jones, C. G., J. H. Lawton, and M. Shachak. 1994. "Organisms as ecosystem engineers." *Oikos* 69: 373–386.

Joosten, H., and D. Clarke. 2002. *Wise use of mires and peatlands: Background and principles including a framework for decision-making*. Edited by International Mire Conservation Group and International Peat Society. Saarijärven Offset Oy, Saarijärven, Finland.

Jung, J. 1987. *Subjektive Ästhetik.* Lang, Frankfurt am Main.

Kadlec, E. 1976. *Realistische Ethik: Verhaltenstheorie und Moral der Arterhaltung.* Duncker & Humblot, Berlin.

Kafka, P. 1989. *Das Grundgesetz vom Aufstieg: Vielfalt, Gemächlichkeit, Selbstorganisation: Wege zum wirklichen Fortschritt.* Carl Hanser, Munich.

Kant, I. [1756] 1985. "Geschichte und Naturbeschreibung des Erdbebens, welches 1755 einen Teil der Erde erschüttert hat." In I. Kant, *Geographische und andere naturwissenschaftliche Schriften,* edited by J. Zehbe. Felix Meiner, Hamburg, pp. 43–80.

———. [1783] 1976. *Prolegomena zu einer jeden zukünftigen Metaphysik, die als Wissenschaft wird auftreten können.* Edited by K. Vorländer. Felix Meiner, Hamburg.

———. [1785] 1965. *Grundlegung zur Metaphysik der Sitten.* Felix Meiner, Hamburg.

———. [1787] 1976. *Kritik der reinen Vernunft* (Critique of Pure Reason). Edited by R. Schmidt. Felix Meiner, Hamburg.

———. [1788] 1974. *Kritik der praktischen Vernunft* (Critique of Practical Reason). Felix Meiner, Hamburg.

———. [1790] 1959. *Kritik der Urteilskraft* (Critique of Judgment). Edited by K. Vorländer. Felix Meiner, Hamburg.

———. [1797] 1990. *Metaphysische Anfangsgründe der Tugendlehre, Metaphysik der Sitten, Teil 2* (The Metaphysical Principles of Virtue, Metaphysic of Morals, Part 2). Edited by B. Ludwig. Felix Meiner, Hamburg.

Kantor, J. E. 1980. "The 'interests' of natural objects." *Environmental Ethics* 2(2): 163–171.

Katz, E. 1979. "Utilitarianism and preservation." *Environmental Ethics* 1: 357–364.

———. 1987. "Searching for intrinsic value: Pragmatism and despair in environmental ethics." *Environmental Ethics* 9: 231–241.

Kaule, G. 1986. *Arten- und Biotopschutz.* Eugen Ulmer, Stuttgart.

Kaule, G., and K. Henle. 1992. "Forschungsdefizite im Aufgabenbereich des Arten- und Biotopschutzes." *Jahrbuch für Naturschutz und Landschaftspflege* 45: 127–136.

Keller, H.-U. 1992. "Der Tod der Dinosaurier—ein kosmischer Treffer?" In H.-U. Keller, *Das Himmelsjahr 1993.* Frankh-Kosmos, Stuttgart, pp. 108–111.

Keller, M. 1995. "Zuviel Wald macht zornig." *Die Zeit* 33/95: 51.

Kellert, S. R., and E. O. Wilson, eds. 1993. *The biophilia hypothesis.* Island Press, Washington, D.C.

Kelman, S. 1981. "Cost-benefit analysis. An ethical critique." *Regulation (AEI Journal on Government and Society)* Jan./Feb. 81: 33–40.

Ketelhodt, F. von. 1992. "Umweltschutz. Schutz des Menschen vor selbst verursachten Naturkatastrophen?" *Criticón* 129(1/2): 13–14.

Kiemstedt, H. 1991. "Leitlinien und Qualitätsziele für Naturschutz und Landschaftspflege." In G. Kaule and K. Henle, eds., *Arten- und Biotopschutzforschung für Deutschland: Berichte aus der Ökologischen Forschung*, pp. 338–342.

King, A. W., and S. L. Pimm. 1983. "Complexity, diversity, and stability: A reconciliation of theoretical and empirical results." *The American Naturalist* 122(2): 229–239.

Kirchner, J. W. 1991. "The Gaia hypotheses: Are they testable? Are they useful?" In S. H. Schneider and P. J. Boston, eds., *Scientists on Gaia*. MIT Press, Cambridge, Mass., pp. 38–46.

Kirk, G. 1991. "Naturschutz—warum?" *Bombina* 1/2: 1–25.

Kirschenmann, P. P. 1978. "Ecology, ethics, science, and the intrinsic value of things." *Proceedings of XVIth World Congress of Philosophy*. Düsseldorf, pp. 366–370.

Kluge, A. G. 1990. "Species as historical individuals." *Biology and Philosophy* 5: 417–431.

Knapp, A. 1986. "Biologie und Moral." *Impulse* 24(5) der Evangelischen Zentralstelle für Weltanschauungsfragen, Stuttgart.

Knauer, R. H. 1992. "Jobs oder Eulen?" *Die Zeit* 20/92: 41.

Köhler, W. R. 1987. "Zur Debatte um reflexive Argumente in der neueren deutschen Philosophie." In Forum für Philosophie Bad Homburg, ed., *Philosophie und Begründung*. Suhrkamp, Frankfurt am Main, pp. 303–333.

Kornwachs, K., and W. von Lucadou. 1984. "Komplexe Systeme." In K. Kornwachs, ed., *Offenheit—Zeitlichkeit—Komplexität: Zur Theorie der offenen Systeme*. Campus, Frankfurt am Main, pp. 110–165.

Kowarik, I., and H. Sukopp. 1986. "Unerwartete Auswirkungen neu eingeführter Pflanzenarten." *Universitas* 41(2): 828–845.

Krämer, H. 1984. "Zum Problem einer hedonistischen Ethik." *Allgemeine Zeitschrift für Philosophie* 9(1): 11–30.

Krebs, C. J. 1985. *Ecology: The experimental analysis of distribution and abundance*. Harper & Row, New York.

Kreeb, K. H. 1979. *Ökologie und menschliche Umwelt: Geschichte, Bedeutung, Zukunftsaspekte*. Gustav Fischer, Stuttgart.

Krieger, M. H. 1973. "What's wrong with plastic trees?" *Science* 179: 446–455.

Kuhlmann, W. 1987. "Was spricht heute für eine Philosophie des kantischen Typs?" In Forum für Philosophie Bad Homburg, ed., *Philosophie und Begründung*. Suhrkamp, Frankfurt am Main, pp. 84–115.

Kuhn, T. S. 1970. *The structure of scientific revolutions*. University of Chicago Press, Chicago.

Küng, H. 1978. *Existiert Gott? Antwort auf die Gottesfrage der Neuzeit*. Piper, Munich.

Küppers, B.-O. 1982. "Der Verlust aller Werte." *Natur* 4/82: 65–79.

Kurt, F. 1977. *Wildtiere in der Kulturlandschaft*. Rentsch, Erlenbach-Zürich.

————. 1982. *Naturschutz—Illusion und Wirklichkeit*. Parey, Hamburg.

Landmann, M. 1976. *Philosophische Anthropologie: Menschliche Selbstdeutung in Geschichte und Gegenwart*. Walter de Gruyter, Berlin.

————. 1981. "Ökologische und anthropologische Verantwortung—eine neue Dimension der Ethik." In A. Eildermuth and A. Jäger, eds., *Gerechtigkeit: Themen der Sozialethik*. Mohr, Tübingen.

Lang, B., C. Lupi, M. Omlin, and I. Reinhardt. 1994. "Wo speist man ökologischer? Möglichkeiten und Grenzen von Ökobilanzen im Restaurant-Vergleich." *Gaia* 3(2): 108–115.

Laws, R. M. 1970. "Elephants as agents of habitat and landscape change in East Africa." *Oikos* 21: 1–15.

Lawton, J. H. 1992. "Feeble links in food webs." *Nature* 355: 19–20.

————. 1994. "What do species do in ecosystems?" *Oikos* 71: 367–374.

Lawton, J. H., and V. K. Brown. 1993. "Redundancy in ecosystems." In E.-D. Schulze and H. A. Mooney, eds., *Biodiversity and ecosystem function*. Springer, Berlin, pp. 255–270.

Lee, K. 1993. "Instrumentalism and the last person argument." *Environmental Ethics* 15: 333–344.

Lehnes, P. 1994. "Zur Problematik von Bewertungen und Werturteilen auf ökologischer Grundlage." *Verhandlungen der Gesellschaft für Ökologie* 23: 421–426.

Leisinger, K. M. 1994. "Bevölkerungsdruck in Entwicklungsländern und Umweltverschleiß in Industrieländern als Haupthindernisse für zukunftsfähige globale Entwicklung." *Gaia* 3(3): 131–143.

Leitzell, T. L. 1986. "Species protection and management decisions in an uncertain world." In B. G. Norton, ed., *The preservation of species: The value of biological diversity*. Princeton University Press, Princeton, N.J., pp. 243–267.

Lenk, H. 1977. "Anforderungen an die Philosophie in der gegenwärtigen Situation." *Universitas* 32: 931–939.

————. 1983a. "Erweiterte Verantwortung. Natur und künftige Generationen als ethische Gegenstände." In D. Mayer-Maly and P. M. Simons, eds., *Das Naturrechtsdenken heute und morgen*. Duncker & Humblot, Berlin, pp. 833–846.

————. 1983b. "Verantwortung für die Natur: Gibt es moralische Quasirechte von oder moralische Pflichten gegenüber nicht-menschlichen Naturwesen?" *Allgemeine Zeitschrift für Philosophie* 8(3): 1–17.

————. 1993. "Über Verantwortungsbegriffe und das Verantwortungsproblem in der Technik." In H. Lenk and G. Ropohl, eds., *Technik und Ethik*. Reclam, Stuttgart, pp. 112–148.

Leopold, A. [1949] 1968. *A Sand County almanac*. Oxford University Press, New York.

Liedke, G. 1981. *Im Bauch des Fisches: Ökologische Theologie*. Kreuz, Stuttgart.

Lippoldmüller, W. 1982. *Schützen und leben lassen: Die in Bayern geschützten Tiere*.

Edited by Bayerisches Staatsministerium für Landesentwicklung und Umweltfragen, Munich.

Locke, J. [1690] 1966. *The second treatise of government*. Barnes & Noble, New York.

Lombardi, L. G. 1983. "Inherent worth, respect, and rights." *Environmental Ethics* 5: 257–270.

Lorenz, K. 1973. *Die Rückseite des Spiegels: Versuch einer Naturgeschichte menschlichen Erkennens*. Piper, Munich.

_____. 1986. *Der Abbau des Menschlichen*. Piper, Munich.

Lovejoy, T. 1976. "We must decide which species will go forever." *Smithsonian* 7(4): 52–59.

Lovelock, J. E. 1988. *The ages of Gaia: A biography of our living earth*. W. W. Norton, London.

_____. 1995. *Gaia: A new look on life on earth*. Oxford University Press, Oxford. (German edition: Lovelock, J. E. 1982. *Unsere Erde wird überleben: Gaia, eine optimistische Ökologie*. Piper, Munich).

Löw, R. 1989. "Philosophische Begründung des Naturschutzes." *Scheidewege* 18 (1988–89): 149–167.

_____. 1990. "Brauchen wir eine neue Ethik?" *Universitas* 3/90: 291–296.

Luhmann, N. 1990. *Ökologische Kommunikation: Kann die moderne Gesellschaft sich auf ökologische Gefährdungen einstellen?* Westdeutscher Verlag, Opladen, Germany.

MacArthur, R. M. 1965. "Patterns of species diversity." *Biological Review* 40: 510–533.

MacIntyre, A. 1966. *A short history of ethics: A history of moral philosophy from the Homeric Age to the twentieth century*. Macmillan, New York.

Mackie, J. L. 1990. *Ethics: Inventing right and wrong*. Penguin Books, London.

Mader, H.-J. 1985. "Welche Bedeutung hat die Vernetzung für den Artenschutz?" *Schriftenreihe des Deutschen Rats für Landespflege* 46: 631–634.

Mannison, D. S. 1980. "A prolegomenon to a human chauvinistic aesthetic: Comments stimulated by Reinhardt's remarks." In D. S. Mannison, M. A. McRobbie, and R. Routley, eds., *Environmental philosophy*. Monograph Series, No. 2, Department of Philosophy, Australian National University, Canberra, pp. 212–216.

Marcus Aurelius. 1990. *The meditations of Marcus Aurelius Antonius*. Translated by A. S. L. Farquharson. Oxford University Press, Oxford.

Marcus, R. B. 1987. "Moral dilemmas and consistency." In C. W. Gowans, ed., *Moral dilemmas*. Oxford University Press, New York, pp. 188–204.

Maren-Grisebach, M. 1982. "Philosophie der Grünen." *Schriftenreihe: Geschichte und Staat—Kritisches Forum*, vol. 267. Günther Olzog, Munich.

Margulis, L., and G. Hinkle. 1991. "The biota and Gaia—150 years of support for environmental sciences." In S. H. Schneider and P. J. Boston, eds., *Scientists on Gaia*. MIT Press, Cambridge, Mass., pp. 11–18.

Marietta, D. E., Jr. 1993. "Environmental holism and individuals." In S. J. Armstrong and R. G. Botzler, eds., *Environmental ethics: Divergence and convergence*. McGraw-Hill, New York, pp. 405–411.

———. 1995. *For people and the planet: Holism and humanism in environmental Ethics*. Temple University Press, Philadelphia.

Markl, H. 1981. "Das Ende der Menschheit—na und?" *Natur* 11/81: 24–31.

———. 1983. "Die Dynamik des Lebens: Entfaltung und Begrenzung biologischer Populationen." In H. Markl, ed., *Natur und Geschichte*. R. Oldenbourg, Munich, pp. 71–100.

———. 1989. "Natur als Kulturaufgabe." In L. Franke, ed., *Wir haben nur eine Erde*. Wissenschaftliche Buchgesellschaft, Darmstadt, Germany, pp. 30–39.

———. 1995. "Pflicht zur Widernatürlichkeit." *Der Spiegel* 48/95: 206–207.

Marsch, W.-D. 1973. "Ethik der Selbstbegrenzung. Theologische Überlegungen zum Umweltschutz." *Evangelische Kommentare* 6: 18–20.

Matthies, D., B. Schmid, and P. Schmid-Hempel. 1995. "The importance of population processes for the maintenance of biological diversity." *Gaia* 4(4): 199–209.

Maurer, R. 1982 "Ökologische Ethik?" *Allgemeine Zeitschrift für Philosophie* 7(1): 17–39.

———. 1984. "Ökologische Ethik." In W. Beer and G. de Haan, eds., *Ökopädagogik: Aufstehen gegen den Untergang der Natur*. Beltz, Weinheim, Germany, pp. 57–68.

May, J. 1979. "Fehlt dem Christentum ein Verhältnis zur Natur?" *Una Sancta* 34(2): 159–171.

May, R. M. 1973. *Stability and complexity in model ecosystems*. Princeton University Press, Princeton, N.J.

———. 1976. *Theoretical ecology: Principles and applications*. W. B. Saunders, Philadelphia.

———. 1988. "How many species are there on earth?" *Science* 241: 1441–1449.

May, R. M., J. R. Beddington, C. W. Clark, S. J. Holt, and R. M. Laws. 1978. "Management of multispecies fisheries." *Science* 205: 267–275.

Mayr, E. 1942. *Systematics and the origin of species*. Columbia University Press, New York.

———. 1982. *The growth of biological thought*. Harvard University Press, Belknap Press, Cambridge, Mass.

———. 1988a. *Toward a new philosophy of biology: Observations of an evolutionist*. Harvard University Press, Belknap Press, Cambridge, Mass.

———. 1988b. "Die Darwinsche Revolution und die Widerstände gegen die Selektionstheorie." In H. Meier, ed., *Die Herausforderung der Evolutionsbiologie*. Piper, Munich, pp. 221–249.

McCloskey, H. J. 1979. "Moral rights and animals." *Inquiry* 22: 23–54.

McKibben, B. 1990. *The end of nature*. Penguin Books, London.

Meadows, D. H., D. L. Meadows, E. Zahn, and P. Milling. 1972. *The limits of growth: A report for the Club of Rome's project on the predicament of mankind*. Universe Books, New York.

Melchart, D., and H. Wagner. 1993. *Naturheilverfahren: Grundlagen einer autoregulativen Medizin*. Schattauer, Stuttgart.

Mesarovic, M., and E. Pestel. 1974. *Mankind at the turning point: The second report to the Club of Rome*. Dutton, New York.

Meyer, T. 1986. "Zur Begründung und Durchsetzung einer neuen Ethik." In T. Meyer and S. Miller, eds., *Zukunftsethik 1 (Zukunftsethik und Industriegesellschaft)*. J. Schweitzer, Munich, pp. 154–156.

Meyer-Abich, K. M 1979. "Kritik und Bildung der Bedürfnisse. Aussichten auf Veränderungen der Nachfrage- und Bedarfsstruktur." In K. M. Meyer-Abich and D. Birnbacher, eds., *Was braucht der Mensch, um glücklich zu sein?* Beck, Munich, pp. 58–77.

————. 1982. "Vom bürgerlichen Rechtsstaat zur Rechtsgemeinschaft der Natur. Bedingungen einer verfassungsmäßigen Ordnung der menschlichen Herrschaft in der Naturgeschichte." *Scheidewege* 12: 581–605.

————. 1984. *Wege zum Frieden mit der Natur: Praktische Naturphilosophie für die Umweltpolitik*. Carl Hanser, Munich.

————. 1987. "Naturphilosophie auf neuen Wegen." In O. Schwemmer, ed., *Über Natur: Philosophische Beiträge zum Naturverständnis*. Vittorio Klostermann, Frankfurt am Main, pp. 63–73.

————. 1989. "Von der Umwelt zur Mitwelt. Unterwegs zu einem neuen Selbstverständnis des Menschen im Ganzen der Natur." *Scheidewege* 18 (1988/89): 128–148.

————. 1990. *Aufstand für die Natur: Von der Umwelt zur Mitwelt*. Carl Hanser, Munich.

————. 1991. "Die holistische Alternative." In R. P. Sieferle, ed., *Natur: Ein Lesebuch*. Beck, Munich, pp. 159–170.

Mill, J. S. [1871] 2000. *Utilitarianism*. Edited by Roger Crisp. Oxford University Press, Oxford.

Mills, L. S., M. E. Soulé, and D. F. Doak. 1993. "The keystone-species concept in ecology and conservation." *BioScience* 43: 219–224.

Mishler, B. D., and R. N. Brandon. 1987. "Individuality, pluralism and the phylogenetic species concept." *Biology and Philosophy* 2: 397–414.

Mittelstraß, J. 1984. "Naturalismus." In J. Mittelstraß, ed., *Enzyklopädie Philosophie und Wissenschaftstheorie*, vol. 2. Bibliographisches Institut, Mannheim, Germany, p. 964.

Mohr, H. 1987. *Natur und Moral: Ethik in der Biologie*. Wissenschaftliche. Buchgesellschaft, Darmstadt, Germany.

Moore, G. E. [1903] 1994. *Principia ethica*. Edited by T. Baldwin. Cambridge University Press, Cambridge.

Morowitz, H. J. 1991. "Balancing species preservation and economic considerations." *Science* 253: 752–754.

Morscher, E. 1986. "Was ist und was soll Evolutionäre Ethik? (Oder: Wie man offene Türen einrennt). Ein Kommentar zu Gerhard Vollmers Programm einer Evolutionären Ethik." *Conceptus* 20(49): 73–77.

Müller, C., and F. Müller. 1992. "Umweltqualitätsziele als Instrumente zur Integration ökologischer Forschung und Anwendung." *Kieler Geographische Schriften* 85: 131–166.

Müller, M. 1974. *Philosophische Anthropologie*. Freiburg, Munich.

Müller-Christ, G. 1995. *Wirtschaft und Naturschutz: Von der technologischen zur humanorientierten Problemsicht*. R.E.A.-Verlag Managementforschung, Bayreuth, Germany.

Müller-Herold, U. 1992. "Umwelthygiene: Gesundheit für die Ethosphäre." *Gaia* 1(1): 26–33.

Müller-Motzfeld, G. 1991. "Artenschwund und Artenschutz bei Insekten." *Mitteilungen des Zoologischen Museums Berlin* 67(1): 195–207.

Mutz, M. 1992. "Genormte Welt. Die abwechslungsreiche Natur wird zum Armenhaus." *Greenpeace Magazin* 2/92: 8–14.

Myers, N. 1976. "An expanded approach to the problem of disappearing species." *Science* 193: 198–202.

_____. 1979. *The sinking ark*. Pergamon Press, Oxford.

Naess, A. 1984. "A defence of the deep ecology movement." *Environmental Ethics* 6: 265–270.

_____. 1986. "The deep ecological movement: some philosophical aspects." *Philosophical Inquiry* 8(1/2): 10–31.

Nagel, T. 1979. "Ethics without biology." In T. Nagel, *Moral questions*. Cambridge University Press, Cambridge, pp. 142–146.

_____. *What does it all mean?* Oxford University Press, Oxford.

Nash, R. 1977. "Do rocks have rights?" *The Center Magazine* 10(6): 2–12.

Nelson, L. [1932] 1970. *System der philosophischen Ethik und Pädagogik* (Gesammelte Schriften, vol. 5). Felix Meiner, Hamburg.

Nielsen, K. 1984. "Why should I be moral? Revisited." *American Philosophical Quarterly* 21(1): 81–91.

Nietzsche. F. [1886] 1990. *Jenseits von Gut und Böse: Vorspiel einer Philosophie der Zukunft*. Goldmann, Augsburg, Germany.

Norberg-Hodge, H. 1991. *Ancient futures: Learning from Ladakh*. Sierra Club Books, San Francisco.

Nordenstam, T. 1982: "Vom 'Sein' zum 'Sollen'—Deduktion oder Artikulation?" In

W. Kuhlmann and D. Böhler, eds., *Kommunikation und Reflexion: Zur Diskussion der Transzendentalpragmatik: Antworten auf Karl-Otto Apel.* Suhrkamp, Frankfurt.

Norton, B. G. 1984. "Environmental ethics and weak anthropocentrism." *Environmental Ethics* 6: 131–148.

———. 1986. "On the inherent danger of undervaluing species." In B. G. Norton, ed., *The preservation of species: The value of biological diversity.* Princeton University Press, Princeton, N.J., pp. 110–137.

———. 1987. *Why preserve natural variety?* Princeton University Press, Princeton, N.J.

———. 1988. "Commodity, amenity, and morality: The limits of quantification in valuing biodiversity." In E. O. Wilson, ed., *Biodiversity.* National Academy Press, Washington, D.C.

Obermann, H. 1992. "Eingreifen oder laufen lassen—was soll der Naturschutz wollen?" *NNA-Berichte* 5(1): 34–36.

Odum, E. P. 1971. *Fundamentals of ecology* (3rd ed.). W. B. Saunders, Philadelphia.

———. 1975. *Ecology* (2nd ed.). Holt, Rinehart and Winston, New York.

———. 1977. "The emergence of ecology as a new integrative discipline." *Science* 195: 1289–1293.

O'Neil, R. V., D. L. DeAngelis, J. B. Waide, and T. F. H. Allen. 1986. *Hierarchical concept of ecosystems.* Princeton University Press, Princeton, N.J.

Osche, G. 1978. *Ökologie: Grundlagen, Erkenntnisse, Entwicklungen der Umweltforschung.* Herder, Freiburg.

Ott, J. A. 1985. "Ökologie und Evolution." In J. A. Ott, G. P. Wagner, and F. M. Wuketits, eds., *Evolution, Ordnung und Erkenntnis.* Parey, Berlin, pp. 47–68.

Page, T. 1978. "A generic view of toxic chemicals and similar risks." *Ecology Law Quarterly* 7(2): 207–244.

Paine, R. T. 1966. "Food web complexity and species diversity." *American Naturalist* 100: 65–75.

———. 1969. "A note on trophic complexity and community stability." *American Naturalist* 103: 91–93.

———. 1980. "Food webs: Linkage, interaction strength and community infrastructure." *Journal of Animal Ecology* 49: 667–685.

Palumbi, S. R., and L. A. Freed. 1988. "Agonistic interactions in a keystone predatory starfish." *Ecology* 69: 1624–1627.

Pascal, B. 1993. *Pensées.* Édition de Ph. Sellier. Classiques Garnier. Bordas, Paris.

Passmore, J. 1974. *Man's responsibility for nature: Ecological problems and Western traditions.* Charles Scribner's Sons, New York.

Pate, J. S., and S. D. Hopper. 1993. "Rare and common plants in ecosystems, with special reference to the south-west Australian flora." In E.-D. Schulze and H. A. Mooney, eds., *Biodiversity and ecosystem function.* Springer, Berlin, pp. 293–325.

Patzig, G. 1983. "Ökologische Ethik." In H. Markl, ed., *Natur und Geschichte*. Oldenbourg, Munich, pp. 329–347.

Pearce, D. 1987. "Foundations of an ecological economics." *Ecological Modelling* 38: 9–18.

Peine, H. G. 1990. "Die Vielfalt der Grenzwerte." In BASF, ed., *Denken Planen, Handeln. Umweltbericht 1990*. BASF, Ludwigshafen, Germany.

Peters, R. H. 1976. "Tautology in evolution and ecology." *American Naturalist* 110: 1–12.

Peters, R. L., and J. D. S. Darling. 1985. "The greenhouse effect and nature reserves." *BioScience* 35(11): 707–717.

Pfordten, D. von der. 1996. *Ökologische Ethik: Zur Rechtfertigung menschlichen Verhaltens gegenüber der Natur*. Rowohlt, Reinbek, Germany.

Pianka, E. 1985. "A wild analogy." *BioScience* 35(11): 685.

Pimm, S. L. 1980. "Food web design and the effect of species deletion." *Oikos* 35: 139–149.

———. 1982. *Food webs*. Chapman & Hall, London.

———. 1984. "The complexity and stability of ecosystems." *Nature* 307: 321–326.

———. 2001. *The world according to Pimm: A scientist audits the earth*. McGraw-Hill, New York.

Pimm, S. L., and P. Raven. 2000. "Extinction by numbers." *Nature* 403: 843-845.

Pippenger, N. 1978. "Complexity theory." *Scientific American* 238: 90–100.

Pister, E. P. 1979. "Endangered species: Costs and benefits." *Environmental Ethics* 1: 341–352.

Pitelka, L. F. 1993. "Biodiversity and policy decisions." In E.-D. Schulze and H. A. Mooney, eds., *Biodiversity and ecosystem function*. Springer, Berlin, pp. 481–493.

Plachter, H. 1990. *Naturschutz*. Fischer, Stuttgart.

Plachter, H., and F. Foeckler. 1991. "Entwicklung von naturschutzfachlichen Analyse- und Bewertungsverfahren." In G. Kaule and K. Henle, eds., *Arten- und Biotopschutzforschung für Deutschland: Berichte aus der Ökologischen Forschung* 4: 323–337.

Platnick, N. I. 1992. "Patterns of biodiversity." In N. Eldredge, ed., *Systematics, ecology, and the biodiversity crisis*. Columbia University Press, New York, pp. 15–24.

Plato. 1990. *Sophistes*. Translated by W. S. Cobb. Rowan & Littlefield, Savage, Md.

Plessner, H. [1928] 1975. *Die Stufen des Organischen und der Mensch: Einleitung in die philosophische Anthropologie*. Walter de Gruyter, Berlin.

Pluhar, E. B. 1983. "The justification of an environmental ethic." *Environmental Ethics* 5: 47–61.

Pojman, L. P. 1994. *Environmental ethics: Readings in theory and application*. Jones & Bartlett Publishers, Boston.

Popper, K. R. 1957. *The poverty of historicism*. Basic Books, New York.

_____. 1959. *The logic of scientific discovery*. Hutchinson, London.

_____. 1972. *Objective knowledge: An evolutionary approach*. Oxford University Press, Oxford.

Popper, K. R., and J. C. Eccles. 1977. *The self and its brain: An argument for interactionism*. Springer, Heidelberg.

Postman, N. 1992. *Technopoly: The surrender of culture to technology*. Alfred A. Knopf, New York.

Prigogine, I., and I. Stengers. 1984. *Order out of chaos: Man's new dialogue with nature*. Random House, New York.

Primack, R. B. 1993. *Essentials of conservation biology*. Sinauer Associates, Sunderland, Mass.

Primas, H. 1992. "Umdenken in der Naturwissenschaft." *Gaia* 1(1): 5–15.

Pschyrembel, W. 1986. *Klinisches Wörterbuch* (255th ed.). Walter de Gruyter, Berlin.

Quine, W. V. 1977. *Ontological relativity and other essays*. Columbia University Press, New York.

Randall, A. 1986. "Human preferences, economics, and the preservation of species." In B. G. Norton, ed., *The preservation of species: The value of biological diversity*. Princeton University Press, Princeton, N.J., pp. 79–109.

_____. 1988. "What mainstream economists have to say about the value of biodiversity." In E. O. Wilson, ed., *Biodiversity*. National Academy Press, Washington, D.C.

Rawls, J. 1973. *A theory of justice*. Oxford University Press, Oxford.

Regan, D. H. 1986. "Duties of preservation." In B. G. Norton, ed., *The preservation of species: The value of biological diversity*. Princeton University Press, Princeton, N.J., pp. 195–220.

Regan, T. 1976. "Feinberg on what sorts of beings can have rights." *The Southern Journal of Philosophy* 14: 485–498.

_____. 1980. "On the connection between environmental science and environmental ethics." *Environmental Ethics* 2(4): 363–367.

_____. 1981. "The nature and possibility of an environmental ethic." *Environmental Ethics* 3: 19–34.

_____. 1993. "The case for animal rights." In S. J. Armstrong and R. G. Botzler, eds., *Environmental ethics: Divergence and convergence*. McGraw-Hill, New York, pp. 321–329.

Reichelt, G. 1979. "Wurzeln der Umweltkrise—ethische Gesichtspunkte zum Umweltschutz." *Veröffentlichungen der Aktionsgemeinschaft Natur- und Umweltschutz Baden-Württemberg* No. 6. Stuttgart.

Reichholf, J. H. 1981. "Verrostendes Wasser." *Nationalpark* 31: 41–43.

_____. 1993. *Comeback der Biber: Ökologische Überraschungen*. Beck, Munich.

———. 1996. "Feuer im Genarchiv: Das Artensterben und seine Folgen." *Bild der Wissenschaft* 2/96: 60–64.

Remmert, H. 1984. *Ökologie: Ein Lehrbuch*. Springer, Berlin.

———. 1990. *Naturschutz: Ein Lesebuch nicht nur für Planer, Politiker, Polizisten, Publizisten und Juristen*. Springer, Berlin.

———, ed. 1991. *The mosaic-cycle concept of ecosystems*. Springer, Berlin.

Rescher, N. 1980. "Why save endangered species?" In N. Rescher, *Unpopular essays on technological progress*. University of Pittsburgh Press, Pittsburgh, pp. 79–92.

Richardson, J. L. 1980. "The organismic community: Resilience of an embattled ecological concept." *BioScience* 30(7): 465–471.

Richter, H. E. 1988. *Der Gotteskomplex: Die Geburt und die Krise des Glaubens an die Allmacht des Menschen*. Rowohlt, Reinbek, Germany.

Ricken, F. 1987. "Anthropozentrismus oder Biozentrismus? Begründungsprobleme der ökologischen Ethik." *Theologie und Philosophie* 62(1): 1–21.

———. 1989. *Allgemeine Ethik*. Kohlhammer, Stuttgart.

Riedl, R. 1980. *Biologie der Erkenntnis: Die stammesgeschichtlichen Grundlagen der Vernunft*. Parey, Berlin.

———. 1985. *Die Spaltung des Weltbildes: Die biologischen Grundlagen des Erklärens und Verstehens*. Parey, Berlin.

Rifkin, J. 1981. *Entropy: A new worldview*. Bantam Books, New York.

Ring, I. 1994. *Marktwirtschaftliche Umweltpolitik aus ökologischer Sicht: Möglichkeiten und Grenzen*. B. G. Teubner Verlagsgesellschaft, Stuttgart.

Rippe, K. P. 1994. "Artenschutz als Problem der praktischen Ethik." In G. Meggle and U. Wessels, eds., *Analyomen 1: Proceedings of the 1st Conference "Perspectives in Analytical Philosophy"*. Walter de Gruyter, Berlin, pp. 805–817.

Roberts, L. 1988. "Hard choices ahead on biodiversity." *Science* 241: 1759–1761.

Robinson, J. V., and W. D. Valentine. 1979. "The concepts of elasticity, invulnerability and invadability." *Journal of Theoretical Biology* 81: 91–104.

Rolston, H., III. 1982. "Are values in nature subjective or objective?" *Environmental Ethics* 4: 125–151.

———. 1985. "Duties to endangered species." *BioScience* 35(11): 718–726.

———. 1988. *Environmental ethics: Duties to and values in the natural world*. Temple University Press, Philadelphia.

———. 1994. "Value in nature and the nature of value." In R. Attfield and A. Belsey, eds., *Philosophy and the natural environment*. Royal Institute of Philosophy Supplement, Cambridge University Press, Cambridge, pp. 13–31.

Routley, R. 1973. "Is there a need for a new, an environmental ethic?" In Bulgarian Organizing Commitee, ed., *Proceedings of XVth World Congress of Philosophy*, vol. 1, Sophia Press, Sophia, pp. 205–210.

Routley, R., and V. Routley. 1979. "Against the inevitability of human chauvinism."

In K. E. Goodpaster and K. M. Sayre, eds., *Ethics and problems of the 21st century*. University of Notre Dame Press, London, pp. 36–59.

———. 1980. "Human chauvinism and environmental ethics." In D. S. Mannison, M. A. McRobbie, and R. Routley, eds., *Environmental Philosophy*. Monograph Series, No. 2, Department of Philosophy, Australian National University, Canberra, pp. 96–189.

Roweck, H. 1993. "Zur Naturverträglichkeit von Naturschutz-Maßnahmen." *Verhandlungen der Gesellschaft für Ökologie* 22: 15–20.

Ruh, H. 1987. "Zur Frage nach der Begründung des Naturschutzes." *Zeitschrift für Evangelische Ethik* 31: 125–133.

Russow, L.-M. 1981. "Why do species matter?" *Environmental Ethics* 3: 101–112.

Sachsse, H. 1967. *Naturerkenntnis und Wirklichkeit*. Vieweg, Braunschweig.

Sagan, C. 1994. *Pale blue dot: A vision of the human future in space*. Random House, New York.

Sagoff, M. 1981. "At the shrine of our Lady of Fatima, or Why political questions are not all economic." *Arizona Law Review* 23(4): 1283–1298.

Salthe, S. N., and B. M. Salthe. 1989. "Ecosystem moral considerability: A reply to Cahen." *Environmental Ethics* 11: 355–361.

Scarre, G. 1981. "On the alleged irrelevance of biology to ethics." *The Journal of Value Inquiry* 15: 243–252.

Schäfer, L. 1987. "Selbstbestimmung und Naturverhältnis des Menschen." In O. Schwemmer, ed., *Über Natur: Philosophische Beiträge zum Naturverständnis*. Vittorio Klostermann, Frankfurt am Main, pp. 15–35.

Schäfer, W. 1982. "Soziale Naturwissenschaft." In G. Michelsen, U. Rühling, and F. Kalberlah, eds., *Der Fischer Öko-Almanach II. Daten, Fakten, Trends der Umweltdiskussion*. Fischer, Frankfurt am Main.

Scheler, M. [1927] 1995. *Die Stellung des Menschen im Kosmos*. Bouvier, Bonn.

Schemel, H.-J. 1984. "Wir brauchen rationale Argumente für den Naturschutz— Versuch einer Entgegnung auf den Beitrag von E. Bierhals." *Landschaft + Stadt* 16(3): 190–191.

Scherzinger, W. 1991. "Biotop-Pflege oder Sukzession?" *Garten + Landschaft* 2/91: 24–28.

Schlick, M. 1984. *Fragen der Ethik*. Suhrkamp, Frankfurt am Main.

Schlitt, M. 1992. *Umweltethik: Philosophisch-ethische Reflexionen—theologische Grundlagen—Kriterien*. Schöningh, Paderborn.

Schmidt, E. 1993. "Der Naturgarten—ein neuer Weg?" In R. Schäfer, ed., *Was heißt denn schon Natur?* Callwey, Munich, pp. 11–24.

Schmidt, S. J. 1975. *Zum Dogma der prinzipiellen Differenz zwischen Natur- und Geisteswissenschaft*. Vandenhoeck & Ruprecht, Göttingen.

Schmidt-Moser, R. 1982. "Süßwasserspeicherbecken im Watt—eine Alternative für den Seevogelschutz? Das Beispiel Hauke-Haien-Koog." *Seevögel* 3(3): 110–111.

Schnädelbach, H. 1987. "Über Rationalität und Begründung." In Forum für Philosophie Bad Homburg, ed., *Philosophie und Begründung*. Suhrkamp, Frankfurt am Main, pp. 67–83.

Schönherr, H.-M. 1985. *Philosophie und Ökologie: Philosophische und politische Essays*. Die Blaue Eule, Essen.

———. 1987. "Ökologie als Hermeneutik. Ein wissenschaftstheoretischer Versuch." *Philosophia Naturalis* 24: 311–332.

———. 1989. *Von der Schwierigkeit, Natur zu verstehen: Entwurf einer negativen Ökologie*. Fischer Taschenbuch Verlag, Frankfurt am Main.

Schopenhauer, A. [1840] 1922. "Preisschrift über die Grundlage der Moral." In A. Schopenhauer, *Sämmtliche Werke in fünf Bänden*, vol. 3. Inselverlag, Leipzig, pp. 493–672.

Schreiber, R. L., ed. 1978. *Rettet die Vögel—wir brauchen sie* (Save the Birds—We Need Them). Herbig, Munich.

Schröder, W. 1978. "Fauna in geänderter Landschaft." *Tutzinger Studien* (Munich) 1: 19–35.

Schuh, H. 1994. "Unbekümmerte Klima-Ingenieure." *Die Zeit* 10/94: 49

———. 1995. "Schöpfer und Zerstörer. Anmerkungen zu E. Wilsons Thesen: Die Rolle des Menschen in der Evolution." *Die Zeit* 26/95: 34.

Schulze, E.-D. 1993. "Ökologie und Ökosystemforschung. Zwei Begriffe im Wandel der Zeit." *Biologie in unserer Zeit* 23(5): 273–275.

Schulze, E.-D., and H. A. Mooney. 1993. "Ecosystem function of biodiversity: A summary." In E.-D. Schulze and H. A. Mooney, eds., *Biodiversity and ecosystem function*. Springer, Berlin, pp. 497–510.

Schulze, E.-D., and H. Zwölfer. 1987. "Synthesis." In E.-D. Schulze and H. Zwölfer, eds., *Potentials and limitations of ecosystem analysis*. Springer, Berlin, pp. 416–423.

Schurz, G. 1986. "Ökologische Ethik. Thesen—Antithesen—Synthesen." In W. Leinfellner and F. M. Wuketits, eds., *Die Aufgaben der Philosophie in der Gegenwart*. Akten des X. internationalen Wittgenstein-Symposiums 1985 in Kirchberg am Wechsel. Hölder-Pichler-Tempsky, Vienna, pp. 247–251.

———. 1991. "How far can Hume's is-ought-thesis be generalized? An investigation in alethic-deontic modal predicate logic." *Journal of Philosophical Logic* 20: 37–95.

Schütze, C. 1989. *Das Grundgesetz vom Niedergang: Arbeit ruiniert die Welt*. Hanser, Munich.

Schüz, M. 1990. "'Ehrfurcht vor dem Leben' in der industriellen Welt. Albert Schweitzers Ethik angesichts der verschärften Risikosituation von heute." In C. Günzler, E. Grässler, B. Christ, and H. H. Eggebrecht, eds., *Albert Schweitzer heute: Brennpunkte seines Denkens*. Katzmann, Tübingen, pp. 125–153.

Schweitzer, A. [1923] 1974. *Gesammelte Werke* (5 vols., quoted as 1974 a–e). Beck, Munich.

———. 1991. *Die Ehrfurcht vor dem Leben: Grundtexte aus fünf Jahrzehnten*. Edited by H. W. Bähr. Beck, Munich.

Schweppe-Kraft, B. 1992. "Bewertung des Arten- und Biotopschutzes mit Hilfe der Zahlungsbereitschaftsanalyse. Ein Beitrag zur fachübergreifenden Abwägung bei Eingriffsregelung und UVP." *Jahrbuch für Naturschutz und Landschaftspflege* 45: 114–126.

Searle, J. R. 1967. "How to derive 'ought' from 'is'." In P. Foot, ed., *Theories of ethics*. Oxford University Press, Oxford, pp. 101–114.

Seel, M. 1991. "Ästhetische Argumente in der Ethik der Natur." *Deutsche Zeitschrift für Philosophie* 39(8): 901–913.

Seneca, L. A. 1969. *Letters from a stoic*. Translation by R. Campbell. Penguin Books, New York.

Sepkoski, J. J., Jr. 1992. "Phylogenetic and ecological patterns in the Phanerozoic history of marine biodiversity." In N. Eldredge, ed., *Systematics, ecology, and the biodiversity crisis*. Columbia University Press, New York, pp. 77–100.

Sepkoski, J. J., Jr., R. K. Bambach, D. M. Raup, and J. W. Valentine. 1981. "Phanerozoic marine diversity and the fossil record." *Nature* 293: 435–437.

Singelmann, A. 1993. "Natur als Symbol auf einem hohen Syntheseniveau." In R. Schäfer, ed., *Was heißt denn schon Natur?* Callwey, Munich, pp. 103–114.

Singer, P. 1979a. "Not for humans only: The place of nonhumans in environmental issues." In K. E. Goodpaster and K. M. Sayre, eds., *Ethics and problems of the 21st century*. University of Notre Dame Press, London, pp. 191–206.

———. 1979b. *Practical ethics*. Cambridge University Press, Cambridge.

———. 1993. *Practical ethics* (2nd ed., revised and expanded). Cambridge University Press, New York.

Sitter-Liver, B. 1994. "Natur als Polis: Vertragstheorie als Weg zu ökologischer Gerechtigkeit." In H. J. Koch et al., eds., *Theorien der Gerechtigkeit*. Steiner, Stuttgart, pp. 139–162.

Skirbekk, G. 1995. "Ethischer Gradualismus: Jenseits von Anthropozentrismus und Biozentrismus?" *Deutsche Zeitschrift für Philosophie* 43(3): 419–434.

Slobodkin, L. B. 1986. "On the susceptibility of different species to extinction: Elementary instructions for owners of a world." In B. G. Norton, ed., *The preservation of species: The value of biological diversity*. Princeton University Press, Princeton, N.J., pp. 226–242.

Smith, F. D. M., R. M. May, R. Pellew, T. H. Johnson, and K. R. Walter. 1993a. "How much do we know about the current extinction rate?" *Trends in Ecology and Evolution* 8(10): 375–378.

Smith, T. B., M. W. Bruford, and R. K. Wayne. 1993b. "The preservation of process: The missing element of conservation programs." *Biodiversity Letters* 1: 164–167.

Sober, E. 1986. "Philosophical problems for environmentalism." In B. G. Norton, ed., *The preservation of species: The value of biological diversity*. Princeton University Press, Princeton, N.J., pp. 173–194.

Soulé, M. E. 1985. "What is conservation biology?" *BioScience* 35(11): 727–734.

Spaemann, R. 1979. "Die christliche Religion und das Ende des modernen Bewußtseins." *Internationale Katholische Zeitschrift Communio* 8(3): 251–270.

_____. 1980. "Technische Eingriffe in die Natur als Problem der politischen Ethik." In D. Birnbacher, ed., *Ökologie und Ethik*. Reclam, Stuttgart, pp. 180–206.

_____. 1984. "Tierschutz und Menschenwürde." In U. M. Händel, ed., *Tierschutz—Testfall unserer Menschlichkeit*. Fischer, Frankfurt am Main, pp. 71–81.

_____. 1986. *Moralische Grundbegriffe*. Beck, Munich.

_____. 1990. *Glück und Wohlwollen: Versuch über Ethik*. Klett-Cotta, Stuttgart.

Sprigge, T. L. S. 1979. "Metaphysics, physicalism, and animal rights." *Inquiry* 22: 101–143.

Stanley, S. M. 1981. *The new evolutionary timetable: Fossils, genes, and the origin of species*. Basic Books, New York.

_____. 1987. *Extinction*. Scientific American Books, New York.

Stegmüller, W. 1969a. *Probleme und Resultate der Wissenschaftstheorie und der analytischen Philosophie*. Springer, Berlin.

_____. 1969b. *Metaphysik, Skepsis, Wissenschaft*. Springer, Berlin.

Steinvorth, U. 1991. "Wie eine Moralbegründung aussehen könnte." *Deutsche Zeitschrift für Philosophie* 39(8): 879–889.

Stenmark, M. 2002. *Environmental ethics and policy-making*. Ashgate Publishing, Burlington, Vt.

Stern, H. 1976. *Mut zum Widerspruch*. Rowohlt, Reinbek, Germany.

Stevenson, C. L. 1937. "The emotive meaning of ethical terms." *Mind* 46: 14–31.

Stöckler, M. 1986. "Über die Schwierigkeiten und Aussichten einer Evolutionären Ethik." *Conceptus* 20(49): 69–72.

Störig, H. J. 1981. *Kleine Weltgeschichte der Philosophie*, vol 1. Fischer, Frankfurt am Main.

Stone, C. D. 1988. *Should trees have standing? Toward legal rights for natural objects*. Tioga Press, Palo Alto, Calif.

_____. 1993. "Moral pluralism and the course of environmental ethics." In S. J. Armstrong and R. G. Botzler, eds., *Environmental ethics: Divergence and convergence*. McGraw-Hill, New York, pp. 76–85.

Stork, N. E. 1993. "How many species are there?" *Biodiversity and Conservation* 2: 215–232.

Strey, G. 1989. *Umweltethik und Evolution: Herkunft und Grenzen moralischen Verhaltens gegenüber der Natur.* Vandenhoeck & Ruprecht, Göttingen.

Stugren, B. 1978. *Grundlagen der Allgemeinen Ökologie.* Gustav Fischer, Stuttgart.

Sukopp, U., and H. Sukopp. 1993. "Das Modell der Einführung und Einbürgerung nicht einheimischer Arten. Ein Beitrag zur Diskussion über die Freisetzung gentechnisch veränderter Kulturpflanzen." *Gaia* 2(5): 267–288.

Swift, M. J., and J. M. Anderson. 1993. "Biodiversity and ecosystem function in agricultural systems." In E.-D. Schulze and H. A. Mooney, eds., *Biodiversity and ecosystem function.* Springer, Berlin, pp. 15–41.

Tahvanainen, J. O., and R. B. Root. 1972. "The influence of vegetational diversity on the population ecology of a specialized herbivore, *Phyllotreta cruciferae* (Coleoptera: Chrysomelidae)." *Oecologia* 10 (4): 321–346.

Taylor, P. W. 1981. "The ethics of respect for nature." *Environmental Ethics* 3(3): 197–218.

———. 1983. "In defense of biocentrism." *Environmental Ethics* 5(3): 237–243.

———. 1984. "Are humans superior to animals and plants?" *Environmental Ethics* 6: 149–160.

———. 1986. *Respect for nature: A theory of environmental ethics.* Princeton University Press, Princeton, N.J.

Teutsch, G. M. 1980. "Umwelt oder Schöpfung? Vom ethischen Aspekt des Umgehens mit der Natur." *Politische Studien,* Sonderheft 1 (Energie, Umwelt, Ernährung): 119–129.

———. 1985. *Lexikon der Umweltethik* (Encyclopedia of Environmental Ethics). Vandenhoeck & Ruprecht, Göttingen.

———. 1987. *Mensch und Tier: Lexikon der Tierschutzethik.* Vandenhoeck & Ruprecht, Göttingen.

———. 1988. "Schöpfung ist mehr als Umwelt." In K. Bayertz, ed., *Ökologische Ethik.* Schnell & Steiner, Munich, pp. 55–65.

———. 1990. "Ehrfurchtsethik und Humanitätsidee. Albert Schweitzer beharrt auf der Gleichwertigkeit alles Lebens." In C. Günzler, E. Grässler, B. Christ, and H.H. Eggebrecht, eds., *Albert Schweitzer heute: Brennpunkte seines Denkens.* Katzmann, Tübingen, pp. 101–109.

———. 1995. *Die "Würde der Kreatur": Erläuterungen zu einem neuen Verfassungsbegriff am Beispiel des Tieres.* Paul Haupt, Bern.

Teilhard de Chardin, P. 1961. *The phenomenon of man.* Harper & Row, New York.

Thielcke, G. 1978. "Aktiver Naturschutz." *BUND-Information* 2/78. BUND Verlagsgesellschaft, Freiburg.

Thiessen, H. 1988. "Naturschutz durch Nichtstun?" *Bauernblatt/Landpost* Jan. 30, pp. 64–65.

Thoreau, H. D. [1854] 1992. *Walden and other writings.* Edited by B. Atkinson. Modern Library, New York.

Tischler, W. 1976. *Einführung in die Ökologie*. Gustav Fischer, Stuttgart.

Töpfer, K. 1991. "Wieviel ungestörte Natur können wir uns leisten?" *Umwelt* 11/91: 494–498.

Tranøy, K. E. 1972. "'Ought' implies 'can': A bridge from fact to norm?" *Ratio* 14(2): 116–130.

Trepl, L. 1983. "Ökologie—eine grüne Leitwissenschaft? Über Grenzen und Perspektiven einer modischen Disziplin." *Kursbuch 74 (Zumutungen an die Grünen)*. Kursbuch Verlag, Berlin, pp. 6–27.

———. 1988. "Gibt es Ökosysteme?" *Landschaft + Stadt* 20(4): 176–185.

———. 1991. "Forschungsdefizite: Naturschutzbegründungen." In K. Henle and G. Kaule, eds., *Arten- und Biotopschutzforschung für Deutschland: Berichte aus der Ökologischen Forschung* 4: 424–432.

———. 1994. *Geschichte der Ökologie: Vom 17. Jahrhundert bis zur Gegenwart: Zehn Vorlesungen*. Beltz Athenäum, Weinheim, Germany.

Tribe, L. H. 1976. "Ways not to think about plastic trees." In L. H. Tribe, C. S. Schelling, and J. Voss, eds., *When values conflict: Essays on environmental analysis, discourse and decision*. Cambridge, Mass., pp. 61–91.

Tügel, H., and C. Fetscher, C. 1988. "Antarktis bald unterm Bohrer?" *Greenpeace-Nachrichten* 1/88: 30–34.

Tugendhat, E. 1984. "Retraktationen." In E. Tugendhat, *Probleme der Ethik*. Reclam, Stuttgart, pp. 132–176.

———. 1989. "Die Hilflosigkeit der Philosophen." In *Die Neue Gesellschaft / Frankfurter Hefte* 10/89: 927–935.

———. 1994. *Vorlesungen über Ethik*. Suhrkamp, Frankfurt am Main.

United Nations. 1992. *Convention on biological diversity*. United Nations, Rio de Janeiro.

Varner, G. E. 1987. "Do species have standing?" *Environmental Ethics* 9: 57–72.

———. 1990. "Biological functions and biological interests." *The Southern Journal of Philosophy* 28(2): 251–270.

Vauk, G., and J. Prüter. 1987. *Möwen: Arten, Bestände, Verbreitung, Probleme*. Jordsand-Buch No. 6, Niederelbe-Verlag, Otterndorf, Germany.

Vermeij, G. J. 1986. "The biology of human-caused extinction." In B. G. Norton, ed., *The Preservation of species: The value of biological diversity*. Princeton University Press, Princeton, N.J., pp. 28–49.

Vester, F. 1980. *Neuland des Denkens: Vom technokratischen zum kybernetischen Zeitalter*. Deutsche Verlags-Anstalt, Stuttgart.

———. 1984. *Der Wert eines Vogels* (The value of a bird). Kösel, Munich.

Vico, G. [1709] 1963. *De nostri temporis studiorum ratione*. Wissenschaftliche Buchgesellschaft, Darmstadt, Germany.

Vitousek, P. M., and D. U. Hooper. 1993. "Biological diversity and terrestrial ecosystem biogeochemistry." In E.-D. Schulze and H. A. Mooney, eds., *Biodiversity and ecosystem function*. Springer, Berlin, pp. 3–14.

Vitousek, P. M., H. A. Mooney, J. Lubchenco, and J. M. Melillo. 1997. "Human domination of earth's ecosystems." *Science* 277(25): 494–499.

Vollmer, G. 1975. *Evolutionäre Erkenntnistheorie*. Hirzel, Stuttgart.

————. 1985. "Über vermeintliche Zirkel in einer empirisch orientierten Erkenntnistheorie." In G. Vollmer, *Was können wir wissen?* (Vol. 1). *Die Natur der Erkenntnis: Beiträge zur Evolutionären Erkenntnistheorie*. Hirzel, Stuttgart, pp. 217–267.

————. 1986. *Was können wir wissen?* (Vol. 2). *Die Erkenntnis der Natur: Beiträge zur modernen Naturphilosophie*. Hirzel, Stuttgart.

————. 1986a. "Kann es von einmaligen Ereignissen eine Wissenschaft geben?" In G. Vollmer, *Was können wir wissen?* (Vol. 2). *Die Erkenntnis der Natur: Beiträge zur modernen Naturphilosophie*. Hirzel, Stuttgart, pp. 53–65.

————. 1986b. "Die Einheit der Wissenschaft in evolutionärer Perspektive." In G. Vollmer, *Was können wir wissen?* (Vol. 2). *Die Erkenntnis der Natur: Beiträge zur modernen Naturphilosophie*. Hirzel, Stuttgart, pp. 163–199.

————. 1986c. "Jenseits des Mesokosmos. Anschaulichkeit in Physik und Didaktik." In G. Vollmer, *Was können wir wissen?* (Vol. 2). *Die Erkenntnis der Natur: Beiträge zur modernen Naturphilosophie*. Hirzel, Stuttgart, pp. 138–162.

————. 1986d. "Über die Möglichkeit einer Evolutionären Ethik." *Conceptus* 20(49): 51–68.

————. 1987. "Über die Chancen einer Evolutionären Ethik. Oder: Wie man Türen zuschlägt." *Conceptus* 21(52): 87–94.

————. 1989. "Von den Grenzen unseres Wissens." *Naturwissenschaftliche Rundschau* 42(10): 387–392.

————. 1990. "Naturwissenschaft Biologie (I)—Aufgaben und Grenzen." *Biologie heute* 371: 3–7.

Vorholz, F. 1995. "Die letzte Party. Ohne Ökoumbau droht der Kollaps." *Die Zeit* 42/95: 42.

Vossenkuhl, W. 1974. *Wahrheit des Handelns: Untersuchungen zum Verhältnis von Wahrheit und Handeln* (Münchner Philosophische Forschungen, vol. 8). Bouvier, Bonn.

————. 1983. "Die Unableitbarkeit der Moral aus der Evolution." In P. Koslowski, P. Kreuzer, and R. Löw, eds., *Die Verführung durch das Machbare: Ethische Konflikte in der modernen Medizin und Biologie* (Civitas Resultate, vol. 3). Hirzel, Stuttgart, pp. 141–154.

————. 1992a. "Vernünftiges und unvernünftiges Wissen." *Naturwissenschaften* 79: 97–102.

————. 1992b. "Moralische Dilemmata." In O. Höffe, M. Forschner, and W. Vossenkuhl, eds., *Lexikon der Ethik*. Beck, Munich, pp. 188–189.

―――. 1992c. "Vernünftige Wahl, rationale Dilemmas und moralische Konflikte." In M. Hollis and W. Vossenkuhl, eds., *Moralische Entscheidung und rationale Wahl*. Oldenbourg, Munich, pp. 153–173.

―――. 1993a. "Normativität und Deskriptivität in der Ethik." In L. H. Eckensberger and U. Gähde, eds., *Ethische Norm und empirische Hypothese*. Suhrkamp, Frankfurt am Main, pp. 133–150.

―――. 1993b. "Ökologische Ethik. Über den moralischen Charakter der Natur." *Information Philosophie* 1/93: 6–19.

Walker, B. H. 1992. "Biodiversity and ecological redundancy." *Conservation Biology* 6(1): 18–23.

Warnock, G. J. 1971. *The object of morality*. Methuen, London.

Watson, R. A. 1983. "A critique of anti-anthropocentric biocentrism." *Environmental Ethics* 5: 245–256.

Weber, J. 1990. *Die Erde ist nicht Untertan: Grundrechte für Tiere und Umwelt*. Eichborn, Frankfurt am Main.

Weber, M. [1919] 1985. "Wissenschaft als Beruf." In M. Weber, *Gesammelte Aufsätze zur Wissenschaftslehre*, edited by J. Winckelmann. Tübingen.

Weber, M., C. Körner, B. Schmid, and W. Arber. 1995. "Diversity of life in a changing world." *Gaia* 4(4): 185–190.

Wehnert, D. 1988. *Noahs letzte Warnung: Artensterben und menschliche Zivilisation*. Ullstein, Frankfurt am Main.

Weikard, H.-P. 1992. *Der Beitrag der Ökonomik zur Begründung von Normen des Tier- und Artenschutzes: Eine Untersuchung zu praktischen und methodologischen Problemen der Wirtschaftsethik*. Duncker & Humblot, Berlin.

Weiss, P. A. 1969. "The living system: Determinism stratified." *Studium Generale* 22: 361–400.

Weissert, H. 1994. "Erdgeschichtliche Treibhausepisoden. Fluchtpunkt Mutter Erde: Die Gaia-Hypothese als Leitbild." *Gaia* 3(1): 25–35.

Weizsäcker, C. von, and E. U. von Weizsäcker. 1984. "Fehlerfreundlichkeit." In K. Kornwachs, ed., *Offenheit, Zeitlichkeit, Komplexität: Zur Theorie der offenen Systeme*. Campus, Frankfurt am Main, pp. 167–201.

―――. 1986. "Fehlerfreundlichkeit als Evolutionsprinzip." *Universitas* 41(483): 791–799.

Weizsäcker, C. F. von. 1960. *Zum Weltbild der Physik*. Hirzel, Stuttgart.

―――. 1977. *Der Garten des Menschlichen*. Hanser, Munich.

―――. 1979. "Modelle des Gesunden und Kranken, Guten und Bösen, Wahren und Falschen." In C. F. von Weizsäcker, *Die Einheit der Natur*. Hanser, Munich, pp. 320–341.

―――. 1991. "Geist und Natur." In H.-P. Dürr and W. C. Zimmerli, eds., *Geist und Natur: Über den Widerspruch zwischen naturwissenschaftlicher Erkenntnis und philosophischer Welterfahrung*. Scherz, Bern, pp. 17–27.

Weizsäcker, E. U. von 1992. *Erdpolitik: Ökologische Realpolitik an der Schwelle zum Jahrhundert der Umwelt*. Wissenschaftliche Buchgesellschaft, Darmstadt, Germany.

Wendnagel, J. 1990. *Ethische Neubesinnung als Ausweg aus der Weltkrise? Ein Gespräch mit dem "Prinzip Verantwortung" von Hans Jonas*. Königshausen & Neumann, Würzburg, Germany.

Wenz, P. S. 1993. "Minimal, moderate, and extreme moral pluralism." *Environmental Ethics* 15: 61–74.

Wenzel, U. J. 1992. *Anthroponomie: Kants Archäologie der Autonomie*. Akademie Verlag, Berlin.

Werner, H. 1978. "Leben auf dem Aussterbe-Etat." In O. Schatz, ed., *Was bleibt den Enkeln? Die Umwelt als politische Herausforderung*. Styria, Graz, pp. 133–164.

West, B. J., and A. L. Goldberger. 1987. "Physiology in fractal dimensions." *American Scientist* 75: 354–365.

Westman, W. E. 1977. "How much are nature's services worth?" *Science* 197: 960–964.

———. 1990. "Managing for biodiversity." *BioScience* 40(1): 26–33.

White, L., Jr. 1967. "The historical roots of our ecological crisis." *Science* 155: 1203-1207.

Whittaker, R. H. 1972. "Evolution and measurement of species diversity." *Taxon* 21(2/3): 213–251.

Wiedmann, F. 1990. "Anstößige Denker. Die Wirklichkeit als Natur und Geschichte in der Sicht von Außenseitern." Fischer, Frankfurt am Main.

Wiens, J. A., J. F. Addicott, T. J. Case, and J. Diamond. 1986. "The importance of spatial and temporal scale in ecological investigations." In J. Diamond and T. J. Case, eds., *Community ecology*. Harper & Row, New York, pp. 145–153.

Wigner, E. P. 1964. "Two kinds of reality." *The Monist* 48: 248–264.

Wiley, E. O. 1980. "Is the evolutionary species fiction? A consideration of classes, individuals and historical entities." *Systematic Zoology* 29(1): 76–80.

Willard, L. D. 1980. "On preserving nature's aesthetic features." *Environmental Ethics* 2: 293–310.

Williams, M. 1980. "Rights, interests, and moral equality." *Environmental Ethics* 2(2): 149–161.

Williamson, M. 1987. "Are communities ever stable?" In A. J. Gray, M. J. Crawley, and P. J. Edwards, eds., *Colonization, succession and stability*. Blackwell Oxford, pp. 353–371.

Willmann, R. 1985. *Die Art in Raum und Zeit: Das Artkonzept in der Biologie und Paläontologie*. Parey, Berlin.

Wilson, E. O. 1985. "The biological diversity crisis." *BioScience* 35(11): 700–706.

———. 1992. *The diversity of life*. Harvard University Press, Belknap Press, Cambridge, Mass.

———. 1995. "Jede Art ein Meisterwerk." *Die Zeit* 26/95: 33.

———, ed. 1988. *Biodiversity*. National Academy Press, Washington, D.C.

Wimmer, R. 1984. "Naturalismus (ethisch)". In J. Mittelstraß, ed., *Enzyklopädie Philosophie und Wissenschaftstheorie*, vol. 2. Bibliographisches Institut, Mannheim, Germany, pp. 965–966.

Wissel, C. 1992. "Mathematik—nur eine andere Sprache?" *Verhandlungen der Gesellschaft für Ökologie* 21: 43–47.

Wissenschaftsrat (Scientific Advisory Committee). 1994. *Stellungnahme zur Umweltforschung in Deutschland: Kurzfassung*. Cologne.

Wittgenstein, L. 1963. *Tractatus logico-philosophicus: Logisch-philosophische Abhandlung*. Suhrkamp, Frankfurt am Main.

Wodzicki, K. 1950. "Introduced mammals of New Zealand: An ecological and economical survey." *Department of Science and Industry Research Bulletin* 98, Wellington.

Wolf, U. 1987. "Brauchen wir eine ökologische Ethik?" *Prokla* 69: 148–173.

———. 1988. "Haben wir moralische Verpflichtungen gegen Tiere?" *Zeitschrift für philosophische Forschung* 42: 222–246.

Wolgast, E. 1981. "The transcendence of ethics." In E. Morscher and R. Stranzinger, eds., *Ethik: Grundlagen, Probleme und Anwendungen*. Akten des 5. internationalen Wittgenstein-Symposiums 1980 in Kirchberg am Wechsel. Hölder-Pichler-Tempsky, Vienna, pp. 144–147.

Wolters, G. 1995. "'Rio' oder die moralische Verpflichtung zum Erhalt der natürlichen Vielfalt. Zur Kritik einer UN-Ethik." *Gaia* 4(4): 244–249.

Woodward, F. I. 1993. "How many species are required for a functional ecosystem?" In E.-D. Schulze and H. A. Mooney, eds., *Biodiversity and ecosystem function*. Springer, Berlin, pp. 271–291.

Worster, D. 1993. "The ecology of order and chaos." In S. J. Armstrong and R. G. Botzler, eds., *Environmental ethics: Divergence and convergence*. McGraw-Hill, New York, pp. 39–51.

Wuketits, F. M. 1981. *Biologie und Kausalität: Biologische Ansätze zur Kausalität, Determination und Freiheit*. Parey, Berlin.

———. 1983. *Biologische Erkenntnis: Grundlagen und Probleme*. Gustav Fischer, Stuttgart.

———. 1988. *Evolutionstheorien: Historische Voraussetzungen, Positionen, Kritik*. Wissenschaftliche Buchgesellschaft, Darmstadt, Germany.

———. 1995. "Evolution ohne Fortschritt." *Kosmos* 11/95: 34–35.

Zahrnt, A. 1993. "Zeitvergessenheit und Zeitbesessenheit der Ökonomie—und ihre ökologischen Folgen." In M. Held and A. Geißler, eds., *Ökologie der Zeit: Vom Finden der rechten Zeitmaße*. Hirzel, Stuttgart, pp. 111–120.

Zimmerli, W. C. 1991. "Technik als Natur des westlichen Geistes." In H.-P. Dürr and W. C. Zimmerli, eds., *Geist und Natur: Über den Widerspruch zwischen natur-*

wissenschaftlicher Erkenntnis und philosophischer Welterfahrung. Scherz, Bern, pp. 389–409.

Ziswiler, V. 1965. *Bedrohte und ausgerottete Tiere: Eine Biologie des Aussterbens und des Überlebens*. Springer, Berlin.

Zwölfer, H. 1978. "Was bedeutet 'Ökologische Stabilität'?" *Bayreuther Hefte für Erwachsenenbildung* 3: 13–33.

_____. 1980. "Artenschutz für unscheinbare Tierarten?" *Schriftenreihe Naturschutz und Landschaftspflege* 12: 81–88.

_____. 1989. "Die menschliche Bevölkerungsentwicklung aus zoologisch-populationsökologischer Sicht." Ruprecht-Karls-Universität Heidelberg, ed., *Bevölkerungsexplosion–Bevölkerungsschwund: Vorträge im Sommersemester 1988*. Heidelberger Verlagsanstalt, Heidelberg, pp. 21–27.

Zwölfer, H., and W. Völkl. 1993. "Artenvielfalt und Evolution." *Biologie in unserer Zeit* 23(5): 308–315.

About the Author

Martin Gorke was born in Stuttgart, Germany, in 1958. He studied biology and philosophy at the University of Bochum and the University of Bayreuth. From 1985 to 1993 he spent eight months each year as the game warden of a bird reserve on an island in the North Sea, Norderoog. He completed his Ph.D. in biology in 1989, specializing in animal ecology. In 1997 he received a second Ph.D., this time in philosophy. Since 1997 he has been an assistant professor in the Department of Environmental Ethics at the University of Greifswald.

INDEX

Island Press Board of Directors

Victor M. Sher, Esq.
Chair
Environmental Lawyer,
Sher & Leff

Dane A. Nichols
Vice-Chair
Environmentalist

Carolyn Peache
Secretary
President,
Campbell, Peachey & Associates

Drummond Pike
Treasurer
President,
The Tides Foundation

Robert E. Baensch
Director, Center for Publishing
New York University

David C. Cole
President, Aquaterra, Inc.

Catherine M. Conover
Chair, Board of Directors, Quercus LLC

Henry Reath
President, Collectors Reprints Inc.

Will Rogers
President, Trust for Public Land

Charles C. Savitt
President, Center for Resource
Economics/Island Press

Susan E. Sechler
Senior Advisor on
Biotechnology Policy
The Rockefeller Foundation

Peter R. Stein
General Partner
The Lyme Timber Company

Diana Wall, Ph.D.
Director and Professor
Natural Resource Ecology Laboratory
Colorado State University

Wren Wirth
President,
The Winslow Foundation